Advances in Applied Mechanics

Volume 11

Contributors to Volume 11

T. K. CAUGHEY

Y. C. FUNG

T. H. LIN

PIOTR PERZYNA

MARTIN SICHEL

TH. YAO-TSU WU

ADVANCES IN
APPLIED MECHANICS

Edited by Chia-Shun Yih

COLLEGE OF ENGINEERING
THE UNIVERSITY OF MICHIGAN
ANN ARBOR, MICHIGAN

VOLUME 11

1971

ACADEMIC PRESS New York and London

ACADEMIC PRESS, INC.
111 Fifth Avenue, New York, New York 10003

United Kingdom Edition published by
ACADEMIC PRESS, INC. (LONDON) LTD.
24/28 Oval Road, London NW1 7DD

LIBRARY OF CONGRESS CATALOG CARD NUMBER: 48-8503

PRINTED IN THE UNITED STATES OF AMERICA

Contents

List of Contributors

Numbers in parentheses indicate the pages on which the authors' contributions begin

T. K. CAUGHEY, California Institute of Technology, Pasadena, California (209)

Y. C. FUNG, Department of Engineering Sciences (including Bioengineering), University of California at San Diego, La Jolla, California (65)

T. H. LIN, Mechanics and Structures Department, University of California, Los Angeles, California (255)

PIOTR PERZYNA, Institute of Fundamental Technical Research, Polish Academy of Sciences, Warsaw, Poland (313)

MARTIN SICHEL, Department of Aerospace Engineering, The University of Michigan, Ann Arbor, Michigan (131)

TH. YAO-TSU WU, California Institute of Technology, Pasadena, California (1)

Preface

Each article appearing in the Advances in Applied Mechanics is intended to be an informative, systematic, and up-to-date account of an area of mechanics, with a view to introducing non-experts to the methods used and results obtained in that area and stimulating experts with the new points of view and the suggestiveness that often come with a systematic treatment of a subject. Our first concern is therefore that the articles be interesting in content and attractively written, so that they may be read with interest, profit, and perhaps even enjoyment.

While we are not neglecting the long cultivated fields of mechanics, from which results of classical beauty have never ceased to flow, our attention is naturally called to new centers of interest in mechanics, where much recent research is concentrated. Two of these are biomechanics and geophysical fluid mechanics. However, it is the tillers rather than the fields they till that are of first importance, and we look for the fields where the plough cuts deep, however solitary the tiller or serried the field.

We take this opportunity to express our gratitude to Professor Gustav Kuerti, not only for his devotion and contributions during the many years when he was Managing Editor, but also for kindly editing one article (by Piotr Perzyna) in this volume.

CHIA-SHUN YIH

Advances in Applied Mechanics

Volume 11

Hydromechanics of Swimming of Fishes and Cetaceans

TH. YAO-TSU WU

California Institute of Technology, Pasadena, California

I. Introduction

Swimming propulsion of aquatic animals in water or in other liquid media covers a wide range of body size and speed. Large cetaceans, such as porpoises and whales, may have lengths from 2 to 30 m and can swim at cruising speeds of 8 to 12 m/sec. Microscopic organisms such as paramecia and spermatozoa, ranging from 300 μ (microns) down to 50 μ in length with length-to-diameter ratios of 20 to 100, can swim at speeds of 80 μ to 1000 μ/sec. Some bacteria, such as *Vibrio comma*, of body length from 1 to 5 μ, have been observed to propel themselves as fast as

200 μ/sec, or about 50 to 150 body lengths per second. In between these two extremities there are various classes of fishes and aquatic animals of all sizes and speeds. Based on the characteristic length l of a body moving at velocity U in a liquid of kinematic viscosity ν, the Reynolds number $\mathbf{R} = Ul/\nu$ is of the order of 10^8 for the most rapid cetaceans, 10^6 for migrating fishes, 10^5–10^3 for a great variety of fishes, about 10^2 for tadpoles, down to about $O(1)$ for turbatrix, 10^{-3} or less for paramecia and spermatozoa, and to the extreme of 10^{-6} or less for bacteria. Thus, the Reynolds number $\mathbf{R}$ covers practically the entire range of interest known to hydrodynamicists. The problems of hydrodynamical interest alone arising from studies of the locomotion of these aquatic animals embrace, in fact, a scope too vast to be attempted in a single survey.

This article is not intended to give a complete and thorough survey of all the hydromechanical aspects underlying the propulsion of aquatic animals, although an attempt will be made to present a discussion, as self-contained as possible, of the basic principles and some recent developments in this field, to describe several points and issues of current interest, to consider a few applications of the theory, and to discuss some experimental results. Due to the limitation of scope and space, the material presented here will be rather selective, leaving some important contributions to the literature. Fortunately, a comprehensive treatment of the biology, physics, and mathematics of the subject, as well as an extensive review of the literature, can be found in the recent treatise by Gray (1968). An excellent survey of the hydromechanics of aquatic animal propulsion has been advanced by Lighthill (1969), giving much needed elucidation to both the zoological and hydromechanical aspects of the subject. That survey includes a discussion of aquatic propulsion in some twenty classes within the animal kingdom; an important concept was centered there about the hydromechanical efficiency η of an animal's propulsive flexural movements. The mathematical theory pertaining to the major ideas expounded in the generally nonmathematical discussion of Lighthill (1969) was subsequently given by Lighthill (1970). This recent, very important study of Lighthill (1970) was graciously made available, prior to its publication, to the author by Professor M. J. Lighthill when this article was in preparation, and has been of valuable help in setting the present discussion in a much more complete and up-to-date form.

Discussion of the hydromechanics of swimming depends on the main methods of propulsion, in terms of the modes of movements employed by the body and appendages, as well as other important flow parameters, which include the Reynolds number $\mathbf{R}$ and the reduced frequency σ. Although $\mathbf{R}$ may vary from case to case, the most effective movements employed for swimming propulsion by aquatic animals of almost all sizes appear to have the body in the formation of a transverse wave progressing along the body, from head to tail. A great majority of many classes of fishes can be singled out as preeminent representatives of this undulatory mode of propulsion. The remarkable performance of some cetaceans (porpoises, whales, etc.) and some well-known percomorph fish families (tuna, wahoo, marlin, swordfish, etc.) using strong tails of large aspect-ratio is but a variation of this basic undulatory mode. In the microorganism world an enormous variety of creatures, ranging from minute bacteria, larger but still primitive protozoa, to higher level spermatozoa, have been observed to employ either uniformly propagating transverse waves, or whiplike waves, or helical waves along slender flagella as the principal means of propulsion. The basic transverse wave mode thus seems to be little affected by the Reynolds number over such a wide range, and will therefore be of main interest in the present discussion, whether the transverse waves be two- or three-dimensional in nature. However, the fundamental principles underlying the hydromechanics of swimming propulsion do become very different for large or small values of the Reynolds number. These different $\mathbf{R}$ regimes will form two major categories of hydromechanical interest.

The undulatory motions that most aquatic animals make to propel themselves through water may be divided into two components: one perpendicular and the other tangential to the animal's instantaneous longitudinal axis of centroid (e.g., fish's spinal column, or the central axis of a flagellum). Where the Reynolds number $\mathbf{R}$ is large, the normal component of the bodily motion gives rise to "reactive" forces between a bulk of surrounding water and the parts of the animal's body surface. These forces, due to the inertia of the water, are proportional to the rate of change of the normal velocity of animal surface relative to the surrounding water. The tangential component of motion, on the other hand, generates (at large $\mathbf{R}$) a thin boundary layer next to the body surface, across which the tangential velocity relative to the body surface varies rapidly from zero at the surface to that of the irrotational flow outside the boundary layer. This tangential component of flow results in a "resistive" force, or skin friction, which can be evaluated separately after the exterior irrotational flow is determined, since the latter depends only on the normal component of motion. The forward component of the reactive force gives the thrust required to overcome the resistive viscous drag.

When the Reynolds number $\mathbf{R}$ is small, however, the situation is very different, since the vorticity now penetrates deeply into the surrounding water and the inertia of the water becomes insignificant compared with the pressure and viscous stresses of the water, except possibly when the reduced frequency based on the lateral dimension of the body is sufficiently high. Because of the widespread vorticity, calculation of the flow, strictly speaking, can no longer be made separately for the normal and tangential components of body motion, suggesting that the rigorous determination of the forces acting between the body and the water is generally very complicated. When the body is very slender, as is the case of flagellated propulsion of a large number of microorganisms, the mathematical analysis can be greatly simplified by adopting the so-called "resistive theory." This is based on the assumption that the force between a small section of the body and the water is resistive and viscous in origin, depending primarily on the instantaneous value of the velocity of that body section relative to the surrounding liquid. Some excellent theoretical studies of the undulatory mode of propulsion based on this type of resistive theory have been developed, notably by Taylor (1952), and by Gray and Hancock (1955). Lighthill (1969) has summarized these in a simplified form. The theory of Gray and Hancock has been recently applied to study the helical movement of a flagellum by Chwang and Wu (1971).

The present discussions will concentrate on the hydromechanics of the aqueous medium—as a study of the external biomechanics. It should be emphasized that much of the area interrelating the biological, physical, and engineering fields urgently remains to be developed, and future endeavor should prove to be most fruitful if based on a close and continuous cooperation between biologists, physicists, engineers, and mathematicians. To this end we quote the strong conviction of Sir James Gray (1968): "There cannot be the slightest doubt that on the borderlines of physics, engineering and biology lie some of the most fascinating and challenging aspects of animal locomotion."

II. Inviscid Flow Theory

For a large Reynolds number, the swimming propulsion, as explained in the previous section, depends primarily on the inertial effect since the flow outside a thin boundary layer next to the body surface is irrotational. Viscosity of the fluid is unimportant except in its role of generating the vorticity shed into the wake, and of producing a thin boundary layer, and hence a skin friction at the body surface.

The basic mechanism of swimming propulsion can also be envisaged by considering the flow momentum at some distance from the swimming animal. As the body performs an undulatory wave motion and attains a forward momentum, the propulsive force pushes the fluid backward with a net total momentum equal and opposite to that of the action, while the frictional resistance of the body gives rise to a forward momentum of the fluid by entraining some of the fluid surrounding the body. The momentum of reaction to the inertia forces is concentrated in the vortex wake due to the small thickness and amplitude of the undulatory trailing vortex sheet; this backward jet of fluid expelled from the body can however, be counterbalanced by the momentum in response to the viscous drag. When a self-propelled body is cruising at a constant speed, the forward and backward momenta exactly balance; they can nevertheless be evaluated separately. This mechanism of swimming motion at large Reynolds numbers has been elucidated by von Kármán and Burgers (1943) for the simple case of a rigid plate in transverse oscillation. This basic principle will be demonstrated for slender fish in Section IV, B, and for the waving motion of a two-dimensional flexible plate in Section IV, C. An alternative approach to evaluating the qualitative motion of a flexible planar body is based on the first principles of energy balance, which is given in Section II, A below.

A. Some First Principles; Energy Balance

To visualize why the motion of a transverse wave progressing along the body is desirable for swimming propulsion, Wu (1966, 1968, 1971a) considered the energy balance for the typical case of a flexible planar body of negligible thickness, performing an arbitrary unsteady motion of small amplitude, achieving in time t a rectilinear forward velocity $U(t)$ through a fluid which is otherwise at rest. Choosing a Cartesian coordinate system (x, y, z) fixed at the mean position of the body, with the stretched plan form S_b of the body lying in the $y = 0$ plane and with the free stream velocity $U(t)$ pointing in the positive x direction, the body motion can be written generally as

$$y = h(x, z, t) \qquad (x, z \in S_b), \tag{2.1}$$

$|\,\partial h/\partial x\,|$ and $|\,\partial h/\partial z\,|$ being assumed small. In the linearized inviscid flow theory, the boundary condition on the normal component of velocity relative to the solid surface reads

$$v(x, \pm 0, z, t) \equiv V(x, z, t) = (\partial h/\partial t) + U(\partial h/\partial x) \qquad (x, z \in S_b). \tag{2.2}$$

The planar body may admit sharp leading edges (physically rounded leading edges of large curvature) and sharp trailing edges. When the latter kind is present, we shall impose there, as usual, the Kutta condition. The following discussion is also applicable to plane flows, say in the xy plane, in which case the dependence on z simply drops out, and all the quantities will then refer to a unit span in the z direction.

The thrust (positive when directed in the negative x direction) acting on the body, based on the inviscid linear theory, is given by the integration of the pressure component in the forward direction:

$$T = T_{\text{p}} + T_{\text{s}} = \int_S (\Delta p) \frac{\partial h}{\partial x} \, dS + \int_{\text{L.E.}} F_{\text{s}}(x, z, t) \, dz, \tag{2.3}$$

where (Δp) denotes the pressure difference across the flexible plate, $\Delta p = p(x, -0, z, t) - p(x, +0, z, t)$, F_s is the singular force per unit arc length along the leading edge due to the leading-edge suction, and the last integral is evaluated along the leading edge $z = b(x)$. The power required to maintain the motion is equal to the rate of work done by the plate against the reaction of the fluid in the direction of the transverse plate motion:

$$P = -\int_S (\Delta p) \frac{\partial h}{\partial t} \, dS. \tag{2.4}$$

The third quantity of interest is the mechanical energy imparted to the fluid per unit time which, in this inviscid flow, is equal to the rate of work done by the pressure over the body surface, or

$$E = -\int_S (\Delta p) \, V(x, z, t) \, dS - T_{\text{s}} U. \tag{2.5}$$

The above three quantities satisfy the principle of conservation of energy which asserts that the power input P is equal to the rate of work done by the thrust, TU, plus the kinetic energy E lost to the fluid in unit time,

$$P = TU + E. \tag{2.6}$$

On physical grounds it can be inferred that the energy loss E is non-negative in several cases of broad interest. One of such cases is the periodic body movement with constant forward velocity,

$$U = \text{const.}, \qquad h(x, z, t) = h_1(x, z) \, e^{i\omega t} \qquad (x, z \in S_{\text{b}}), \tag{2.7}$$

where $i = (-1)^{1/2}$ is the imaginary unit for the periodic time motion, $h_1(x, z)$ may generally be complex, and h is to be interpreted by its real part. After the transient stage is over, the kinetic energy imparted to the fluid will be largely confined to the wake which contains the trailing vortex sheet and is lengthening at the rate U. Therefore E, or its time average at least, cannot be negative. (A mathematical proof of this statement has been given for slender bodies by Lighthill (1960a) and for waving plates in plane flows by Wu (1961); see also (2.25) and (2.69) in the following text.)

Another example is when the body starts to swim from a state of rest.

$$U = U(t), \qquad h = h(x, z, t) \qquad (t > 0), \tag{2.8}$$

while U, h, and the components of the perturbation velocity (u, v, w) all vanish for $t < 0$. In this case any disturbance generated in the flow must correspond to a gain of kinetic energy of the fluid (see Section II, C, 2).

The following discussion will be based on the presumption that $E \geqslant 0$. Under this condition we have, by (2.6),

$$P \geqslant TU \qquad \text{if} \quad E \geqslant 0. \tag{2.9}$$

P, however, need not be positive definite. When P is negative, energy is transferred out of the fluid (like a turbine); then $T < 0$ according to (2.9), indicating that there must be an inertial drag acting on the body. (This also explains why an aeroplane wing cannot keep fluttering, once started, without the external thrust supplied by its engine.) Forward swimming is possible only when the thrust $T > 0$, large enough to overcome the viscous drag; then $P > 0$, and hence a power is required to maintain the motion. Now, from (2.3) it is seen that a sufficient condition for producing a positive thrust is satisfied if (Δp) and $(\partial h/\partial x)$ are everywhere of the same sign, for the suction force F_s in forward movement is never negative. In view of the inequality (2.9) and the expression (2.4) for P, (Δp) and $(\partial h/\partial t)$ cannot also be positively correlated everywhere on S_b. Suppose, as a qualitative picture, $(\partial h/\partial x)$ and $(\partial h/\partial t)$ are everywhere opposite in sign, then clearly h represents a transverse wave propagating towards the tail (see Fig. 1).

To investigate further the qualitative features of such periodic waving motions it suffices to consider the case of simple harmonic form (2.7) since for arbitrary time dependence all linear effects such as the pressure, lift, moment, can be obtained by the Fourier synthesis and as for the quadratic effects such as T, P, and E, it can be seen that in their time averages, the components with different multiple frequencies are not coupled. In fact, consider two functions:

$$g(\mathbf{x}, t) = \operatorname{Re}\left[\sum_n g_n(\mathbf{x}) \exp(i\omega_n t)\right], \qquad h(\mathbf{x}, t) = \operatorname{Re}\left[\sum_n h_n(\mathbf{x}) \exp(i\omega_n t)\right], \tag{2.10a}$$

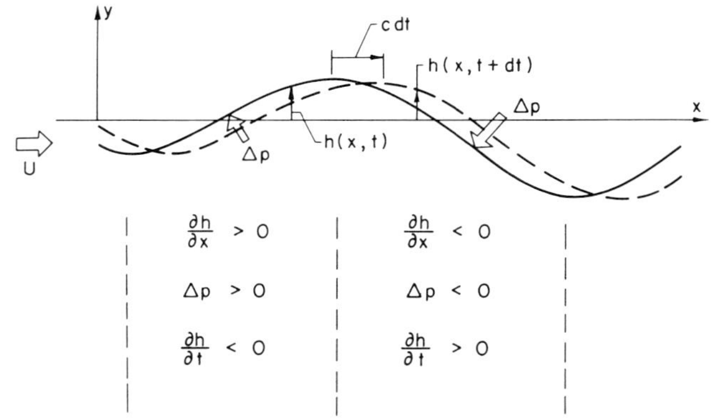

FIG. 1. A consideration of energy conservation indicates that in forward swimming, transverse movements of the body wave propagate not only backward (from head to tail) with velocity c, but also backward relative to the fluid, since $c \geqslant U$.

where Re denotes the real part, then the time average of gh is

$$\overline{gh} = \lim_{T\to\infty} \frac{1}{T}\int_0^T g(\mathbf{x}, t)\, h(\mathbf{x}, t)\, dt = \frac{1}{2}\,\mathrm{Re}\Big[\sum_n g_n(\mathbf{x})\, h_n{}^*(\mathbf{x})\Big], \tag{2.10b}$$

where h^* is the complex conjugate of h. This result is readily extended to the integral form when g, h are expressed by integrals over a continuous spectrum.

Returning to the waving motion, we consider the fundamental form:

$$h = \mathrm{Re}[h_1(x, z)\, e^{i(\omega t - kx)}] \qquad (x, z \in S_b), \tag{2.11}$$

which represents a simple wave propagating streamwise along the planar body, with phase velocity $c = \omega/k$ and amplitude $|\, h_1 \,|$. Substituting (2.11) in (2.3) and (2.4), and taking the time average, we obtain

$$\bar{T}_p = \frac{k}{2}\,\mathrm{Re}\int_S (\Delta p_1)\left[ih_1{}^* + \frac{1}{k}\frac{\partial h_1{}^*}{\partial x}\right] e^{ikx}\, dS, \tag{2.12a}$$

$$\bar{P} = \frac{\omega}{2}\,\mathrm{Re}\int_S (\Delta p_1)(ih_1{}^*)\, e^{ikx}\, dS, \tag{2.12b}$$

where $(\Delta p_1) = (\Delta p) \exp(-i\omega t)$, being independent of t as a result of the linearized theory. Since the thrust T_s due to the leading edge suction is always nonnegative, it follows from inequality (2.9) that

$$\bar{P} \geqslant U\bar{T} \geqslant U\bar{T}_P \tag{2.13}$$

provided $\bar{E} \geqslant 0$. Consequently, if $\partial h_1/\partial x = 0$, or when $|\partial h_1/\partial x| \ll |kh_1|$, then from (2.12) and (2.13) we immediately have

$$c = \omega/k \geqslant U. \tag{2.14}$$

This result shows that not only a progressive transverse wave is desirable, but also its phase velocity must be greater than U (under the stated conditions) in order to achieve a given swimming velocity U. This qualitative feature has been explained in another simple elegant manner by Lighthill (1969); it remains true for a wide class of amplitude functions $h_1(x, z)$.

B. Swimming of Slender Fish

In a very valuable paper, Breder (1926) showed that the patterns of bodily movements exhibited by a wide variety of species of fishes are related to the relative length and flexibility of the tail. Three main types (see also Gray, 1968) can be recognized: (1) *Anguilliform*—fish with long thin and flexible tails exhibiting movements similar to those of an eel; (2) *Carangiform*—fish with tapering tails of medium length, ending in a well-defined caudal fin, exhibiting movements similar to those of the salmon, trout, and most other pelagic fish; (3) *Ostraciiform*—fish with relatively large and rigid bodies and very short tails which oscillate almost symmetrically about the tail-base neck and exhibit very little bending. Between these three main types lies a wide range of intermediate patterns, as the relative length and flexibility of the tail decreases throughout a series of different species, so a continuous variation from anguilliform motion through carangiform to ostraciiform can be found in nature. In terms of the hydromechanical form parameter called the slenderness ratio, δ, defined as the ratio of the maximum body width b_0, as a measure of the transverse dimension of the body, to the body length l, or $\delta = b_0/l$, both anguilliform and carangiform may be regarded as slender ($\delta \ll 1$) whereas ostraciiform is mostly not. This classification is useful for identifying different categories of hydromechanical problems to be discussed below.

In a pioneering paper which opened the field of study on fish locomotion at high Reynolds numbers, Lighthill (1960a) investigated the inviscid flow around a slender fish which makes swimming movements in a direction transverse to its direction of locomotion, while its cross section varies along it only gradually. Based on the slender body theory, Lighthill obtained the result for thrust produced by the fish, time rate of work done by it, and the rate of shedding of energy, showing that the mean values of these quantities all depend on the movement and body shape at the tail-end section only, and that they will vanish with the "virtual mass" of the tail. What has primarily been implied here is that the body cross section varies so gradually and its cross-sectional shape is so well-rounded (no sharp edges) that the cross flow remains attached to the body, leaving no vortex sheet until the tail-end section is reached. This situation may represent, quite accurately, several wide classes of aquatic animals of anguilliform, such as the eel *Anquilla vulgaris*, eel-like fishes of the order Heteromi, ribbon fishes of the order Allotriognathi, and members of the order Anacanthini, like cod and similar fishes, as have been discussed in a review article by Lighthill (1969).

There also exist other classes of fishes of more advanced orders which are known to be strong active swimmers, capable of putting on impressive performances. This group of fish orders contains members of the order Isospondyli (such as salmon, trout, etc.) and those of the large order Ostariophysi, to which belong most of the successful freshwater fishes. As an indication of the over-all performance, the swimming speeds U, when expressed in units of body length/sec (or $n = U/l$), of pike (*Esox lucius*), salmon (*Salmo salar*), trout (*Salmo gairdneri*), dace (*Leuciscus leuciscus*), and herring (*Culpea harengus*) have been observed to vary from $n = 6$/sec to 12/sec (see Hertel, 1963, Chap. G; Gray, 1968, Chap. 3). Speeds considerably greater than 10/sec, as high as 20/sec, have also been recorded by Walters and Fiersteine (1964) for tunny (*Thunnas albacores*) and wahoo in open water. By and large, the slenderness parameter, δ, of this general group of fishes is moderate to small (from about 0.4 down to 0.18 based on body depth, with dorsal and ventral fins extended). Furthermore, they have developed transverse compression [or flattening of the cross section broadwise to the lateral (side-to-side) undulations] which enhances the virtual-mass effect to improve propulsive efficiency. As a general feature, the transverse crosssections of these fishes have rather rounded edges anterior to the section of maximum depth, changing to a more-or-less lenticular shape with increasing transverse compression, to fairly pointed edges at the rear part of the body in which may be found a great variety of dorsal, ventral, pectoral, anal, and possibly other smaller fins, to be followed eventually by the caudal fin.

As for the detailed fin shapes and locations, there are perhaps as many configurations as the number of species. However, it may be useful for hydromechanical reasons to classify crudely the dorsal and ventral fins into two main types: the elongated "ribbon fin" and the triangular or trapezoidal "sail-shaped fin." Some preeminent families of fast fishes that have developed conspicuous ribbon fins are dolphin fish *Coryphaena hippurus*, which have a dorsal ribbon fin all the way from head to caudal peduncle and a shorter ventro-anal one, yellowtail *Seriola quinqueradiata*, atka mackerel *Pleurogrammus azonus*, and porgies and breams (see Fig. 2).

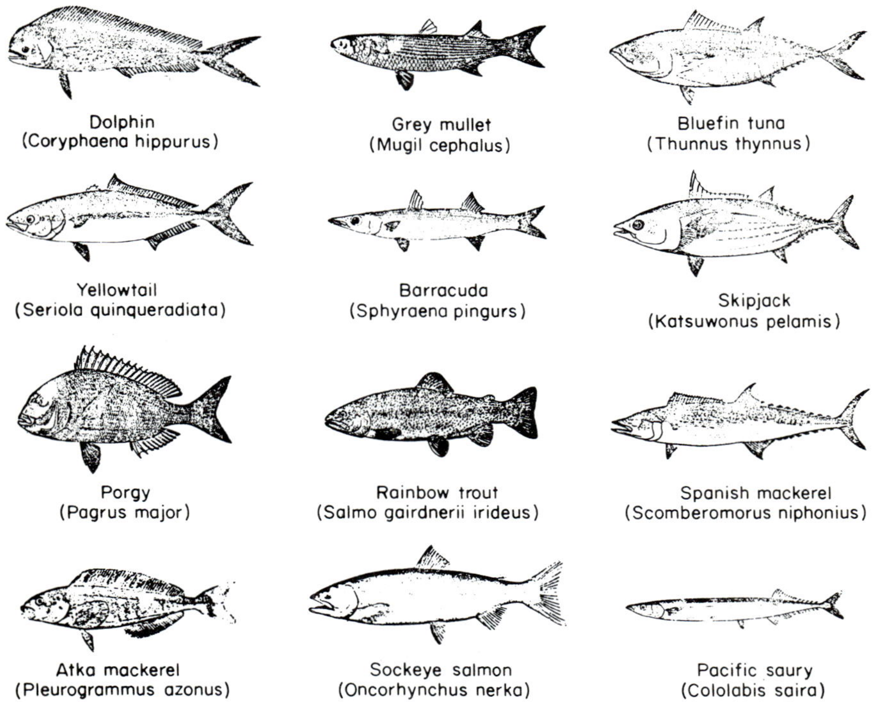

FIG. 2. A hydromechanical classification of fish fins. In the left column are examples of "ribbon-type" ventral and dorsal fins; the central column represents some triangular or trapezoidal "sail-shaped" fins; the right column contains specimens having a combination of both types of fins, and a series of auxiliary fins. The caudal fins of dolphin, tuna, yellowtail, skipjack and mackerel, when fully expanded, are some variants of the "lunate tail" as characterized by the crescent shape of rather large aspect-ratio.

Examples of fast fishes having sail-shaped dorsal and ventral fins are mullet *Mugil cephalus*, which is very fast and migrating widely, barracuda *Sphyraena pinguis*, trout, salmon, and herring. Also, many families of fishes have both types of fins, notably bluefin tuna *Thunnus thynnus*, skipjack, and some mackerels. Furthermore, several fishes (wahoo, tuna, skipjack, mackerel, and saury) have behind the main fins a series of small fins, extruding out with alternate angles of yaw, which are probably developed for boundary layer control.

In most of these cases the trailing side edges, in some parts terminating with extruded spiny ventro-dorsal fins, are sharp enough to have an oscillating vortex sheet shed from the body in an undulatory swimming motion. In order to determine the effect of shedding of a vortex sheet on the swimming performance, Lighthill (1970) examined the case of discrete dorsal (or ventral) fins with unswept straight trailing edges, assuming that the body sections behind the dorsal and ventral fins are smoothly shaped and do not shed any further vorticity. This is reasonable for many fishes including the freshwater Ostariophysi, such as dace [for its shape, see Bainbridge (1963)]. When a series of such discrete fins is present, the vortex sheets filling the gaps between them are shown to maintain continuity rather effectively, avoiding thrust reduction and permitting a slight decrease in drag. The ribbon-fin-type problem has been investigated by Wu (1971c), the sharp trailing side edges of these ribbon fins being assumed to have a gradual change in slope. When both types of fins are present, these two theories can be combined to form an all-embracing theory.

From the physical point of view, the main difference between the fins with straight trailing edges and those with slanted ones is that the strength of the vortex sheet shed from a fin of the former type is known when the flow is determined up to that trailing-edge station, whereas in the latter case the vortex sheet strength remains unknown as a part of the problem. However, it has been found that in both cases the resultant propulsive thrust depends not only on the virtual mass of the tail-end section of the caudal fin, but also on an integral effect of variations of the virtual mass along the entire body segment containing the trailing vortex sheets until the caudal fin is reached, and that this latter effect can greatly enhance the thrust-making. Both the original slender-fish theory of Lighthill (1960a) and Wu's extension can be conveniently discussed together, as will be presented below. The new theory of Lighthill (1970) on vortex sheets shed by fins with straight trailing edges will be included in a later section. Also on this subject Gadd (1952) considered a waving long thin body, and later (1963) discussed the effects of vortex shedding and flow separation related to the leg action in human swimming.

1. *Slender Fish with Ribbon-Shaped Fins*

As depicted in Fig. 3, a Cartesian coordinate system (x, y, z) is fixed at the mean position of the body, with the stretched body plan form S_b lying in the $z = 0$ plane, the free stream velocity being U, pointing in the x direction. The body configuration will be assumed to be symmetrical with respect to y, so that the side edges are given by $y = \pm b(x)$, and to have its thickness symmetrical in z. [The more

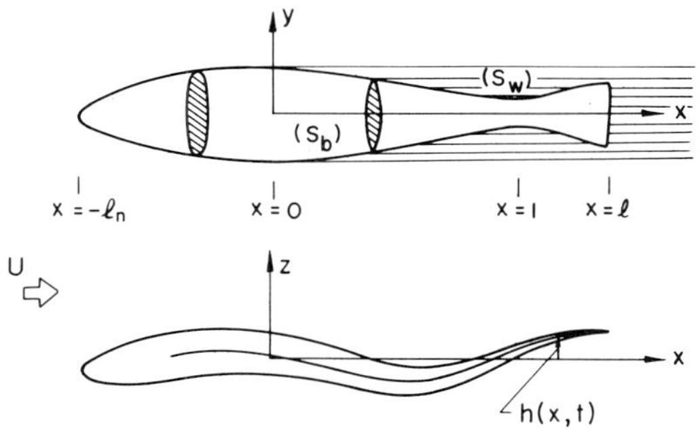

FIG. 3. Slender fish with sharp trailing edges (in $0 < x < 1$), from which a vortex sheet is shed to form the wake plane form S_w.

general case of asymmetrical shape, in y, has no effect on the final result. See Wu (1971c).] The origin is at the midpoint of the maximum span so that $y = b(0) = b_0$ is the maximum of $b(x)$, and the body extends from its nose at $x = -l_n$ to its tail end at $x = l$. The tail-base neck, just in front of the caudal fin, is at $x = l_m$, which will be normalized to be of unit length. The body motion is given by (assumed to be independent of y)

$$z = h(x, t) \qquad (x, y \in S_b). \tag{2.15}$$

The cross flow velocity $V(x, t)$ at the body plan form S_b, now in the z direction, is again given by (2.2) (with z in (2.2) changed to y).

The "unsteady elongated-body theory" [a generalized version of the classical slender-body theory; for an earlier review of the classical theory, see Lighthill (1960b)] for such body shapes and movements as are considered here has been under critical study by Lighthill (1970). Provided that the longitudinal variation of cross-sectional properties is slight on a scale of the maximum cross-sectional depth (or span as it is usually called in the "wing theory") b_0, and *if* the wavelength of significant harmonic components of the undulatory motions [that may be included in the general expression $h(x, t)$] exceeds $5b_0$, the basic approach of the classical theory is applicable. That this general theory is still valid for the case when the body has fins with rather unswept trailing edges which give abrupt changes in cross-sectional dimensions is because the vortex sheets shed from these trailing edges provide a "smoothing effect" to maintain continuity of cross-flow properties.

In the framework of this generalized slender-body theory, the entire flow has two components: one is the steady flow around the stretched straight body, which gives no resultant force or moment for symmetric bodies, and the other is the cross flow due to the displacement $h(x, t)$, which generates in the cross flow plane the velocity $V(x, t)$ at the body plan form S_b. This latter component alone determines the lift, moment, thrust, and other relevant quantities. The cross flow momentum is $\rho A(x) V(x, t)$, where ρ is the density of the fluid and $\rho A(x)$ is the "virtual mass" of the longitudinal segment of body per unit length at x, for motions in the z direction. Thus, $A(x)$ has the dimension of area and can be determined once the cross-sectional shape is given. For example, if the body thickness is zero, or when the cross section is an ellipse with minor axis in the z direction, $A(x) = \pi b^2(x)$. For various fishes whose cross sections are such that a fraction of the total local depth $2b(x)$ is occupied by the fish body, which may be approximated by an ellipse or a convex lens, and the remainder by dorsal and/or ventral fins, the virtual mass $\rho A(x)$ varies little from $\rho \pi b^2$ (see Lighthill, 1970, Fig. 1). [For the general formula of $A(x)$ applicable to slender bodies, see Wu (1971c), Appendix.]

In the front section ($-l_n < x < 0$) and the tail section* ($1 < x < l$), in which the side edges are leading edges and hence $dA/dx > 0$, the instantaneous lift (or the force in the z direction) per unit length of fish, $\mathscr{L}(x, t)$, is equal and opposite to the rate of change of the z component of momentum of the fluid passing station x; that is, for $-l_n < x < 0$ and $1 < x < l$,

$$\mathscr{L}(x, t) = -\rho \left(\frac{\partial}{\partial t} + U \frac{\partial}{\partial x}\right) [A(x)\, V(x, t)]. \tag{2.16a}$$

In the trailing-edge section ($0 < x < 1$) with sharp side edges from which a vortex sheet is continually shed into the wake, the instantaneous lift per unit length has been shown (see Wu, 1971c), under the assumption that the flow at side trailing edges satisfies the Kutta condition, to assume the value

$$\mathscr{L}(x, t) = -\rho A(x) \left(\frac{\partial}{\partial t} + U \frac{\partial}{\partial x}\right) V(x, t) \qquad (0 < x < 1), \tag{2.16b}$$

in which the virtual mass $\rho A(x)$ refers to the fish body cross section alone, not including the cross section of the vortex sheet as a part of an effective "virtual body." Thus the effect of vortex shedding is equivalent to treating $A(x)$ as if it is a constant in calculating the changes of the cross flow momentum. Alternatively, this effect may be considered to contribute a correction term $\rho V U\, dA/dx$ to the $\mathscr{L}$ based on the calculation without accounting for the effect of vortex shedding, i.e., the $\mathscr{L}$ given by (2.16a). We note that $dA/dx < 0$ in the trailing-edge section, $0 < x < 1$.

* For the tail section Eq. (2.16a) serves only as a crude approximation since the effect of the trailing vortex sheet on the flow around the caudal fin was largely neglected.

The total lift in the z direction is given by the integration of $\mathscr{L}$,

$$L(t) = -\rho \int_{-l_n}^{l} D[A(x)\, V(x, t)]\, dx + \rho U \int_0^1 V(x, t) \frac{dA}{dx}\, dx, \tag{2.17}$$

where

$$D \equiv \partial/\partial t + U(\partial/\partial x). \tag{2.18}$$

The moment of force about the origin is

$$M(t) = \rho \int_{-l_n}^{l} xD[A(x)\, V(x, t)]\, dx - \rho U \int_0^1 xV(x, t) \frac{dA}{dx}\, dx. \tag{2.19}$$

The power P expended by the fish in making a displacement $h(x, t)$ is equal to the rate of work done by the body against the reaction of the cross flow, $-\mathscr{L}(x, t)$, at the rate of the body transverse velocity h_t, or

$$\begin{aligned} P &= -\int_{-l_n}^{l} h_t \mathscr{L}(x, t)\, dx, \\ &= \rho \int_{-l_n}^{l} [D(AVh_t) - AVV_t]\, dx - \rho U \int_0^1 h_t V A_x\, dx, \\ &= \rho \frac{\partial}{\partial t} \int_{-l_n}^{l} AV \left(h_t - \frac{V}{2}\right) dx + \rho U [AVh_t]_{x=l} - \rho U \int_0^1 h_t V A_x\, dx, \end{aligned} \tag{2.20}$$

in which $A(x)$ is assumed to vanish at the nose, $A(-l_n) = 0$.

Since the kinetic energy of lateral fluid motion per unit length of a body is $\frac{1}{2}\rho AV^2$, the rate of imparting this energy is clearly equal to $D(\frac{1}{2}\rho AV^2)$ in the leading edge sections ($-l_n < x < 0$ and $1 < x < l$) where no vortex sheet is shed. In the trailing-edge section ($0 < x < 1$), the rate of shedding of the flow energy becomes $A(x)D(\frac{1}{2}\rho V^2)$ since the effect of vortex shedding is equivalent to regarding $A(x)$ as a constant. The total kinetic energy E imparted to the fluid in unit time is therefore given by

$$\begin{aligned} E &= \int_{-l_n}^{l} D(\tfrac{1}{2}\rho AV^2)\, dx - \int_0^1 \tfrac{1}{2}\rho V^2 U A_x\, dx, \\ &= \frac{1}{2}\rho \frac{\partial}{\partial t} \int_{-l_n}^{l} AV^2\, dx + \frac{1}{2}\rho U [AV^2]_{x=l} - \frac{1}{2}\rho U \int_0^1 V^2 A_x\, dx. \end{aligned} \tag{2.21}$$

Finally, by (2.6) the instantaneous thrust is $T = (P - E)/U$, or

$$T = -\rho \frac{\partial}{\partial t} \int_{-l_n}^{l} AVh_x \, dx + \rho \left[A \left(\frac{1}{2} V^2 - UVh_x \right) \right]_{x=l} - \rho \int_0^1 \left(\frac{1}{2} V^2 - UVh_x \right) \frac{dA}{dx} \, dx. \tag{2.22}$$

The mean values of the thrust, power required, and energy loss, obtained by averaging T, P, and E over a long time for periodic or almost-periodic motions are

$$\bar{T} = \frac{1}{2} \rho A(l) \overline{[h_t^2 - U^2 h_x^2]}_{x=l} - \frac{1}{2} \rho \int_0^1 \overline{(h_t^2 - U^2 h_x^2)} \frac{dA}{dx} \, dx, \tag{2.23}$$

$$\bar{P} = \rho U A(l) \overline{[h_t V]}_{x=l} - \rho U \int_0^1 \overline{h_t V} \frac{dA}{dx} \, dx, \tag{2.24}$$

$$\bar{E} = \frac{1}{2} \rho U A(l) \overline{[V^2]}_{x=l} - \frac{1}{2} \rho U \int_0^1 \overline{V^2} \frac{dA}{dx} \, dx, \tag{2.25}$$

in which the bar denotes the time average. The first terms on the right-hand side of these three equations, all involving quantities at the tail-end section only, were first given by Lighthill (1960a), whereas the second terms in the integral form, representing contribution due to the effect of vortex shedding along the entire trailing side edges, are according to Wu (1971c).

The physical significance of these results is clear. The tail-end contribution to P is equal to the product of the body lateral velocity h_t with the rate of shedding of lateral flow momentum $(\rho AV)U$ at the tail trailing edge. It is also clear that according to the previous interpretation of the effect of vortex sheet formation, the rate of shedding of lateral momentum from the trailing side edges between sections $A(x)$ and $A(x + dx) = A(x) + dA$ is $(-\rho V \, dA)U$ (the negative sign being taken since $dA = (dA/dx) \, dx < 0$ for $0 < x < 1$), and hence its product with the local lateral velocity h_t of body cross section integrated over the entire trailing side edges provides the total contribution due to the effect of vortex shedding. The first term on the right of (2.20), after its integrand is recast, using (2.2), as $(\frac{1}{2}\rho AV^2 - U\rho AVh_x)$, can be interpreted, after Lighthill (1970), to represent the rate of work involving an exchange of energy of water that has not yet reached the wake. In this integrand the first term represents the energy of the cross flow, $\frac{1}{2}\rho AV^2$, per unit length; the second term stands for the loss by convection, at velocity U, of the x component flow momentum ρAVh_x, again per unit length. The instantaneous rate of change of these two quantities of the water ahead of the trailing edge must be supplied by part of the power expended by the fish, although their mean rate of change vanishes in undulatory motions.

As for the kinetic energy imparted by the fish to the surrounding water, as given by (2.21), the tail-end contribution to E is equal to the shedding of the kinetic energy $\frac{1}{2}\rho AV^2$ (per unit length of fish) at rate U, whereas the interpretation of the contribution due to the vortex sheet formation ahead of the tail end is entirely parallel to that for P. The first term in (2.21) represents the rate of change of the flow energy in the water ahead of the tail end.

Although the instantaneous thrust T in (2.22) was deduced from the energy equation (2.6), a direct physical interpretation of the result is possible. We note, after Lighthill (1970), that the first term on the right in (2.22) is the rate of change of the x component of the water momentum in the region anterior to a plane Π through the tail trailing edge perpendicular to the x direction. The term $-\rho AUVh_x$ in square brackets again accounts for the loss by convection at rate U of the x component momentum ρAVh_x, as already noted earlier. Based on pressure considerations and further implemented by vorticity considerations, Lighthill (1970) identified the term $\frac{1}{2}\rho AV^2$ in square brackets as the resultant pressure force acting over the control plane Π by the water behind Π. Finally, the last term in (2.22) represents the contribution by the vortex sheets shed from the side trailing edges, whose physical significance can be interpreted in a manner similar to that for P and E. This contribution now contains a part due to the transport of momentum, $U\rho Vh_xA_x$, into the side vortex sheet, and another part, given by $-\frac{1}{2}\rho V^2A_x$, due to the local pressure force associated with the rate of change of momentum in the side vortex wake, both parts being based on unit length of the fish. Lighthill's interpretation of the term $\frac{1}{2}\rho AV^2$ as the net pressure force acting over a transverse plane has a far-reaching significance. With this verification the concept of bulk momentum and energy is then complete, which enables Lighthill (1971) to develop a more general theory for large amplitude motions. Such a theory is desirable because amplitudes used by fishes swimming at their top speed and in the accelerating stage are in reality quite large; without it little can be done to analyze turning maneuvers.

It may be noted that wherever the time-average quantities in the integrands of (2.23) and (2.24) are positive, the local contribution to $\overline{T}$ and $\overline{P}$ due to the effect of vortex shedding is also positive since $dA/dx < 0$ for $0 < x < 1$.

Another point of significance is that by (2.25), $\overline{E} \geqslant 0$ in this case, as should be expected on the physical grounds stated before. When $\bar{T}$ is positive, the hydrodynamic efficiency of swimming will be defined as

$$\eta = U\bar{T}/\bar{P} = 1 - (\bar{E}/\bar{P}). \tag{2.26}$$

Lighthill reasoned that high efficiency can be achieved if $V \ll h_t$ but V and h_t must be positively correlated, i.e., $\overline{h_t V} > 0$ (for otherwise, $\bar{P} < 0$; hence $\bar{T}$ is also negative by (2.9)). The problem of optimum shape for maximum efficiency under certain side conditions will be further discussed in Section IV. In concluding this section, we mention two specific examples given by Lighthill (1960a). When the body motion is a standing wave, the efficiency η is always <0.5. If the body motion is a progressive wave with phase velocity c towards the tail, an estimate shows that η can be as high as 0.9 at $c = 1.25U$ provided the variation of the wave amplitude is small at the tail. These general features still remain true in the presence of vortex sheet formation, as will be shown in Section IV.A.

2. *Slender Fish with Sail-Shaped Fins*

In contrast with the ribbon-type fins, there exists, as discussed earlier in this section, another type, called "sail-shaped fins," which can be characterized by their two salient features that the main dorsal (or ventral) fin terminates with an almost unswept (or nearly vertical) trailing edge; and that there is a large gap between the main dorsal fin and the caudal fin. Several notable examples have been given earlier, a few of them being shown in Fig. 2. Analysis of their motion must take into account the extensive region of vortex sheet between the dorsal and caudal fins, where the differences, in both phase and amplitude, between the motion of the body and of the vortex sheet, at the same longitudinal station, may be sufficiently important to modify both thrust and power required. Lighthill (1970) treated this problem with the assumption that the body sections behind the dorsal and ventral fins are smoothly shaped and do not shed any further vorticity, although the depth of these body sections may continue to decrease gradually towards the caudal peduncle.

To see how the vortex sheet shed from such a fin is determined, we suppose that the unswept trailing edge of the fin is at $x = x_F$, where the body cross section, including the fin, is a thin strip of depth $2\tilde{b}$. The velocity potential φ for motion of such a strip with velocity $V(x_F, t)$ in the z direction takes at $z = \pm 0$ values

$$\varphi^{\pm}(x_F, y, t) = \pm V(x_F, t)(\tilde{b}^2 - y^2)^{1/2} \qquad (|y| < \tilde{b}). \tag{2.27}$$

This velocity potential, which already exhibits an elliptical distribution, generates at the trailing edge, due to a discontinuity in velocity there, a vortex sheet to trail the fin. The condition of continuity of pressure across the vortex sheet, however, requires that in the body frame of reference, $\varphi_t^{\pm} + U\varphi_x^{\pm} = 0$ within the framework of the linearized theory. Therefore, for (x, y) lying in the vortex wake S_W,

$$\varphi^{\pm}(x, y, t) = \varphi^{\pm}(x_F, y, t - (x - x_F)/U), \tag{2.28}$$

where $\varphi^{\pm}(x_F, y, t)$ is given by (2.27). The velocity potentials $\varphi^{\pm}$ on two sides of S_W are thus determined everywhere on S_W. The vorticity vector $\boldsymbol{\gamma}$ in the thin layer of this vortex sheet has only two components, $\boldsymbol{\gamma} = (\gamma_1, \gamma_2, 0)$, with its x and y components given by Lighthill (1970) and Wu (1971c) [for (x, y) lying in S_W] as

$$\gamma_1(x, y, t) = -2\varphi_y^{+}(x, y, t), \qquad \gamma_2(x, y, t) = -(2/U)\,\varphi_t^{+}(x, y, t). \tag{2.29}$$

Equation (2.27), and its use in (2.28) and (2.29), would give a good approximation if the body were thin relative to its depth; corresponding results for thicker bodies can, however, be calculated by complex-variable theory.

With the strength of the vortex sheet determined, the velocity potential of the cross flow at any station $x > x_F$ can be calculated, following Lighthill (1970), by decomposing the potential into two parts, $\varphi + \tilde{\varphi}$, which satisfy separately the boundary conditions

$$\partial\varphi/\partial n = V(x, t)\, n_z \quad \text{on} \quad C_x, \qquad \varphi^{\pm} = 0 \quad \text{on} \quad S_W; \tag{2.30a}$$

$$\partial\tilde{\varphi}/\partial n = 0 \quad \text{on} \quad C_x, \qquad \tilde{\varphi}^{\pm} = \text{given on} \quad S_W, \tag{2.30b}$$

where n_z is the z component of a unit normal $\mathbf{n}$ to the body surface C_x. The given value of $\tilde{\varphi}^{\pm}$, from (2.27) and (2.28), is proportional to

$$V_F(x, t) = V(x_F, t - (x - x_F)/U). \tag{2.31}$$

The z component of the cross-flow momentum is therefore

$$m(x)\, V(x, t) + \tilde{m}(x)\, V_F(x, t); \tag{2.32a}$$

and its kinetic energy (since φ, $\tilde{\varphi}$ are orthogonal in a space with kinetic energy as norm) is

$$\tfrac{1}{2}m(x)\, V^2 + \tfrac{1}{2}\tilde{m}(x)\, V_F^2, \tag{2.32b}$$

where $m(x) = \rho A(x)$, $\tilde{m}(x) = \rho\tilde{A}(x)$ are virtual masses, per unit length, associated respectively with φ and $\tilde{\varphi}$. Lighthill (1970) gave (in the present notation)

$$m(x) = \rho\pi b^2, \qquad \tilde{m}(x) = \rho\pi(\tilde{b}^2 - b^2). \tag{2.32c}$$

The z component of flow force acting on the body per unit length is therefore

$$\mathscr{L} = -D[m(x)\, V(x, t)] - U\tilde{m}'(x)\, V_F(x, t). \tag{2.33}$$

The corresponding power expended by the fish is given by Lighthill [1970, Eq. (26)] as (again partly converted to the present notation)

$$P = \frac{\partial}{\partial t} \int_{-l_n}^{l} \left(mVh_t - \frac{1}{2} mV^2 \right) dx + U[mVh_t]_{x=l} + U\tilde{m}(x_F)\, V(x_F\,, t)\, h_t(x_F\,, t)$$

$$+ U \int_{x_F}^{l} \tilde{m}'(x)\, V_F(x, t)\, h_t(x, t)\, dx. \tag{2.34}$$

Lighthill further reasoned that the effect of the vortex sheet, as given by the last two terms of (2.34) can improve the performance by increasing the total power output without an appreciable increase in the energy wasted.

At this stage it is of interest to apply the theory discussed in Section II, B, 1 to the present case and to compare it with Lighthill's theory. First, we note that (2.16b), after being recast as

$$\mathscr{L} = -D[m(x)\, V(x, t)] + Um'(x)\, V(x, t) \qquad (x > x_F), \tag{2.16b$'$}$$

differs from (2.33) only in that V in the last term of (2.16b)$'$ is the local body section velocity whereas the corresponding V_F in (2.33) is the velocity of the trailing-edge section (at $x = x_F$) evaluated at the retarded time (see (2.31); also note that $m'(x) = \rho A'(x) = -\tilde{m}'(x)$). For the integrated quantities, we write for the "sail-shaped fin" the value of $A(x)$ with a jump (represented by the Heaviside step function) at $x = x_F$ as

$$A(x) = A_c(x) - (\Delta A_F)\, H(x - x_F), \qquad \Delta A_F = A(x_F^-) - A(x_F^+), \tag{2.35}$$

where $A_c(x)$ is the continuous part of $A(x)$. That $\rho A(x)$, the virtual mass per unit length of the body alone, may admit jumps, such as above, is still valid because the vortex sheets shed from these jumps produce a far greater continuity in the cross flow than might be suggested by the geometry. With dA from (2.35) regarded as in the Stieltjes' sense, or as having a delta function, (2.20) yields

$$P = \rho \frac{\partial}{\partial t} \int_{-l_n}^{l} AV \left(h_t - \frac{1}{2} V \right) dx + \rho U[AVh_t]_{x=l} + \rho U(\Delta A_F)[Vh_t]_{x=x_F}$$

$$- \rho U \int_{x_F}^{1} V(x, t)\, h_t(x, t) \frac{dA_c}{dx}\, dx. \tag{2.36}$$

Noting that $\tilde{m}(x_F) = \rho(\Delta A_F)$, the only difference between (2.34) and (2.36) is in the last integrals where V_F in (2.34) is again replaced by V in (2.36) and where the flow at the caudal-fin leading edge is treated differently. This difference is a result of the different cases being considered; the body behind the main fin continues to shed a vortex sheet in one, but not in the other. Both cases may arise in reality, thus dictating which theory to apply. In practice, however, these two theories come to much the same thing, because (i) the wave velocity of undulatory body motion is only slightly greater than U in efficient swimming (see Section IV, A), thus diminishing the difference between V and V_F, and (ii) the extra term for the caudal-fin leading edge [see the last term of (2.34)] has in effect only a small mean value owing to phase shifts.

C. General Theory of Two-Dimensional Swimming Motion

Although the flow around swimming fish is certainly three-dimensional, the theory of two-dimensional swimming motion is still of considerable interest since it can be applied to estimate the propulsion of a tail of large aspect ratio of some species of cetaceans and pelagic fishes, or even the propulsion of flapping wings of migrating birds. The main drawback of two-dimensional theory is of course its inability to provide significant modifications to the two-dimensional flow, such as the unsteady downwash induced at the lifting surface by the finite-aspect-ratio effects. Its significant strength, however, outweighs any crude three-dimensional theory in the capacity of providing an accurate account of the singular leading-edge thrust, an aspect which turns out to be so important in undulatory propulsion (see Section IV, C) as to mark the crucial difference between whether or not the assumed all-attached flow configuration, particularly near the leading edge, can be physically realized.

Basically, the theory of two-dimensional swimming motion is a further development of the oscillating airfoil theory which dates back to Theodorson (1934), Küssner and Schwarz (1940), and Schwarz (1940). The basic mechanism of swimming propulsion has been elucidated by von Kármán and Burgers (1943) for the simple case of a rigid plate in transverse oscillation. Further studies of drag on oscillating airfoils in incompressible flow have been given by Jones (1957) with the aid of the Mellin transform, and by Veltkamp (1958) using the velocity potential in terms of a Riemann–Hilbert problem. With the specific purpose of application to swimming propulsion in mind, Wu (1961) treated the waving motion of a two-dimensional flexible plate in an incompressible fluid, using the Fourier series method for the acceleration potential. Subsequently, Siekmann (1962, 1963) considered the same problem of the waving plate, using a distribution of vortices along the plate and its wake, and yielding the same result as that of Wu. Using a mechanical model, Kelly (1961) carried out a series of experiments and demonstrated that the linearized theory is found to be in good agreement with the experimental results. This steady swimming problem has since been extended by Wu (1962) to the more general case of time-dependent forward velocity. The effect of finite body thickness has been discussed by Uldrick and Siekmann (1964), but classical aerodynamics already tells us that the effect of airfoil thickness based on the linearized theory has no influence upon the lift on the one hand, and that on the other, this effect based on the exact inviscid flow theory, such as the Joukowsky airfoil, is at variance with the experimental results of lift, a fact generally attributed to the boundary layer effect.

Another point that should be mentioned concerns the accuracy of the quasi-steady theory. The trailing vortex sheet behind an oscillating airfoil has a significant effect on the pressure distribution at the airfoil, particularly near the trailing edge and at high reduced frequencies, as demonstrated by von Kármán and Sears (1938). Consequently, the quasi-steady theory becomes increasingly poorer the higher the frequency of oscillation. Smith and Stone (1961) also discussed the flow around an oscillating flexible plate, but they neglected the effect of the vortex wake. As a result, their theory considerably overestimates the thrust, particularly at high frequencies, as noted by Kelly (1960) and again by Pao and Siekmann (1964).

Practical applications of the theory may require, on occasion, that the forward velocity be arbitrary. For instance, this is necessary for accelerating swim. In the flapping flight of birds, it is more realistic to consider the forward velocity of the wing to be nonuniform since the wing movements, in up-and-down strokes, are known to have very large backward-and-forward components (see Gray, 1968). Furthermore, such a general theory may also have applications to artificial propulsive devices such as the vertical-axis propeller (Voith–Schneider type, whose blades move relative to the fluid with variable velocity and pitch angle), and may be particularly useful in the control theory for hydrofoils in waves and other devices when the transient behavior is of essence. With these prospects in view, we present here the inviscid flow theory of swimming of a two-dimensional flexible plate at variable forward speed in terms of Prandtl's acceleration potential formulated as a Riemann-Hilbert problem. (For the omitted details, see Wu, 1962, 1971a.)

Consider the incompressible, inviscid, plane flow past a flexible plate of zero thickness, performing a waving motion of the general form

$$y = h(x, t) \qquad (-1 < x < 1, \quad t > 0), \tag{2.37}$$

h being again an arbitrary continuous function of x, t and assumed to be always small. The motion starts at $t = 0$ from a uniform state, and the free stream velocity $U(t)$ may depend on t. Let u and v again denote, respectively, the x and y components of the perturbation velocity. We introduce the Prandtl acceleration potential

$$\phi(x, y, t) = (p_\infty - p)/\rho \tag{2.38}$$

where p_∞ is the pressure at infinity. In the linear theory of this incompressible irrotational flow, p, and hence also ϕ, is a harmonic function of x, y for all t. A harmonic function $\psi(x, y, t)$ conjugate to ϕ may be defined by $\phi_x = \psi_y$, $\phi_y = -\psi_x$. Hence the complex acceleration potential $f = \phi + i\psi$ and the complex velocity $w = u - iv$ are analytic functions of the complex variable $z = x + iy$ for all real t. (We borrow the notation w and z for this different purpose in this section.) f and w are related to the linearized equation of motion

$$\partial f/\partial z = \partial w/\partial t + U(t)\, \partial w/\partial z. \tag{2.39}$$

The linearized boundary conditions are

$$v(x, 0\pm, t) \equiv V(x, t) = h_t + Uh_x \qquad -1 < x < 1, \tag{2.40}$$

$$-\frac{\partial \psi}{\partial x} = \left(\frac{\partial}{\partial t} + U\frac{\partial}{\partial x}\right) V \qquad y = 0\pm,\ |x| < 1 \tag{2.41}$$

$$\phi(x, 0, t) = 0 \qquad |x| > 1, \tag{2.42}$$

$$|f(1, t)| < \infty \qquad \text{for all } t, \tag{2.43}$$

$$f(z, t) \to 0 \quad \text{as} \quad |z| \to \infty; \qquad w(z, t) \to 0 \quad \text{as} \quad z \to -\infty. \tag{2.44}$$

Here, condition (2.41) follows from the imaginary part of (2.39) and condition (2.40); condition (2.42) follows from the fact that ψ is even, and hence ϕ is odd in y; (2.43) is the Kutta condition for the flow at the trailing edge $z = 1$. Condition (2.44) for w may also be specified as $|z| \to \infty$, $|\arg z| > 0$, i.e., as $z \to \infty$ in the region excluding the trailing vortex sheet.

A remark is in order here concerning the validity of the Kutta-Joukowsky hypothesis. Greidanus *et al.* (1952) measured the oscillatory forces on two-dimensional airfoils with the reduced frequency (based on half-chord) σ up to $\sigma = 1$. Ransleben and Abramson (1962) measured the oscillatory lift and moment on hydrofoils of aspect ratio 5, fully submerged in water, up to high values of σ. These experimental results of amplitudes of force and moment fall below the linearized theory for $\sigma \gtrsim 0.5$ in the two-dimensional case and for $\sigma \gtrsim 1$ in the hydrofoil case. These authors attributed the discrepancy to a lag in fulfilling the Kutta condition at the trailing edge and/or to an appreciable lateral displacement of the vortex sheet. However, some uncertainties in the experimental procedures have been observed by Ashley *et al.* (1965) that may weaken the explanations suggested. In view of the fact that the reduced frequency σ used in the swimming of cetaceans and fishes with tails of large aspect ratio and in the flapping flight of birds is considerably smaller than 1, we shall uphold the Kutta condition as valid.

For the case of variable $U(t)$ it is convenient to make use of the Laplace transform method with respect to the new variable

$$\tau(t) = \int_0^t U(t)\, dt \qquad (t > 0). \tag{2.45}$$

The inverse function $t = t(\tau)$ is assumed to be unique so that $U = U(t(\tau))$ is a one-valued function of τ, this being the case so long as the swimming proceeds in one direction. Regarding w and f as functions of z and τ, (2.39) becomes

$$\partial F/\partial z = \partial w/\partial \tau + \partial w/\partial z, \tag{2.46}$$

where

$$F(z, \tau) = f(z, \tau)/U(\tau) = \Phi(x, y, \tau) + i\Psi(x, y, \tau). \tag{2.47}$$

Application of the Laplace transform

$$\tilde{F}(z, s) = \int_0^\infty e^{-s\tau} F(z, \tau)\, d\tau \qquad (\operatorname{Re} s > 0) \tag{2.48}$$

to (2.46), under zero initial conditions, yields

$$d\tilde{F}/dz = (d/dz + s)\tilde{w}. \tag{2.49}$$

Integrating (2.49) from $z = -\infty$ under conditions (2.44), and expressing $\tilde{F}$ in terms of $\tilde{w}$, and vice versa, we can apply condition (2.40) to obtain

$$\tilde{\Psi}(x, 0\pm, s) = \tilde{\Psi}_1(x, s) + \tilde{A}_0(s) \qquad (|\, x \,| < 1), \tag{2.50a}$$

where

$$\tilde{\Psi}_1(x, s) = -\left(\frac{\partial}{\partial x} + s\right) \int_{-1}^{x} \tilde{V}(x_1, s)\, dx_1 \qquad (|\, x \,| < 1), \tag{2.50b}$$

$$\tilde{A}_0(s) = -s \int_{-\infty}^{-1} \tilde{v}(x, 0\pm, s)\, dx = s \int_{-\infty}^{-1} e^{s(x+1)} \tilde{\Psi}(x, 0, s)\, dx. \tag{2.50c}$$

Thus $\tilde{\Psi}$ is known for $|x| < 1$ except for an additive constant term $\tilde{A}_0(s)$. Furthermore, by (2.42) and (2.47),

$$\tilde{\Phi}(x, 0\pm, s) = \operatorname{Re} \tilde{F}(x \pm i0, s) = 0 \qquad (|x| > 1). \tag{2.51}$$

Conditions (2.50) and (2.51) together with (2.43) and (2.44) constitute a Riemann–Hilbert problem for $\tilde{F} = \tilde{\Phi} + i\tilde{\Psi}$, which can be readily solved (for the general method, see, e.g., Muskhelishvili, 1953, pp. 235–238), giving, after taking the inverse transform, the solution of f as

$$f(z, t) = iU(t)\left[A_0(\tau) - \frac{1}{2}a_0(\tau)\left(\frac{z-1}{z+1}\right)^{1/2}\right] + \frac{1}{\pi i}\int_{-1}^{1}\left(\frac{z^2-1}{1-\xi^2}\right)^{1/2}\frac{\psi_1(\xi, t)}{\xi - z}\,d\xi, \tag{2.52}$$

$$a_0(\tau) = b_1(\tau) - \int_0^{\tau}[b_0(\tau') + b_1(\tau')]\,H(\tau - \tau')\,d\tau', \tag{2.53}$$

$$b_n(\tau) = \frac{2}{\pi}\int_0^{\pi} V(x, t(\tau))\cos n\theta\,d\theta \qquad (x = \cos\theta,\ n = 0, 1, 2, \ldots,) \tag{2.54}$$

$$H(\tau) = \frac{1}{2\pi i}\int_{\epsilon - i\infty}^{\epsilon + i\infty} e^{s\tau}\frac{K_1(s)}{K_0(s) + K_1(s)}\,ds \qquad \epsilon > 0, \tag{2.55}$$

$$\psi_1(x, t) = -\left(\frac{\partial}{\partial t} + U\frac{\partial}{\partial x}\right)\int_{-1}^{x} V(\xi, t)\,d\xi, \tag{2.56}$$

$$A_0(\tau) = \frac{1}{2}a_0(\tau) - \frac{1}{\pi U(t)}\int_{-1}^{1}\psi_1(\xi, t)(1 - \xi^2)^{-1/2}\,d\xi. \tag{2.57}$$

In (2.52) the function $(z-1)^{1/2}(z+1)^{-1/2}$ is defined with a branch cut from $z = -1$ to $z = 1$ so that this function tends to 1 as $|z| \to \infty$. In (2.55), K_0, K_1 are the modified Bessel functions of the second kind.

In particular, the value of ϕ on the body surface is given by

$$\phi^+(x, t) = \phi(x, 0+, t) = -\phi(x, 0-, t) = -\phi^-(x, t), \tag{2.58}$$

$$\phi^+(x, t) = \frac{1}{2}U(t)\,a_0(\tau)\left(\frac{1-x}{1+x}\right)^{1/2} + \frac{1}{\pi}\oint_{-1}^{1}\left(\frac{1-x^2}{1-\xi^2}\right)^{1/2}\frac{\psi_1(\xi, t)}{\xi - x}\,d\xi, \quad |x| < 1,$$

in which C over the integral sign denotes its Cauchy principal value. The first term in (2.58) gives the leading edge singularity, whereas the integral term is regular wherever ψ_1 is continuous. Furthermore, $a_0(\tau)$ is the only quantity in the solution that is influenced by the history [see (2.53) and (2.55)] and requires expression in terms of τ. The pressure difference across the plate is, by (2.38),

$$\Delta p = p^-(x, t) - p^+(x, t) = 2\rho\phi^+(x, t) \qquad (|x| < 1). \tag{2.59}$$

The lift L acting on the plate and the moment of force M about the midchord (positive in the nose-up sense) are readily obtained by straightforward integrations as

$$L = \int_{-1}^{1} \Delta p \, dx = \pi\rho \left\{ U(t)[a_0(\tau) - b_1(t)] - \frac{1}{2}\frac{d}{dt}[b_0(t) - b_2(t)] \right\}, \tag{2.60}$$

$$M = -\int_{-1}^{1} (\Delta p)x \, dx = \frac{\pi}{2}\rho \left\{ U(t)[a_0(\tau) + b_2(t)] + \frac{1}{4}\frac{d}{dt}[b_1(t) - b_3(t)] \right\}. \tag{2.61}$$

In calculating the thrust T and energy loss E we must also determine the leading edge suction T_s. From (2.46) it follows obviously that w has the same singularity strength at the leading edge as that of F, namely,

$$w(z, \tau) \sim a_0(\tau)[2(z+1)]^{-1/2} + O(1) \qquad \text{as} \quad |z+1| \to 0.$$

This singularity of w is known in aerodynamic theory to give rise to a leading edge suction (directed upstream)

$$T_s(t) = (\pi/8)\,\rho[a_0(\tau) + a_0{}^*(\tau)]^2, \tag{2.62}$$

in which $a_0{}^*$ stands for the complex conjugate of a_0.

Finally, the thrust T, power P, and energy loss E can be determined in terms of $h(x, t)$ by substituting (2.58), (2.59), and (2.62) in (2.3)–(2.5), with ϕ, h_t, and h_x in (2.3)–(2.5) all assuming their real values with respect to t. The final result is

$$T = (\pi/2)\,\rho\{[a_0 + b_0 - \dot\beta_0][a_0 - b_1 + \dot\beta_1] + [\dot\beta_0][\dot\beta_1]\}$$
$$- \frac{2}{\pi}\rho\frac{d}{dt}\int_{-1}^{1}\oint_{-1}^{1}\left(\frac{1-x^2}{1-\xi^2}\right)^{1/2}[V(x,t)][h(\xi,t)]\frac{d\xi\,dx}{\xi - x}, \tag{2.63}$$

$$E = (\pi/2)\,\rho U(t)[a_0 + b_0][b_1 - a_0]$$
$$+ \frac{1}{\pi}\rho\frac{d}{dt}\int_{-1}^{1}\oint_{-1}^{1}\left(\frac{1-x^2}{1-\xi^2}\right)^{1/2}[V(x,t)]\left[\int_{-1}^{\xi} V(\eta,t)\,d\eta\right]\frac{d\xi\,dx}{\xi - x}, \tag{2.64}$$

where $\dot\beta_n = d\beta_n(t)/dt$, and

$$\beta_n(t) = \frac{2}{\pi}\int_0^{\pi} h(x,t)\cos n\theta \, d\theta \qquad (x = \cos\theta,\ n = 0, 1, 2, \ldots.) \tag{2.65}$$

In (2.63) and (2.64), the quantities in square brackets are understood to assume their real values. The power P follows simply from (2.6); $P = TU + E$.

As an alternative expression the integrals in (2.63) and (2.64) can be converted into a series representation upon substituting for V and h their Fourier series, with their respective Fourier coefficients given by (2.54) and (2.65), into the integrand and carrying out the double integration, yielding

$$\frac{2T}{\pi\rho} = (a_0 + b_0 - \dot{\beta}_0)(a_0 - b_1 + \dot{\beta}_1) + \dot{\beta}_0\dot{\beta}_1 - \frac{d}{dt}\sum_{n=1}^{\infty} \beta_n(b_{n-1} - b_{n+1}), \tag{2.63'}$$

$$\frac{2E}{\pi\rho} = U(t)(a_0 + b_0)(b_1 - a_0) + \frac{d}{dt}\sum_{n=1}^{\infty} \frac{1}{4n}(b_{n-1} - b_{n+1})^2, \tag{2.64'}$$

in which the real value of each of a_0, b_n, β_n is implied.

Other flow quantities of related interest are the vorticity distribution along the trailing vortex sheet and the circulation around the plate. The strength of the vorticity in dx of the vortex sheet at a point $(x, 0)$ of the wake $(x > 1)$ is $\gamma(x, t)\,dx$, positive in the counterclockwise sense, where

$$\gamma(x, t) = -2u^+(x, t),$$

which, by virtue of (2.39) and (2.42), satisfies the equation

$$\gamma_t + U(t)\,\gamma_x = 0, \qquad \text{or} \qquad \gamma_\tau + \gamma_x = 0.$$

It therefore follows that

$$\gamma(x, \tau) = \gamma(1, \tau - x + 1) = -2u^+(1, \tau - x + 1) \qquad (x > 1). \tag{2.66}$$

By Kelvin's circulation theorem, the circulation (positive in the clockwise sense) around the plate, $\Gamma(t)$, varies at the rate

$$d\Gamma/dt = U\gamma(1, t),$$

which gives, upon integration,

$$\Gamma(\tau) = \int_0^\tau \gamma(1, \tau)\,d\tau = -2\int_0^t U(t)\,u^+(1, t)\,dt. \tag{2.67}$$

Thus, both $\gamma(x, \tau)$ and $\Gamma(\tau)$ are determined once $u^+(1, \tau)$ is found. This can be done by evaluating first its Laplace transform $\tilde{u}^+(1, s)$, which can be deduced from the solution (2.52) as

$$\tilde{u}^+(1, s) = (\pi/2)\,e^{-s}[\tilde{b}_0(s) + \tilde{b}_1(s)]/[K_0(s) + K_1(s)]. \tag{2.68}$$

$u^+(1, \tau)$ is then given by the inverse transform, which can be written as a convolution integral.

A particularly significant feature of the general solution is noteworthy at this point. If for all t

$$b_0(t) + b_1(t) = 0, \tag{2.69}$$

or equivalently, if $V(x, t)$ assumes the following Fourier expansion [see (2.54)]

$$V(x, t) = b_0(t)(\tfrac{1}{2} - \cos\theta) + \sum_{n=2}^{\infty} b_n(t)\cos n\theta \qquad (x = \cos\theta), \tag{2.70}$$

then, according to (2.68), $u^+(1, t) = 0$, and hence the circulation Γ remains constant (see (2.67)) and the plate sheds no vorticity into the wake regardless of what values $b_0(t)$, $b_2(t)$,... may take. Thus, there are an infinite number of such modes of unsteady motion that will leave no trailing vortex sheet. Furthermore, by (2.53), condition (2.69) also implies that $a_0(t) = b_1(t) = -b_0(t)$. It therefore follows from (2.63) and (2.64) that the first terms in the expression for T and E vanish, thus leaving T, E, and P to vary as the total time derivative of certain functions of t. Since no vortex sheet is shed under condition (2.69), those values of T, E, and P can arise only from the effect of the virtual masses of the fluid. In particular, when the motion is periodic in t, the average values of T, E, and P must all vanish under condition (2.69), implying no net transfer of momentum, no net energy loss, nor any net power being required over each cycle. This last property was observed earlier by Wu (1961) and will be seen later in Section IV of this article to play a particularly significant role in the problem of the optimum shape of $h(x, t)$ for the maximum swimming efficiency.

1. *Simple Harmonic Motion at Constant Forward Speed*

The motion is described by

$$h(x, t) = h_1(x)\, e^{j\omega t} \qquad (-1 \leqslant x \leqslant 1,\ t > 0), \tag{2.71}$$

and $h = 0$ for $t < 0$. $U =$ const.; hence $\tau = Ut$. Then

$$V(x, t) = V_1(x)\, e^{j\omega t} \qquad (-1 \leqslant x \leqslant 1,\ t > 0), \tag{2.72}$$

$$V_1(x) = U[(d/dx) + j\sigma]\, h_1(x), \qquad \sigma = \omega/U, \tag{2.73}$$

and $V = 0$ for $t < 0$. σ is the reduced frequency based on the half-chord, which is of unit length. The asymptotic behavior of the solution involves primarily the values of $a_0(t)$ for large and small t, since $a_0(t)$ is the only history-dependent term. The following results are of interest.

For large τ (actually large Ut/l, l being the half-chord which is being taken to be unity here),

$$a_0(t) = b_1 - (b_0 + b_1)\,\Theta(\sigma) + \frac{(b_0 + b_1)}{j\sigma\tau^2}\,e^{-j\omega t}\left\{1 + O\left(\frac{1}{\tau}\log\tau\right)\right\} \qquad (\tau = Ut/l \gg 1), \quad (2.74)$$

$$\Theta(\sigma) \equiv \frac{K_1(j\sigma)}{K_0(j\sigma) + K_1(j\sigma)} = \mathscr{F}(\sigma) + j\mathscr{G}(\sigma) \qquad (\sigma = \omega l/U). \quad (2.75)$$

$\Theta(\sigma)$ is the Theodorsen function, $\mathscr{F}$ and $\mathscr{G}$ being its real and imaginary parts, and σ is the reduced frequency based on the half-chord l. The last term in (2.74) diminishes monotonically like t^{-2} (noting that the harmonic time factors cancel out) as $t \to \infty$, yielding the steady-state solution of a_0 as

$$a_0(t) = b_1 - (b_0 + b_1)\,\Theta(\sigma). \quad (2.76)$$

For small τ, the asymptotic expansion of $a_0(t)$ is

$$a_0(t) = b_1 - \tfrac{1}{2}(b_0 + b_1)[1 + \tfrac{1}{4}\tau - \tfrac{5}{128}\tau^2 + O(\tau^3)] \qquad (\tau = Ut/l \ll 1). \quad (2.77)$$

This result shows that immediately after the motion starts, the coefficient of the term $(b_0 + b_1)$ is $\frac{1}{2}$, which changes over to $\Theta(\sigma)$ as $t \to \infty$. This feature is quite similar to the Wagner function for the sharp-edged gust effect. The above asymptotic expressions for a_0, (2.74), (2.77), can be directly used to determine T, E, and P at large or small values of t. It is obvious that for t large, T, E, and P differ from their respective steady-state value by a term of $O(\tau^{-2})$, which becomes negligible for $\tau > 10$, or after the body traveled over five chord lengths.

After the transient motion falls off, the time averages of T, E, and P follow immediately from (2.63), (2.64) by applying the averaging formula (2.10). Thus

$$\begin{aligned}\bar{T} &= \tfrac{1}{4}\pi\rho\,\mathrm{Re}[(a_0 + b_0 - \dot{\beta}_0)(a_0{}^* - b_1{}^* + \dot{\beta}_1{}^*) + \dot{\beta}_0\dot{\beta}_1{}^*] \\ &= \tfrac{1}{4}\pi\rho\{|\, b_0 + b_1\,|^2(\mathscr{F}^2 + \mathscr{G}^2 - \mathscr{F}) \\ &\quad - \mathrm{Re}(jU\sigma(b_0 + b_1)[(\beta_0{}^* - \beta_1{}^*)\,\Theta(\sigma) + \beta_1{}^*])\},\end{aligned} \quad (2.78)$$

$$\bar{E} = \tfrac{1}{4}\pi\rho U\,\mathrm{Re}[(a_0 + b_0)(b_1{}^* - a_0{}^*)] = \tfrac{1}{4}\pi\rho U\,|\, b_0 + b_1\,|^2[\mathscr{F} - (\mathscr{F}^2 + \mathscr{G}^2)], \quad (2.79)$$

$$\bar{P} = U\bar{T} + \bar{E} = \tfrac{1}{4}\pi\rho U^2\,\mathrm{Re}\{-j\sigma(b_0 + b_1)[(\beta_0{}^* - \beta_1{}^*)\,\Theta(\sigma) + \beta_1{}^*]\}, \quad (2.80)$$

the final expressions being obtained upon using (2.75) and (2.76). This was the result first given by Wu (1961) and Siekmann (1962), with some differences in notations. The above expression for $\bar{E}$ shows that $\bar{E} \geqslant 0$ since it is known from Theodorsen's function $\Theta(\sigma) = \mathscr{F} + i\mathscr{G}$ that $\mathscr{F} \geqslant (\mathscr{F}^2 + \mathscr{G}^2)$ for $\sigma \geqslant 0$ and the equality holds only when $\sigma = 0$. Therefore $\bar{E} \geqslant 0$ in general; $\bar{E} = 0$ either when $\sigma = 0$, the trivial case of steady motion, or when $b_0 + b_1 = 0$, a special case which, as already discussed in the sequel to (2.69), corresponds to the condition at which no trailing vortex sheet is shed from the body. In fact, the strength of the vortex sheet at the trailing edge is given by

$$\gamma(1, t) = \pi \operatorname{Re} \left\{ \frac{b_0 + b_1}{K_0(j\sigma) + K_1(j\sigma)} e^{j(\omega t - \sigma)} \right\}. \tag{2.81}$$

The main features of the solution may be seen from the following specific example:

$$h(x, t) = \tfrac{1}{16}(x + 1) \cos(kx - \omega t) \qquad (|\, x \,| < 1). \tag{2.82}$$

The thrust coefficient $C_{\mathrm{T}} = \bar{T}/(\frac{1}{4}\pi\rho U^2 l)$ is plotted versus the reduced frequency $\sigma = \omega l/U$ ($l = 1$ being the half-chord) for $k = \pi$ in Fig. 4, in which the experimental results of Kelly (1961) are also shown for comparison (these data include the skin-friction drag). While the experimental error was not precisely known, the agreement between theory and experiment may still be regarded as satisfactory. The theoretical result also shows that $C_{\mathrm{T}} \lesseqgtr 0$ according as $\sigma \lesseqgtr k$, or as the wave velocity $c \lesseqgtr U$. This qualitative feature has already been predicted earlier in Section II, A.

2. *Starting Stage with Constant Acceleration*

A typical starting motion has been considered by Wu (1971c), with the plate starting with a constant acceleration from rest,

$$U(t) = \alpha t \qquad (\alpha > 0,\ t > 0),$$

and with $h(x, t)$ assuming a polynomial of degree 3 in x. The small time behavior of the solution has been evaluated assuming small lift and moment in order to minimize the body recoil in lateral and spinning motions. The result shows that the thrust is generated at a time of the order of t^2, whereas the power is already required at a time $O(t)$, the initial power being positive definite for arbitrary transverse motion $h(x, t)$. When a high efficiency is required in addition, the body profile appears in an S-shape, with a maximum and minimum of h at $x = -0.564$ and $x = 0.295$ approximately.

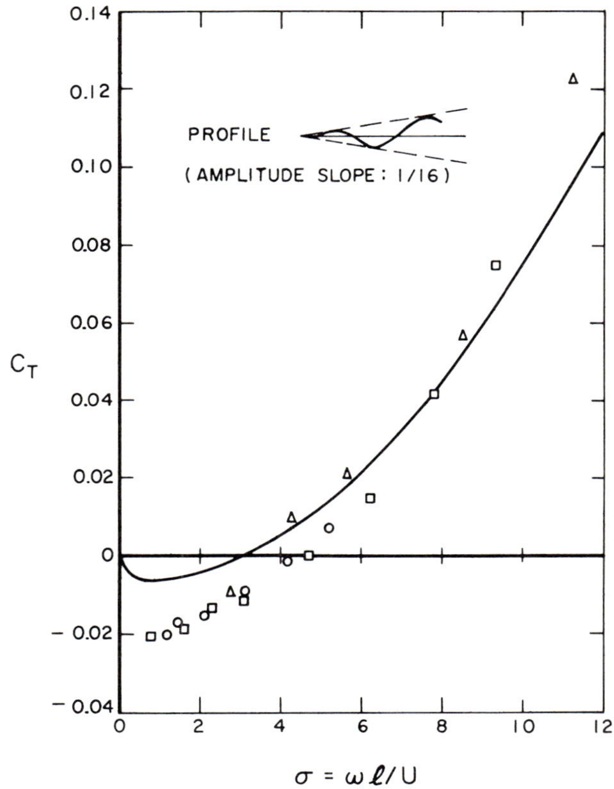

FIG. 4. Variation of thrust coefficient C_T versus the reduced frequency $\sigma = \omega l/U$, the wave number k being π per unit distance (equal to half-chord).—Theory. Kelly's experiment $\triangle$, $U = 1$ ft/sec; $\square$, $U = 2$ ft/sec; $\circ$, $U = 3$ ft/sec.

D. Balance of Recoil of Self-Propelling Bodies

When an aquatic animal propels itself along a rectilinear path, the total force and the moment of the force must balance the time rate of change of their corresponding momenta. Considering the typical case of a three-dimensional planar (or slender) fish, and not taking into account the secondary details such as the movement of pedal fins, it is reasonable to assume that the motive power will come only from the pure moment of the internal forces that can be produced by alternating muscular contractions and relaxations. This moment is analogous to the applied bending moment in the theory of elastic beams. Whether it is possible for aquatic animals to be actually represented by a linear elastic body is of course an open question, since the elasticity of living soft tissues has been found by Fung (1967) to be strongly nonlinear. How much this will be affected by the vertebral column is still not known. We shall however assume that the reactions of the flexible body to the applied bending moment and hydrodynamic forces satisfy the linear elastic relationships. We shall further adopt the elementary beam theory which is considered to be adequate here.

Taking the free-body diagram of a longitudinal section of length dx of the body (see Fig. 5), we obtain the equations governing the motion of

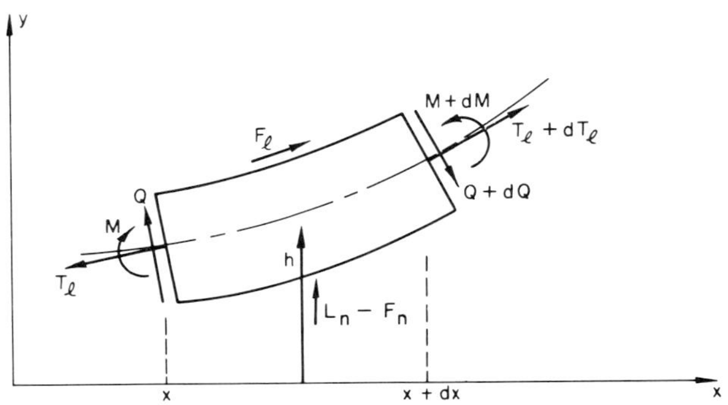

FIG. 5. Hydrodynamic and elastic forces and moments acting on a longitudinal element of a flexible body in transverse movements. The bending moment M includes the active moment M_a due to asymmetrical muscular contractions and relaxations.

a flexible body:

$$\partial T_l/\partial x + F_l = 0, \tag{2.83}$$

$$L_n - F_n = m(x)\frac{\partial^2 h}{\partial t^2} + \frac{\partial Q}{\partial x} - \frac{\partial}{\partial x}\left(T_l \frac{\partial h}{\partial x}\right), \tag{2.84}$$

$$Q = \frac{\partial M_a}{\partial x} + \frac{\partial}{\partial x}\left(E_y I \frac{\partial^2 h}{\partial x^2}\right), \tag{2.85}$$

where $T_l(x, t)$ is the longitudinal tension induced by F_l which represents the longitudinal component of hydrodynamic shear and pressure forces per unit length, L_n is the lift per unit length arising from the pressure, F_n denotes the transverse hydrodynamic viscous drag per unit length, m is the mass of the body per unit length, Q is the elastic shear force in the cross-sectional plane, M_a represents the applied bending moment due to muscular contractions. The quantity (E_yI) stands for the effective bending rigidity, E_y being the effective Young's modulus and I the moment of inertia about the bending axis. From (2.83)–(2.85) one can evaluate the applied moment M_a that must be required for generating the prescribed body motion h, and vice versa, with suitable end conditions (e.g., $Q = M_a = 0$ at the ends). In this respect this set of equations may be useful for biological studies of the activating couple M_a.

Qualitatively speaking, if the thrust and viscous drag are about uniformly distributed along a self-propelling body, F_l and T_l should be everywhere small. Furthermore, F_n is generally small compared with L_n at large Reynolds numbers if the cross flow does not separate. Under these presumptions, we integrate (2.83)–(2.85) along a slender or planar body, giving

$$L(t) = \int_{-1}^{1} m(x)\, h_{tt}(x, t)\, dx, \tag{2.86}$$

$$M(t) = -\int_{-1}^{1} xm(x)\, h_{tt}(x, t)\, dx, \tag{2.87}$$

where L is the total lift and M is the hydrodynamic moment (see (2.17), (2.19) for slender fish and (2.60), (2.61) for the two-dimensional plate). Strictly speaking, the right side of (2.87) further contains an additive term representing the elastic moment $[E_y I h_{xx}]$ evaluated at the two ends of the body at $x = \pm 1$, these two terms being assumed to vanish with the bending moments of inertia there. These two integral conditions were given earlier by Lighthill (1960a).

It may be noted that these two "recoil conditions," if not satisfied by a specific body motion $h(x, t)$, may require extra "rigid-body" motions of sideslip and yaw to be superimposed on h. These recoil movements have been calculated by Lighthill (1970, Section 4) for the carangiform mode, showing that they can be minimized with the right distribution of total inertia. However, when the two-dimensional theory is applied to evaluate the propulsion of the lunate tail of large aspect ratio, or of the wing of some birds, (2.86), (2.87) need not be considered separately since the recoil balance must involve the motion of the entire body.

E. Self-Propulsion in a Perfect Fluid

The previous theories were concerned with the swimming of bodies in fluids of small, but not zero viscosity. Recently, Saffman (1967) raised the interesting hypothetical question: can a fish swim in a perfect fluid whose viscosity is identically zero (as in a superfluid)? It has been shown that the classical paradox of D'Alembert for steady flows of a perfect fluid does not apply to the general unsteady flows past a deformable body, and that a deformable body can move persistently from rest through a perfect fluid without having to produce vorticity in the fluid.

The momentum equation for the rectilinear motion of a deformable body in a perfect fluid can be written

$$[M + m(t)]W = -MU(t) - I_{\mathrm{D}}(t), \tag{2.88}$$

where M is the mass and $m(t)$ the virtual mass of the body, $W(t)$ is the velocity of the geometric centroid $\mathscr{C}$, $U(t)$ is the velocity of the center of mass of the body relative to $\mathscr{C}$, and I_{D} is the component of the fluid impulse due to the change in body shape relative to an instantaneously identical rigid body moving with velocity W. The quantities m, U, and I_{D} are functions only of the shape and structure of the deformable body and are independent of W. It is clear that an arbitrary displacement can be effected without a permanent or net deformation of the body if m, U and I_{D} can be made to vary periodically with t in such a way that W has a nonzero time average

$$\int_0^t W(t')\,dt' \to \overline{W}t \qquad \text{as} \quad t \to \infty \qquad (\overline{W} \neq 0). \tag{2.89}$$

Saffman described two different ways in which this can be accomplished, one for a heterogeneous and other for a homogeneous body.

For a heterogeneous body, we can have $U \neq 0$, and it is simplest to suppose that the surface deformation has fore and aft symmetry so that $I_{\mathrm{D}} = 0$. Then $\overline{W}$ is positive if $U(t)$ and $m(t)$—$m(0)$ oscillate periodically in phase or with an in-phase component. The physical explanation of the propulsion mechanism in this case is clear. There is a hydrodynamic force on a body whenever the body accelerates, which is described by the virtual mass. Now if the center of mass is moved backwards, the recoil will send the shell forward. If then the resistance or virtual mass is less when the shell goes forward than it is when the reverse recoil is moving the shell backwards, the distance covered during the forward motion exceeds that covered during the backwards motion and there is a net forward displacement during each cycle. Note that there is no continuing transfer of momentum between the body and the fluid; the momentum of the body oscillates about a nonzero mean while the oscillating deformation continues. There is of course a transfer of energy between body and fluid, but this is loss-free.

III. Skin-Frictional Resistance of Fish and Cetacean

Hydrodynamics of swimming is still only a part of the whole problem. From a complete bioengineering point of view, the entire process begins with the biochemical energy stored in the swimming body, which can be converted, with efficiency η_1, into mechanical energy for maintaining the body movements. The latter is in turn transformed, with efficiency η_2, into hydrodynamic energy for swimming. For aquatic animals moving at large Reynolds numbers, a part (fraction η_3 say) of the hydrodynamic energy is expended as the useful work done by the thrust, which balances the work done by frictional drag, and the remaining part becomes the energy lost, or dissipated, in the flow wake.

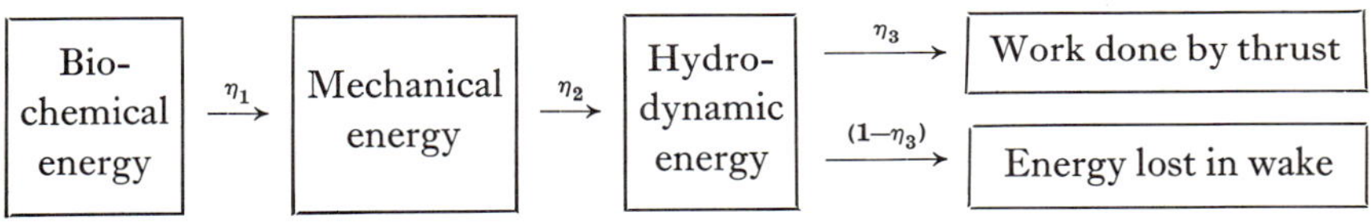

In an effort to ascertain the skin-frictional drag of fish and cetacean in order to make a self-contained account of energy balance, some reported observations appear difficult to explain. For example, Johannessen and Harder (1960) reported several impressively high speeds (about 20 to 22 knots) attained by porpoises, killer whales, and black whales. The boundary layer over a rigid, smooth surface of a similar body in this Reynolds number range is definitely turbulent. If the skin friction is evaluated on this basis, then the power required to maintain such high speeds would violate many times over the rule of thumb in biology that a pound of strong muscle can deliver only up to 0.01 horsepower. Even the latter figure can only be sustained for a few seconds. Another interesting study is that of migratory salmon by Osborne (1960). According to this careful investigation, a detailed estimate again led to one of the two conclusions: either (1) these creatures have a much smaller drag than could be achieved with similar, rigid bodies, or (2) the power output per gram of muscle is much larger than observed from physiological experiments on warm-blooded animals—this being known as the paradox of Gray (1948, 1949). These puzzling conclusions have stimulated investigators to eliminate any uncertainty in the experimental determination of the viscous resistance of a fish on the one hand, and to explore various other possibilities, such as the effect of compliant skin and the effects of mucous surface and additives on frictional drag.

The viscous resistance of a fish, which it must overcome to maintain steady swimming, has been measured by several methods (deceleration in glide, terminal velocity of a dead fish made rigid, models made of paraffin, etc.); these results have been reviewed by Gray (1968, p. 50 ff.). The general picture is that the resistance measured is close to that of a dead fish or rigid models of the same shape, albeit there are also reports (cf. Osborne, 1960) indicating more turbulent skin friction on rigid shapes at the same Reynolds number tested. Typical drag coefficients C_D based on the frontal area are about 0.25.

Recently a series of hydrodynamic experiments with several different species of porpoises (*Tursiops gilli*, *Stenella attenuata*) was performed by Lang and co-workers (Lang, 1962, 1966; Lang and Daybell, 1963; Lang and Norris, 1966; Lang and Pryor, 1966) under more carefully controlled conditions. The test results with a Pacific bottlenose porpoise (*Tursiops gilli*) compare closely with the highest predictions based upon rigid body drag calculations, the same power output per unit body weight as for athletes, and a propulsive efficiency of 85%. The maximum power output of *Stenella attenuata* per unit body weight was, however, 50% greater than for human athletes; and the measured drag coefficient was approximately the same as that of an equivalent rigid body with a near turbulent boundary layer. Thus, in general, no unusual hydrodynamic or physiological performance was observed. Also, it has been pointed out that Gray's paradox can be largely resolved by consideration of duration; Gray's original analysis was based on the power output of humans for a 15 min period and this figure can be raised several times if based on a shorter period, such as a few seconds.

While these results have improved the aspect of the whole picture, a closer examination of the experimental data of Lang and Daybell (1963), as shown in Fig. 6, indicates that there are still cases in which the laminar flow was maintained over a considerably greater percentage of the porpoise skin than for an equivalent rigid body. A number of hypotheses have been proposed in an effort to explain the observed low drag. One of the likely effects is attributed to a favorable pressure gradient over a well-shaped streamline body, as indicated by Van Driest and Blumer (1963) for laminar flows up to $\mathbf{R} = 10^8$. Another possibility is by means of boundary layer control, such as the compliant skin discovered by Kramer (1960). Subsequent theoretical studies of this effect by Betchov (1959), Benjamin (1960), and Landahl (1961) have indicated that the increases in critical Reynolds number obtainable with passive flexible surfaces are too modest to support this effect on the basis of simple stability theory alone. Even though the possibility of activated flexible surfaces has been proposed, the structural complexity of such skin seems to be biologically unfeasible.

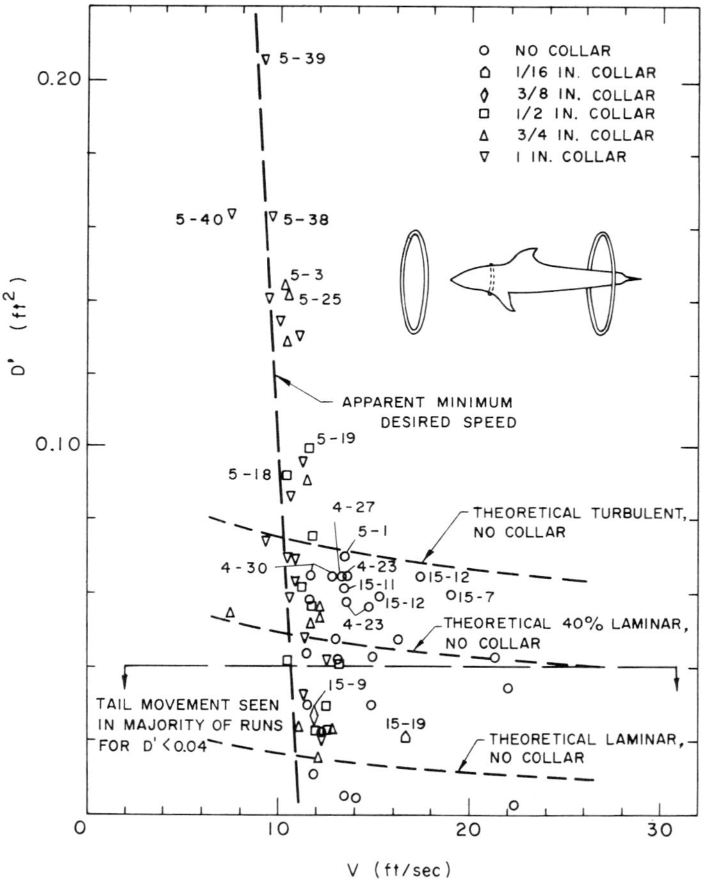

FIG. 6. Experimental results (Lang, 1966) of the "drag area" (Drag/$\frac{1}{2}\rho U^2$) versus swimming speed. (Courtesy of Dr. T. G. Lang.)

A fairly accurate explanation for low drag on fish is the effect on the boundary layer produced by the addition of long-chain molecules, as reported by Fabula, Hoyt, and Crawford (1963). The mucous exuded by fish is composed of a similar type of long-chain molecule and has been found by Hoyt (1970) to bear significantly the same effect. Still another possible explanation, which seems likely to be a potential cause to this author, may be related to the unsteady flow effects, due to body undulations, on the hydrodynamic stability.

IV. Optimum Shape Problems in Swimming Propulsion

Some of the most intriguing problems concerning the phenomena of aquatic animal propulsion and the flapping flights of birds and insects are invariably connected with the highest possible hydrodynamic efficiency. These problems merit study because, as noted by Lighthill (1969), "about 10^9 years of animal evolution in an aqueous environment, by preferential retention of specific variations that increase ability to survive and produce fertile offspring, have inevitably produced rather refined means of generating fast movement at low energy cost." In this section we shall discuss a few problems concerning the optimum shape of body movements for two main categories of body configurations, one being the slender fish, and the other a two-dimensional thin plate as a first approximation for lifting surfaces of large aspect ratio, such as the tails of some cetacean and pelagic fishes and the wings of some birds.

A. Optimum Movements of Slender Fish

The optimum shape problem considered here involves the determination of the transverse oscillatory movements which a slender fish can make, which will produce a prescribed thrust, so as to overcome the frictional drag, at the expense of the minimum work done in maintaining the motion. The solution is for the fish to send a wave down its body at a phase velocity c somewhat greater than the desired swimming speed U, the amplitude being nearly uniform from the maximum span section to the tail. The ratio U/c is found to depend upon the thrust coefficient, reduced wave frequency, and the wave amplitude.

Earlier, Lighthill (1960a) considered a traveling wave which moves down the fish's body with velocity c, and found that an efficiency close to 1 can be realized if c is only slightly greater than the desired swimming speed U (see also Section II,B). Based on the extended slender-body theory as already discussed in Section II,B, Wu (1971c) tried to determine quantitatively the optimum shape under appropriate constraints. This will be discussed below.

Of fundamental importance is the simple harmonic motion

$$h(x, t) = \mathrm{Re}[h_1(x)\, e^{i\omega t}] = \mathrm{Re}[h_1{}^*(x)\, e^{-i\omega t}] \tag{4.1}$$

where $h_1(x)$ is an arbitrary function of x (it may be complex) and $h_1{}^*$ stands for its complex conjugate. It is convenient to use the coefficient form of mean thrust, power required, and energy loss [see (2.23)–(2.25)] as

$$C_T \equiv \bar{T}\Big/\left(\frac{1}{4}\rho U^2 l_{\mathrm{m}}^2\right) = Q_T(l)\,A(l) - \int_0^1 Q_T(x)\,\frac{dA}{dx}\,dx, \tag{4.2}$$

$$C_P \equiv \bar{P}\Big/\left(\frac{1}{4}\rho U^3 l_{\mathrm{m}}^2\right) = Q_P(l)\,A(l) - \int_0^1 Q_P(x)\,\frac{dA}{dx}\,dx, \tag{4.3}$$

$$C_E \equiv \bar{E}\Big/\left(\frac{1}{4}\rho U^3 l_{\mathrm{m}}^2\right) = Q_E(l)\,A(l) - \int_0^1 Q_E(x)\,\frac{dA}{dx}\,dx, \tag{4.4}$$

where $l_{\mathrm{m}} = 1$ is the unit length from the maximum span section to tail base,

$$\sigma = \omega l_{\mathrm{m}}/U \tag{4.5}$$

being the reduced frequency, and

$$Q_T(x) = \sigma^2 hh^* - h_x h_x^*, \tag{4.6}$$

$$Q_P(x) = 2\sigma^2 hh^* + i\sigma(hh_x^* - h^* h_x), \tag{4.7}$$

$$Q_E(x) = Q_P - Q_T = (h_x + i\sigma h)(h_x^* - i\sigma h^*). \tag{4.8}$$

The optimum shape problem at hand can be stated as follows. Within the class of shape function h of (4.1), required to be continuous in x and to satisfy $|\,\partial h/\partial x\,| \ll 1$, find the optimum one which will minimize C_P under the condition of fixed thrust coefficient

$$C_T = C_{T_0} > 0. \tag{4.9}$$

Aside from this isoperimetric condition for optimization, we shall elect not to enforce the recoil conditions (2.86), (2.87), since this practice would be acceptable if they could be satisfied afterward by other means. This is possible in this case by properly adjusting the body motion h in the front part ($-l_n < x < 0$) since h in this range is not subjected to variation.

The optimization under constraint (4.9) is equivalent to minimizing a new functional

$$I[h(x), h_l, \dot{h}_l] = \lambda C_P - (C_T - C_{T_0}) = I_1(h_l, \dot{h}_l) + I_2[h(x)], \tag{4.10}$$

$$I_1(h_l, \dot{h}_l) = [\lambda Q_P(l) - Q_T(l)]\,A_l, \tag{4.11}$$

$$I_2[h] = -\int_0^1 G(h, \dot{h})\,\alpha(x)\,dx, \qquad G(h, \dot{h}) = \lambda Q_P(h, \dot{h}) - Q_T(h, \dot{h}), \tag{4.12}$$

where

$$\alpha(x) = dA/dx, \qquad \dot{h} = dh/dx, \qquad h_l = h(l), \qquad \dot{h}_l = \dot{h}(l), \tag{4.13}$$

and λ is an undetermined multiplier. In the above, the dependence of the fundamental function G on h, h_x as well as their complex conjugate is understood. In the next step, when the shape function h is given an arbitrary variation $\delta h(x)$ in $0 < x < 1$ and a variation at the tail $x = l$, it is of importance to realize that in this extremization the amplitude of h at some point must be left free as a reference amplitude, since the theory, being a linearized one, will not be altered when the solution is uniformly magnified. Therfore, in the variation of I_1, h_l may be held free, or in other words, the extremum of I_1/h_l^2 is to be calculated for an arbitrary variation of $\dot{h}_l/h_l$. For the same reason, the amplitudes $h_0 = h(0)$ and $h_{\mathrm{m}} = h(l_{\mathrm{m}}) = h(1)$ will be left free in the variation of I_2. To fail to observe these free end conditions will lead again to difficulties which will be discussed in Section IV,C.

With the end conditions so clarified, we proceed to impose on h an arbitrary variation $\delta\dot{h}_l$ at $x = l$ and a variation $\delta h(x)$ in $0 < x < 1$, yielding the corresponding first variations of I_1 and I_2 as

$$\delta I_1 = A_l[h_x + i\lambda\sigma h]_{x=l}\,\delta\dot{h}_l^* + \text{complex conjugate}, \tag{4.14}$$

$$\begin{aligned}\delta I_2 &= -\int_0^1 [(\partial G/\partial h^*)\,\delta h^* + (\partial G/\partial \dot{h}^*)\,\delta\dot{h}^*]\,\alpha(x)\,dx + \text{complex conjugate},\\ &= -\frac{\partial G}{\partial \dot{h}^*}\,\alpha(x)\,\delta h^*\Big|_{x=0}^{1} + \int_0^1 \left\{\frac{d}{dx}\left[\alpha\frac{\partial G}{\partial \dot{h}^*}\right] - \frac{\partial G}{\partial h^*}\,\alpha\right\}\delta h^* dx + \text{c.c.}\end{aligned} \tag{4.15}$$

In order that $I[h(x), h_l, \dot{h}_l]$ be extremum, δI_1 and δI_2 must vanish separately for arbitrary $\delta\dot{h}_l$ and $\delta h(x)$. This yields from (4.14) the end condition

$$h_x + i\lambda\sigma h = 0 \qquad (x = l), \tag{4.16}$$

and from (4.15) the Euler–Lagrange equation

$$(d/dx)[(h_x + i\lambda\sigma h)\,\alpha(x)] + [i\lambda\sigma h_x - (2\lambda - 1)\,\sigma^2 h]\,\alpha(x) = 0 \qquad (0 < x < 1) \tag{4.17}$$

and the transversality conditions*

$$(h_x + i\lambda\sigma h)\,\alpha(x) = 0 \qquad (x = 0, 1). \tag{4.18}$$

* These conditions must still be satisfied even when $\alpha(0) \neq 0$, $\alpha(1) \neq 0$.

For slender bodies having a maximum span at $x = 0$ and a minimum at $x = 1$, it is typical that $\alpha(x)$ has simple zeros at $x = 0, 1$, and $\alpha < 0$ for $0 < x < 1$. Then (4.17) is an ordinary differential equation, $x = 0, 1$ being its two regular singularities. Now, with change of variable

$$h(x, t) = g(x)\, e^{i(\omega t - \lambda\sigma x)}, \tag{4.19}$$

Eq. (4.17) and the transversality condition (4.18) become

$$[\alpha g_x]_x + \mu\alpha g = 0 \qquad (0 \leqslant x \leqslant 1, \quad \mu = (1 - \lambda)^2\, \sigma^2). \tag{4.20}$$

$$\alpha(x)\, g_x = 0 \qquad (x = 0, 1). \tag{4.21}$$

This problem of $g(x)$ may manifest itself in a number of ways. (i) $\alpha(x)$ is negative and does not vanish in $0 \leqslant x \leqslant 1$ (e.g., when $b'(x)$ has a jump at $x = 0, 1$)—then (4.20), (4.21) belong to the standard Sturm–Liouville system. (ii) $\alpha < 0$ in $0 < x < 1$, α has simple zeros at $x = 0, 1$ and by choice, $g_x = 0$ at $x = 0, 1$. This is a singular Sturm–Liouville problem. (iii) $\alpha(x)$ has the same properties as in (ii), and the boundary conditions merely require $g(x)$ to remain finite so that (4.21) is still satisfied. This case is no longer an eigenvalue problem. Case (i) does not have general interest, and in case (ii), the eigenfunctions, except possibly for the lowest eigenvalue $\mu_0 = 0$, with corresponding eigenfunction $g_0(x) = \text{const.}$, are seen to bear no physical relevance to high efficiency.

Case (iii) is of general interest. To obtain an approximate solution for this case, we note that at high efficiencies $(1 - \lambda)$ will be small, so will be μ if σ is not too large. Consequently, up to a term of order $(\mu \log \mu)$ the only regular solution of (4.20) is $g_0(x) = \text{const.}$, which completely satisfies all three conditions in (4.16) and (4.21). Thus we have, as the first order solution, the optimum shape given by

$$h(x, t) = \xi_0 e^{i(\omega t - \lambda\sigma x)} \qquad (\xi_0 = \text{const.}, \quad 0 < x < 1). \tag{4.22}$$

The Lagrange multiplier λ can now be determined by applying condition (4.9), yielding

$$C_T = (\xi_0\sigma)^2(1 - \lambda^2)(A_l + A_0 - A_{\mathrm{m}}),$$

which gives for λ two real roots, λ_1 and λ_2,

$$\lambda_1 = -\lambda_2 = [1 - \tilde{C}_{T_0}/\sigma^2]^{1/2} \qquad (\sigma \geqslant \sigma_{\mathrm{c}} \equiv \tilde{C}_{T_0}^{1/2}), \tag{4.23}$$

where

$$\tilde{C}_{T_0} = C_{T_0}/[\xi_0{}^2(A_l + A_0 - A_{\mathrm{m}})]. \tag{4.24}$$

The corresponding optimum mean power coefficient is

$$C_P = 2(\xi_0\sigma)^2(1 - \lambda)(A_l + A_0 - A_{\mathrm{m}}). \tag{4.25}$$

The maximum and minimum hydrodynamic efficiency, under condition (4.9), are therefore

$$\eta_{\max} = \tfrac{1}{2}(1 + \lambda_1) = \tfrac{1}{2}[1 + (1 - \tilde{C}_{T_0}/\sigma^2)^{1/2}], \tag{4.26}$$

$$\eta_{\min} = 1 - \eta_{\max}. \tag{4.27}$$

Although the above solution of optimum efficiency appears to depend mathematically on a single parameter $(\tilde{C}_{T_0}/\sigma^2)$, it is more desirable for the convenience of subsequent discussion of other flow quantities, analyzing experimental studies, as well as for facilitating comparison with the two-dimensional case of Section IV,C, to keep σ and $\tilde{C}_{T_0}$ as two separate parameters. The optimum efficiency is real when σ is greater than the critical frequency $\sigma_{\mathrm{c}} = (\tilde{C}_{T_0})^{1/2}$. As σ increases from σ_{c}, $\eta_{\max}$ increases towards unity, whereas $\eta_{\min}$ decreases towards zero, both rather rapidly from $\eta = 0.5$, as shown in Fig. 7. For $\sigma \gg \sigma_{\mathrm{c}}$,

$$\eta_{\max} \simeq 1 - \tilde{C}_{T_0}/(2\sigma)^2.$$

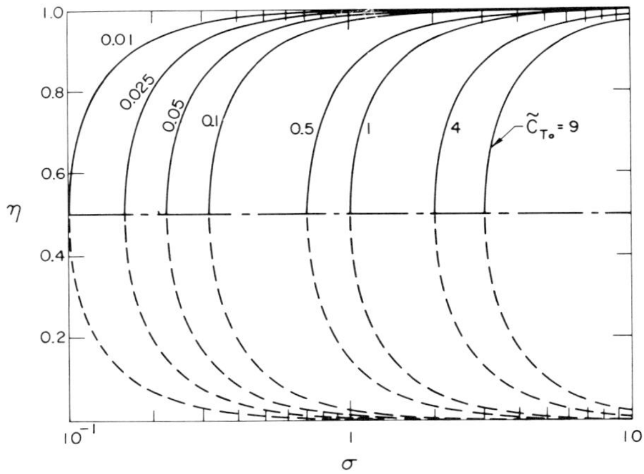

FIG. 7. Maximum and minimum efficiencies of slender fish for fixed $\tilde{C}_{T_0}$.

Another quantity of importance is the phase velocity c of body wave form relative to the swimming speed U. For the optimum shape (4.22), the phase velocity is $c = \omega/\lambda_1\sigma = U/\lambda_1$, or

$$U/c = (1 - \tilde{C}_{T_0}/\sigma^2)^{1/2}. \tag{4.28}$$

For $\sigma \gg \sigma_{\mathrm{c}}$, U is therefore only very slightly less than c.

To estimate the order of magnitude of σ_c, we equate the thrust T and the frictional drag, assuming no flow separation, so that

$$D = \tfrac{1}{2}\rho U^2 C_D S_f = T = \tfrac{1}{4}\rho U^2 l_m{}^2 C_{T_0},$$

where C_D is the skin-frictional drag coefficient based on the total wetted surface S_f. This leads to

$$\tilde{C}_{T_0} = (C_D/\xi_0{}^2)[2S_f/(A_0 - A_m + A_l)]. \tag{4.29}$$

The factor with S_f is clearly of order $O(\delta^{-1})$ for a small slenderness parameter δ.

To give some general idea about the magnitudes of these terms, we take a typical case of an experiment using trout, from the most comprehensive studies of Bainbridge (1960, 1961, and 1963). (Without having the original data, some of the following values are estimated from the figure plots and body geometry provided, but they are thought to be fairly reliable.) The specific case is a trout of total body length $l_0 = 20$ cm, swimming at a speed $U = 1.5$ meters/sec, with tail-beat frequency $f \simeq 17$/second. The Reynolds number based on l_0 is about 3×10^5, at which the turbulent drag coefficient is about $C_D \simeq 0.006$, which is a conservative estimate. From the geometry, the value of l_m (distance between the maximum cross section to caudal peduncle) is $\simeq 8.75$ cm, based on which (for unit length) the following are estimated: $S_f \simeq 2.58$, $A_0 \simeq 0.28$, $A_m \simeq 0.04$, $A_l \simeq 0.21$, $\xi_0 \simeq 0.12$, and the reduced frequency $\sigma \simeq 6.24$. Hence, by (4.29), $\tilde{C}_{T_0} \simeq 4.8$ and $\tilde{C}_{T_0}/\sigma^2 \simeq 0.123$. Consequently,

$$U/c \simeq 0.94 \qquad \eta \simeq 0.97.$$

This estimation shows how close to 1 are the theoretically predicted values of U/c and η. Furthermore, using $c = f\lambda_0$ (f being the beat frequency and λ_0 the wavelength), we can write the swimming speed based on body length as

$$U/l_0 = (U/c)(\lambda_0/l_0)f = (\lambda_0/l_0)f(1 - \tilde{C}_T/\sigma^2)^{1/2}. \tag{4.30a}$$

This expression should be compared with the empirical relationship based on the studies of Bainbridge (1961; see also Gray, 1968, p. 48)

$$U/l_0 = (0.75)f - 1. \tag{4.30b}$$

A comparison between these two equations indicates that, as an approximate rule, the wavelength λ_0 is about $\frac{3}{4}$ of the body length l_0, and U/l_0 is a linear function of beat frequency f, noting that $U/c \simeq 1$.

The result of a nearly uniform amplitude of transverse motion along the slender fish from the maximum span to the tail is another important feature of the solution. With no such requirement for the front part of body, it seems that motions with a somewhat reduced amplitude in the front part help keep the recoil small.

B. Vortex Wake in the Optimum Motion; Momentum Balance

It is instructive to investigate the vortex wake generated by the optimum body motion (4.22). As has been explained earlier the vorticity $\boldsymbol{\gamma}$ of this flow is confined entirely to the body surface S_b and the trailing vortex sheet S_w, and hence has only two components. Within the framework of the linearized theory, $\boldsymbol{\gamma} = (\gamma_1, \gamma_2, 0)$. Since $\boldsymbol{\gamma}$ is the curl of the velocity vector, div $\boldsymbol{\gamma} = 0$, which becomes in this case

$$\partial\gamma_1/\partial x + \partial\gamma_2/\partial y = 0. \tag{4.31}$$

Corresponding to the optimum body motion (4.22) the vorticity component γ_2 of the wake has been determined (see Wu, 1971c) as

$$\gamma_2(x, y, t) = -2U\xi_0\sigma^2(1 - \lambda_1)\, e^{-i\sigma(x-Ut)}\{(b_0{}^2 - y^2)^{1/2} + O(\sigma(1 - \lambda_1))\}, \tag{4.32}$$

which is valid for $(x, y) \in S_w$, $0 < x < 1$, and γ_1 can be obtained by integration from (4.31). γ_2 has an elliptical spanwise distribution and its magnitude is approximately proportional to C_{T_0}/ξ_0 [upon using (4.23)]. From the above solution some of the important features of the vortex wake can be demonstrated.

For a fixed station x, write the phase angle of the transverse motion h as $\theta = \omega t - \lambda_1\sigma x$ so that θ increases by 2π as t varies over a period $2\pi/\omega$. Taking the real part of (4.22) and (4.32) for physical interpretation, we obtain the qualitative picture of $h(x, t)$ and $\gamma_2(x, 0, t)$ as shown in Fig. 8. As γ_2 leads h in phase by $\pi - \sigma(1 - \lambda_1)x$, they are nearly opposite in phase since the phase angle $\Delta\theta = \sigma(1 - \lambda_1)x$ is small, but increases distally. In a reference frame fixed relative to the undisturbed fluid, the section x will traverse a sinusoidal trajectory, shedding in the course of time a trail of vortices γ_2 as shown qualitatively in Fig. 8. The velocity of the fluid particles along the x axis induced by this system of γ_2 vortex sheet is clearly seen to move backwards from the body, resulting in a "jet stream," to which the forward thrust T, regarded as the reaction to this jet flow momentum generated at this optimum operational state, can be wholly attributed. When the viscous wake produced by the diffusion of the vorticity left behind from the boundary layer flow is considered in

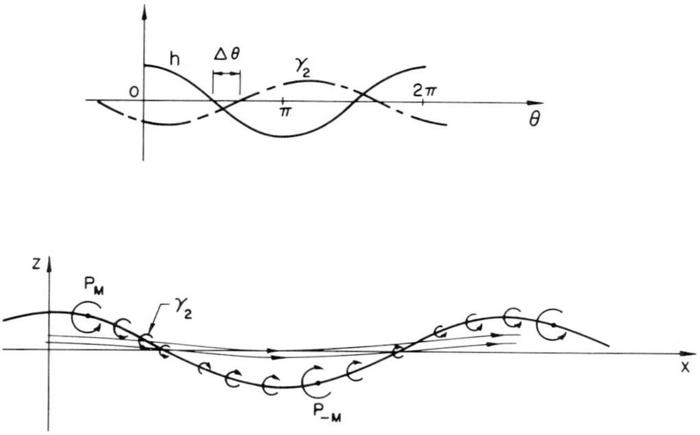

FIG. 8. The vortex wake of the optimum movements. γ_2 reaches maximum at P_M and minimum at P_{-M}.

addition, the entrainment of fluid due to the viscous wake is opposite in direction to that caused by γ_2, thus leaving no net flow momentum at large distances behind the body, if it is self-propelling. This explains the basic mechanism of swimming of a slender fish.

Hertel (1963, p. 157) carried out a series of experimental studies of caudal fin strokes of trout, sturgeon, and whiting. From the recorded movement of the trout, with the transverse (side-to-side) displacement (h_c, say) of the midway section of the caudal fin plotted together with the yawing angle (β, say) of the slightly bent caudal fin, it was found that both variations are almost exactly sinusoidal and that the yawing motion β leads the "heaving" motion h_c by a phase difference of 72°. Putting it somewhat differently, we may also say that the slope dh_c/dx of the path $h_c(x)$ traversed by the caudal fin midpoint leads the yawing angle β by 18° since dh_c/dx leads h_c by 90°. This implies that the trout's caudal fin does not move tangentially along its own trajectory, but sustains a "feathering angle" of 18° to this trajectory. This rather small feathering angle can be seen to further enhance the thrust because it is associated with a forward momentum, and it is in the same direction of phase shift as the y component of tail vorticity γ_2, since we have seen that γ_2 reaches its peak shortly after each peak of lateral displacement h_c of the caudal fin is reached.

To account for the motion with this additional feathering of the caudal fin, a new term representing this extra degree of freedom of yawing (pivoted at the base of caudal fin) will have to be added to h of (4.22). This will cause some modification of the final result, but amounting only to replacing V, h_x, h_t at the tail-end section ($x = l$) by their new values, with the contribution of the feathering motion included, in the previous results.

C. Optimum Movement of a Rigid-Plate Wing

Insofar as the two-dimensional theory provides a first approximation to the motion of lunate tails of large aspect ratio and the flapping motion of bird wings, it is of interest to study the problem of their optimum movement for producing high hydromechanical efficiency. Since in reality these lifting surfaces have very little flexural deviations, it suffices to regard them as rigid, thin, airfoil-shaped bodies. Lighthill (1970), considering the hydromechanical efficiency η of a rigid plate in undulatory propulsion, found a well-defined range of the motion variables in which the maximum η lies, and interpreted the results with clear physical significance. Wu (1971b) investigated the optimum movement of a rigid plate that will minimize the energy loss under the condition of fixed thrust (to overcome the viscous drag) and possibly also under other constraints. This study of Wu's has been further extended to the general case of optimum shape of a two-dimensional flexible plate of a large or infinite number of degrees of freedom which may have other practical interest.

The lateral displacement of a rigid, thin, airfoil-shaped plate may be written, after Wu (1971b), as

$$h(x, t) = [\tfrac{1}{2}\xi_0 + (\xi_1 + i\xi_2)x]\, e^{i\omega t} \qquad (|\, x\,| \leqslant 1), \tag{4.33}$$

where ξ_0, ξ_1, ξ_2 are real. The above h represents a rigid plate performing a heaving with amplitude $\xi_0/2$ (taken positive as the reference phase) and a pitching about the midchord with amplitude $|\, \xi_1 + i\xi_2 \,|$, at a phase angle $\tan^{-1}(\xi_2/\xi_1)$ leading the heaving motion. Now, by (2.65)

$$\beta_0 = \xi_0 e^{i\omega t}, \qquad \beta_1 = (\xi_1 + i\xi_2)\, e^{i\omega t}, \tag{4.34}$$

and by (2.40) and (2.54),

$$b_0 + b_1 = Ue^{i\omega t}[i\sigma\xi_0 + (2 + i\sigma)(\xi_1 + i\xi_2)]. \tag{4.35}$$

Upon substituting (4.34) and (4.35) in (2.78)–(2.80), the coefficients of $\bar{T}$, $\bar{E}$, $\bar{P}$ assume the quadratic forms

$$C_E \equiv \bar{E}/(\tfrac{1}{4}\pi\rho U^3 l) = B(\sigma)(\boldsymbol{\xi}, \mathbb{Q}\boldsymbol{\xi}), \tag{4.36}$$

$$C_P \equiv \bar{P}/(\tfrac{1}{4}\pi\rho U^3 l) = \sigma(\boldsymbol{\xi}, \mathbb{P}\boldsymbol{\xi}), \tag{4.37}$$

$$C_T \equiv \bar{T}/(\tfrac{1}{4}\pi\rho U^2 l) = C_P - C_E\,, \tag{4.38}$$

where

$$B(\sigma) = \mathscr{F} - (\mathscr{F}^2 + \mathscr{G}^2), \tag{4.39}$$

$\boldsymbol{\xi} = (\xi_0, \xi_1, \xi_2)$ is a vector in a three-dimensional vector space, $(\boldsymbol{\xi}, \boldsymbol{\zeta})$ denotes the inner product of $\boldsymbol{\xi}$ and $\boldsymbol{\zeta}$, or $\xi_0\zeta_0 + \xi_1\zeta_1 + \xi_2\zeta_2$, and $\mathbb{Q}$ and $\mathbb{P}$ are 3×3 symmetric matrices with elements

$$\begin{gathered} Q_{11} = Q_{12} = \sigma^2, \qquad Q_{13} = 2\sigma, \qquad Q_{22} = Q_{33} = 4 + \sigma^2, \qquad Q_{23} = 0, \\ P_{11} = \sigma\mathscr{F}, \qquad P_{12} = \mathscr{G} + \tfrac{1}{2}\sigma, \qquad P_{13} = \mathscr{F} - \sigma\mathscr{G}, \\ P_{22} = P_{33} = \sigma(1 - \mathscr{F}) - 2\mathscr{G}, \qquad P_{23} = 0. \end{gathered} \tag{4.40}$$

It can be shown from the properties of $\mathscr{F}$ and $\mathscr{G}$ that $\mathbb{P}$ is nonsingular for $\sigma > 0$ since none of the three eigenvalues of $\mathbb{P}$ vanishes for $\sigma > 0$. However, $\mathbb{Q}$ has the eigenvalues 0, $(\sigma^2 + 4)$, $(2\sigma^2 + 4)$, and hence $\mathbb{Q}$ is singular in the third order, but nonsingular in the second order.

Before we proceed further, we list here two fundamental cases:

(i) *Heaving* only, so that $\xi_1 = \xi_2 = 0$ and only $\xi_0 \neq 0$. The corresponding C_T, C_E, C_P are

$$C_E = \sigma^2 B(\sigma)\, \xi_0^2, \qquad C_P = \sigma^2 \mathscr{F}(\sigma)\, \xi_0^2, \qquad C_T = \sigma^2(\mathscr{F}^2 + \mathscr{G}^2)\, \xi_0^2. \tag{4.41}$$

The hydrodynamic efficiency of heaving propulsion,

$$\eta_{\text{heav}}(\sigma) = C_T/C_P = (\mathscr{F}^2 + \mathscr{G}^2)/\mathscr{F}, \tag{4.42}$$

is seen to depend on σ only, decreasing monotonically from $\eta_h(0) = 1$ to $\eta_h(\infty) = 0.5$, as shown in Fig. 9. The propulsive thrust in this case comes entirely from the leading-edge suction.

(ii) *Pitching* only, so that $\xi_0 = 0$ and we may also set $\xi_2 = 0$ as a reference phase. Then, clearly,

$$C_E = B(\sigma)\, Q_{22}\xi_1^2, \qquad C_P = \sigma P_{22}\xi_1^2, \qquad C_T = T_{22}\xi_1^2, \tag{4.43a}$$

where

$$T_{22} = \sigma P_{22} - B Q_{22}. \tag{4.43b}$$

In this pure pitching mode, both C_E and C_P are positive definite for $\sigma > 0$. But $T_{22}(\sigma) = 0$ has one real root,

$$T_{22}(\sigma_0) = 0 \qquad \text{for} \quad \sigma_0 = 1.781, \tag{4.44}$$

and $T_{22} \lesseqgtr 0$ according as $\sigma \lesseqgtr \sigma_0$. When $C_T > 0$, the hydrodynamic efficiency (2.26) is given by

$$\eta_{\text{pitch}}(\sigma) = C_T/C_P = T_{22}(\sigma)/\sigma P_{22}(\sigma) \qquad (\sigma > \sigma_0), \tag{4.45}$$

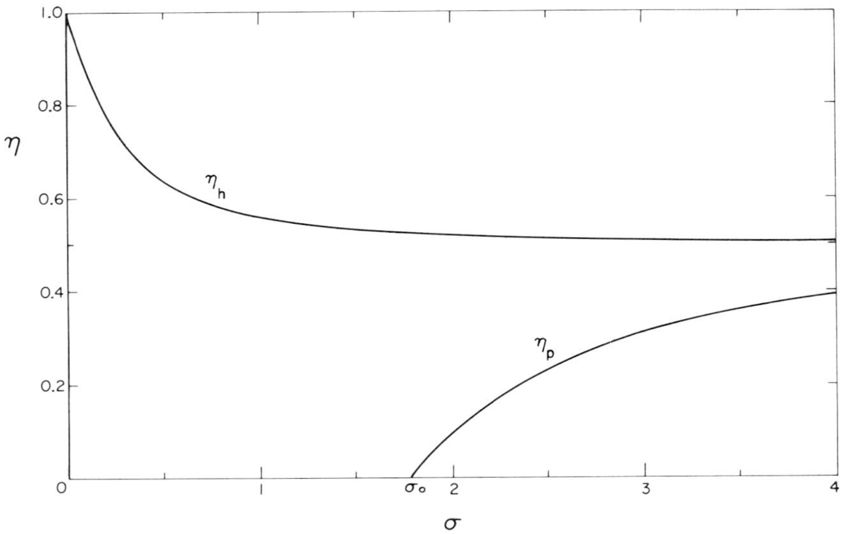

FIG. 9. Hydrodynamic efficiency of heaving propulsion, $\eta_h(\sigma)$, and of pitching propulsion, $\eta_p(\sigma)$, the latter being defined for the reduced frequency $\sigma \geqslant \sigma_0 = 1.781$.

which is found to increase monotonically from $\eta_p(\sigma_0) = 0$ to $\eta_p(\infty) = 0.5$, as shown in Fig. 9. We further note that in both cases (i) and (ii), power must be supplied to maintain the motion; consequently it is impossible to extract energy from the fluid in so far as either (i) or (ii) is concerned. The main objective of optimization is then to determine if the efficiency can be greatly improved when both heaving and pitching modes are admitted.

The optimum problem at hand is to minimize the quadratic form C_E of (4.36) again under the constraint (4.9), while the recoil conditions (2.85), (2.87) are relaxed for the reason already stated. This constrained optimization is equivalent to minimizing a new function

$$C_E' = C_E - \lambda'(C_T - C_{T_0}) = (1 + \lambda')\, C_E - \lambda' C_P + \lambda' C_{T_0}\,, \qquad (4.46)$$

λ' being a Lagrange multiplier.

It has been pointed out by Wu (1971b) that this is not an eigenvalue problem on account of the singular behavior of $\mathbb{Q}$. (The solution of Wang (1966) treating this as an eigenvalue problem is erroneous.) In fact, we note here that $b_0 + b_1 = 0$ when

$$\xi_1 = \hat{\xi}_1 \equiv -\sigma^2(\sigma^2 + 4)^{-1}\, \xi_0\,, \qquad \xi_2 = \hat{\xi}_2 \equiv -2\sigma(\sigma^2 + 4)^{-1}\, \xi_0\,. \qquad (4.47)$$

At the same time, C_T, C_P, and C_E all vanish with $(b_0 + b_1)$, according to (2.78)–(2.80). This particular set of values $(\xi_0, \hat{\xi}_1, \hat{\xi}_2)$ will be seen to play a significant role in the optimum solution.

The correct approach is found by noting that since $\mathbb{Q}$ is singular, but nonsingular in the second order, the quadratic form C_E can always be reduced to a nonsingular form in two variables. In fact, in terms of the new variables

$$\zeta_0 = \xi_0/(4 + \sigma^2), \qquad \zeta_1 = \xi_1 - \hat{\xi}_1, \qquad \zeta_2 = \xi_2 - \hat{\xi}_2, \tag{4.48}$$

with $\hat{\xi}_1$, $\hat{\xi}_2$ given by (4.47), C_E and C_P in (4.36), (4.37) reduce to

$$C_E = B(\sigma) Q_{22}(\zeta_1^2 + \zeta_2^2), \tag{4.49}$$

$$C_P = \sigma\{P_{22}(\zeta_1^2 + \zeta_2^2) + 2A_1\zeta_0\zeta_1 + 2A_2\zeta_0\zeta_2\}, \tag{4.50}$$

$$A_1 = P_{12}Q_{22} - Q_{12}P_{22}, \qquad A_2 = P_{13}Q_{33} - Q_{13}P_{33}. \tag{4.51}$$

Now it is clear that while C_P spans the whole vector space $(\zeta_0, \zeta_1, \zeta_2)$ C_E spans only its subspace (ζ_1, ζ_2). Obviously, the surface $C_E =$ const. $= C_{E_0} > 0$ is a circular cylinder with its central axis along the ζ_0 axis. The quadric $C_P =$ const. $= C_{P_0} > 0$ is seen to be an oblique hyperboloid of one sheet since its intersection with the plane $\zeta_0 =$ const. is a circle centered at $(-A_1\zeta_0/P_{22}, -A_2\zeta_0/P_{22})$, of radius

$$[(A_1^2 + A_2^2)(\zeta_0/P_{22})^2 + C_{P_0}/(\sigma P_{22})]^{1/2}.$$

The extremal solutions under condition (4.9) are therefore given by the points in the subspace (ζ_1, ζ_2) at which (grad C_E) is proportional to (grad C_P). Thus, after setting the derivatives of $C_E'' \equiv (C_E - \lambda'' C_P)$ with respect to ζ_1 and ζ_2 to zero, we obtain

$$\zeta_1 = \lambda A_1\zeta_0, \qquad \zeta_2 = \lambda A_2\zeta_0, \tag{4.52}$$

where λ is a Lagrange multiplier. Upon substituting (4.52) in (4.49), (4.50), we have

$$C_E = BQ_{22}\lambda^2(A\zeta_0)^2, \qquad A^2 = A_1^2 + A_2^2, \tag{4.53}$$

$$C_P = \sigma(P_{22}\lambda^2 + 2\lambda)(A\zeta_0)^2. \tag{4.54}$$

Now, the application of condition (4.9), or $C_P - C_E = C_{T_0}$, results in a quadratic equation for λ:

$$T_{22}(\sigma)\,\lambda^2 + 2\sigma\lambda = \bar{C}_{T_0}(4 + \sigma^2)^2/A^2, \tag{4.55a}$$

where

$$\bar{C}_{T_0} = C_{T_0}/\xi_0^2, \qquad A^2 = A_1^2 + A_2^2, \tag{4.55b}$$

and $T_{22}(\sigma)$ is given by (4.43b). The multiplier λ therefore has two solutions

$$\left.\begin{matrix}\lambda_1\\ \lambda_2\end{matrix}\right\} = \frac{\sigma}{T_{22}}\{-1 \pm (1 + \Lambda)^{1/2}\}, \qquad \Lambda = \bar{C}_{T_0} T_{22} \left(\frac{\sigma^2 + 4}{\sigma A}\right)^2. \tag{4.56}$$

λ_1, λ_2 depend on two parameters: σ and $\bar{C}_{T_0}$. By virtue of the behavior $T_{22}(\sigma) \gtreqless 0$ according as $\sigma \gtreqless \sigma_0 = 1.781$, it follows that for fixed $\bar{C}_{T_0} > 0$, $\Lambda(\sigma, \bar{C}_T)$ increases monotonically from $-\infty$ to 0 as σ moves from $\sigma = 0$ to σ_0. Consequently λ_1, λ_2 will be real (as is required to be physically meaningful) if $\Lambda \geqslant -1$; or equivalently,

$$\sigma \geqslant \sigma_c(\bar{C}_{T_0})$$

where

$$\Lambda(\sigma_c, \bar{C}_{T_0}) = -1. \tag{4.57}$$

The solution $\sigma_c = \sigma_c(\bar{C}_{T_0})$ lies between $\sigma_c(0) = 0$ and $\sigma_c(\infty) = \sigma_0$. For given $\bar{C}_{T_0} > 0$, the real optimum solution therefore exists only for $\sigma \geqslant \sigma_c$. Within this range λ_1 is positive, numerically smaller than λ_2, and corresponds to the highest efficiency attainable under condition (4.9),

$$\eta_{\max} = \frac{C_{T_0}}{C_P} = 1 - \frac{C_E}{C_P} = 1 - \frac{BQ_{22}\lambda_1}{\sigma(P_{22}\lambda_1 + 2)} \qquad (\sigma \geqslant \sigma_c). \tag{4.58}$$

The lowest efficiency $\eta_{\min}$ that can be realized under the same condition (4.9) is given by the last expression of (4.58) with λ_1 replaced by λ_2. For any combination of ζ_0, ζ_1, ζ_2 different from (4.52), the efficiency η is given by $\eta_{\min} < \eta < \eta_{\max}$ so long as $\bar{C}_{T_0}$ is kept fixed. The numerical results of $\eta_{\max}$ and $\eta_{\min}$ are plotted in Fig. 10 for several values of $\bar{C}_{T_0}$. It is of interest to note that this two-dimensional optimum efficiency behaves quite similar to Fig. 7 for slender fish.

The optimum motion of the plate is given by (4.48), (4.52), (4.47), and (4.56),

$$\xi_1/\xi_0 = (\lambda_1 A_1 - \sigma^2)/(\sigma^2 + 4), \qquad \xi_2/\xi_0 = (\lambda_1 A_2 - 2\sigma)/(\sigma^2 + 4). \tag{4.59}$$

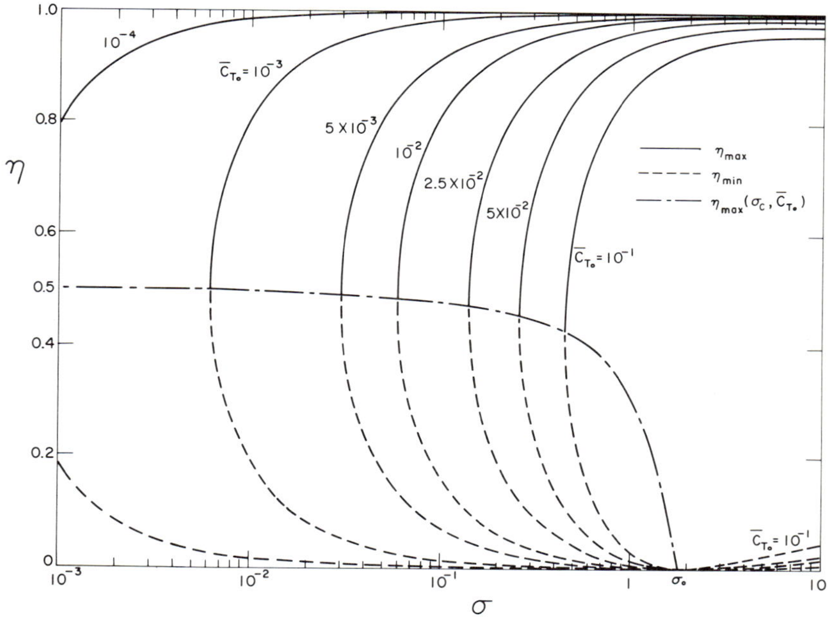

FIG. 10. Maximum and minimum efficiencies of a rigid-plate wing at fixed $\bar{C}_{T_0}$.

Hence the amplitude ratio and the phase advance of pitching relative to the heaving mode are

$$Z_p = (\xi_1^2 + \xi_2^2)^{1/2}/\xi_0 = (\sigma^2 + 4)^{-1}[(\lambda_1 A_1 - \sigma^2)^2 + (\lambda_1 A_2 - 2\sigma)^2]^{1/2}, \tag{4.60a}$$

$$\alpha_p = \tan^{-1} \xi_2/\xi_1 = \tan^{-1}[(\lambda_1 A_2 - 2\sigma)/(\lambda_1 A_1 - \sigma^2)]. \tag{4.60b}$$

These results are shown in Figs. 11 and 12 for several values of $\bar{C}_{T_0}$.

In summary, we first notice the advantage of operating at small values of $\bar{C}_{T_0}$, corresponding to a sufficiently large heaving amplitude. A smaller $\bar{C}_{T_0}$ renders the optimum solution valid to lower frequencies σ and makes η_{max} greater at the same σ. As σ increases from σ_c, the pitching–heaving amplitude ratio Z_p first decreases to a minimum, then increases steadily to a common asymptote. Over the same range of σ, the phase difference α_p changes very rapidly at first, followed by a much slower variation at higher σ. A rather sophisticated control would therefore be necessary if the operating range of σ is chosen in which fast variations of Z_p and α_p may take place. It is a remarkable result that in the higher range of σ, very high efficiencies can be realized with an appropriate interplay between the heaving and pitching motions. This

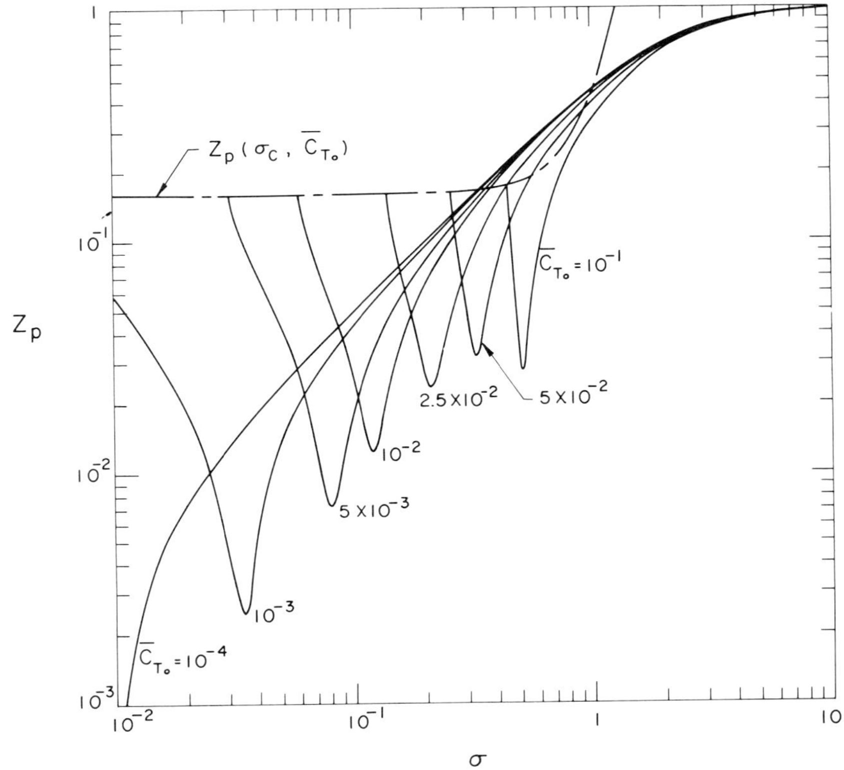

FIG. 11. The amplitude ratio (pitching/heaving) $Z_P(\sigma, \bar{C}_{T_0})$.

effect is exhibited in the result with the pitching amplitude as low as only a small fraction ($\sim$0.1 or less) of the heaving motion, provided the phase difference is correctly observed.

Whether this very attractive operation can be actually achieved must depend on the leading-edge suction being reasonably small, as was emphasized by Lighthill (1970). Although this suction force has been simplified, for mathematical convenience, as a singular force acting on a pointed leading edge, it can be realized physically only when the thin section's leading edge is sufficiently rounded. Large magnitudes of this suction force often indicate that the flow would separate from the airfoil near its leading edge, causing drastic changes in forces. The time average of the leading-edge suction in periodic motion, by (2.62), is

$$\bar{T}_S = \tfrac{1}{4}\pi\rho a_0 a_0^*, \qquad a_0 = (b_0 + b_1)\,\Theta(\sigma) - b_1 . \tag{4.61}$$

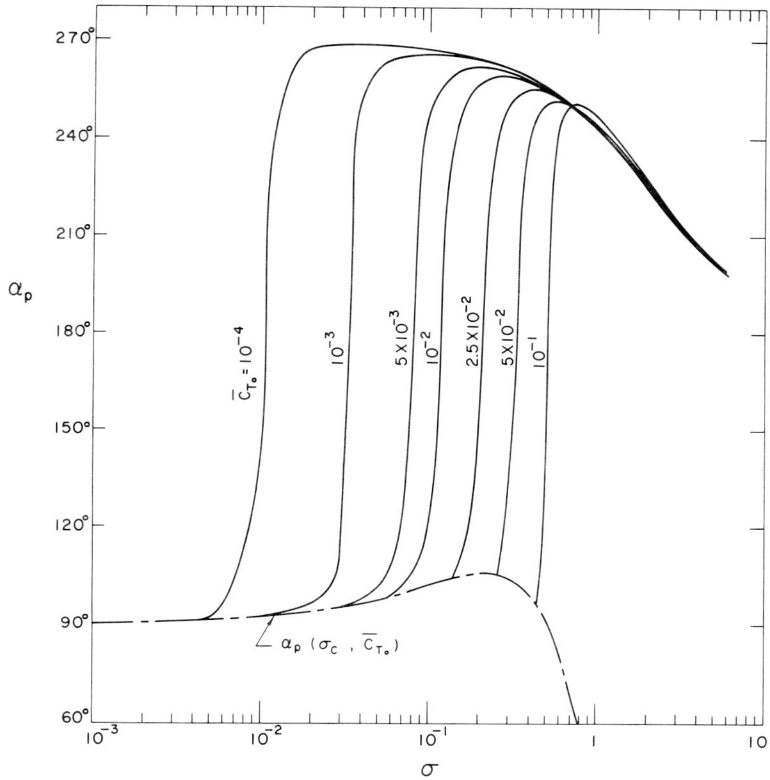

FIG. 12. The phase advance angle $\alpha_p(\sigma, \bar{C}_{T_0})$ of the pitching mode.

The ratio of the mean suction coefficient C_S (nondimensionalized on the same base as C_T) to the total thrust coefficient C_T has been calculated by Wu (1971b) for optimum ξ_1/ξ_0 and ξ_2/ξ_0; the final result is shown in Fig. 13. It is of great interest to note that for a specified "proportional-loading" parameter $\bar{C}_{T_0}$, the ratio C_S/C_T has a minimum at $\sigma = \sigma_m(\bar{C}_{T_0})$ say, and is relatively small only in a short stretch of σ about σ_m. Outside of this range, C_S/C_T increases rapidly beyond 1 and becomes so large (the complementary thrust delivered by the plate surface is then negative) as to be too difficult to realize in practice without leading-edge stalling. It is also noteworthy that $\sigma = \sigma_m$ coincides almost exactly with the value of σ of the maximum of the corresponding $\alpha_p(\sigma)$ curve (see Fig. 12), about which the phase angle α_p of pitching varies relatively slowly with σ. It is thus convincing that the optimum range of operating σ in practice must be somewhere very near σ_m, most likely a little greater than

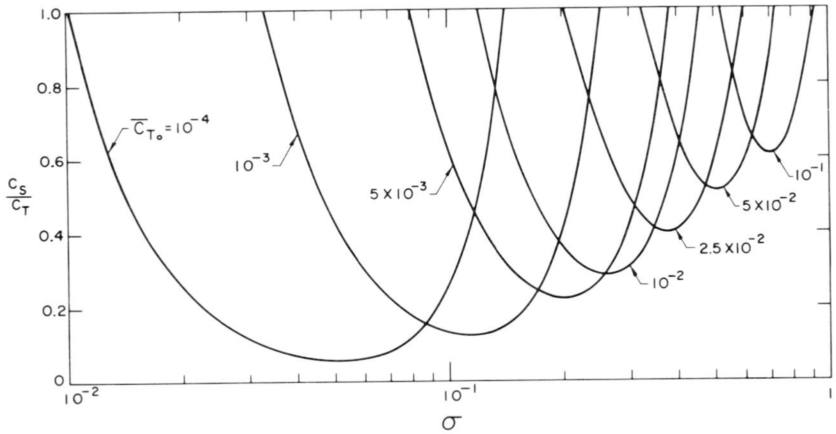

FIG. 13. The ratio of the leading-edge suction coefficient C_S to the total thrust coefficient $C_T = C_{T_0}$.

σ_m before C_S/C_T rises sharply so that a slightly improved efficiency can be achieved without risking leading-edge stall.

The problem of optimum efficiency of a rigid plate in undulatory propulsion has been discussed by Lighthill (1970) by taking the section's lateral displacement in the form

$$y = [h - i\alpha(x - b)]\, e^{i\omega t} \qquad (-l < x < l), \tag{4.62}$$

where h and α are real numbers denoting the amplitude of the heaving and pitching motions, respectively, and $x = b$, $y = 0$ is the axis of pitch. Lighthill's adoption of a fixed phase difference of 90° but generalizing the axis of pitch is of course equivalent to adopting a general phase difference between heaving and pitching about the mid-chord axis. In fact, this equivalence is complete by introducing a reference phase γ to (4.33), and recovering the half-chord length l,

$$y = [\tfrac{1}{2}l\xi_0 + (\xi_1 + i\xi_2)x]\, e^{i(\omega t+\gamma)} \qquad (-l < x < l).$$

Then the above two expressions for y are equivalent if

$$b/l = -\tfrac{1}{2}\xi_0\xi_1(\xi_1^2 + \xi_2^2)^{-1} = -\tfrac{1}{2}Z_p^{-1}\cos\alpha_p\,, \tag{4.63}$$

$$h/l = -\tfrac{1}{2}\xi_0\xi_2(\xi_1^2 + \xi_2^2)^{-1/2} = -\tfrac{1}{2}\xi_0\sin\alpha_p\,, \tag{4.64}$$

$$\alpha = (\xi_1^2 + \xi_2^2)^{1/2} = \xi_0 Z_p\,. \tag{4.65}$$

To serve as a useful measure of the relative magnitudes of pitching and heaving, Lighthill (1969, 1970) further introduced a "proportional-feathering parameter," $\theta = U\alpha/\omega h$. Physically, this parameter is the ratio of the plate slope to the slope of the path traversed in the space by the axis of pitch. Since this path is sinusoidal, the largest value α can assume for positive thrust is the maximum slope of the trajectory, which is $\omega h/U$. Thus, θ is usually less than 1, and $\theta = 1$ corresponds to geometrically complete feathering of the fin. In terms of the other set of notation, θ can also be written as

$$\theta \equiv U\alpha/\omega h = -(2/\sigma)(\xi_1^2 + \xi_2^2)/\xi_0\xi_2 = -(2/\sigma)\, Z_p \csc \alpha_p \,. \tag{4.66}$$

The advantage of Lighthill's form (4.62) first appears in the result where the energy lost in the wake has a sharp minimum when $b = l/2$, and $\theta = 1$, whereas the rate of working increases somewhat for axis position b distal to that and for smaller values of θ. Consequently, an optimum axis position from thrust considerations as well as from efficiency considerations lies somewhere between $b = l/2$ and $b = l$ (i.e., for the pitching axis to lie between the $\frac{3}{4}$-chord point and the trailing edge).

The optimum values of b/l and θ corresponding to the optimum solution (4.59) are plotted versus σ for several values of $\bar{C}_{T_0}$ in Figs. 14 and 15, in which the dotted chain lines correspond to $\sigma = \sigma_m$, along which the leading-edge suction is the smallest possible. Along this line,

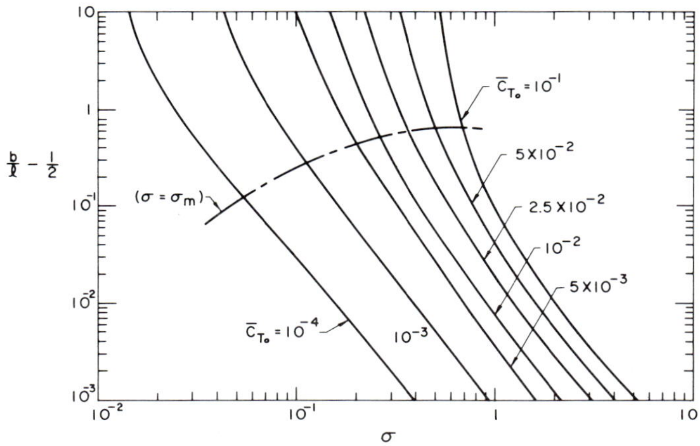

FIG. 14. The optimum location of pitching axis $x = b$ when heaving is taken to lead pitching by 90° in phase. The dotted-chain line corresponds to $\sigma = \sigma_m$ along which the leading-edge suction is minimum.

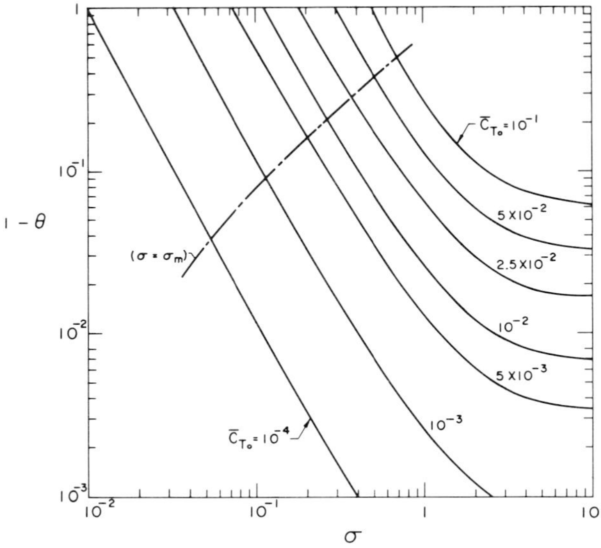

FIG. 15. Variation of the feathering parameter θ with the reduced frequency σ. The dotted-chain line corresponds to $\sigma = \sigma_m$.

the pitch axis b increases from $l/2$ to about $1.2l$ whereas the feathering parameter θ falls off from 1 to 0 as $\bar{C}_{T_0}$ increases. These general features of the optimum movement have been predicted by Lighthill (1970).

The optimum shape of a flexible plate has also been analyzed by Wu (1971b) for the most general case of infinite degrees of freedom. It has been shown that the solution can be determined to a certain extent, but is not always uniquely determinate.

D. Movements of Porpoise Tail

Lang and Daybell (1963) reported a series of experiments dealing with the swimming performance of a porpoise (of genus *Lagenorhyncus obliquidens*, or the Pacific Whitesided Dolphin) which was trained to swim and glide along an almost straight course in a long towing tank. This was perhaps one of very few of the most exhaustive and carefully conducted tests of a live cetacean under a well-controlled condition.

The porpoise tail, nearly triangular but slightly crescent in shape, has an aspect ratio of about 5.4, which may be considered large. It is therefore reasonable to adopt the strip theory, using the local two-

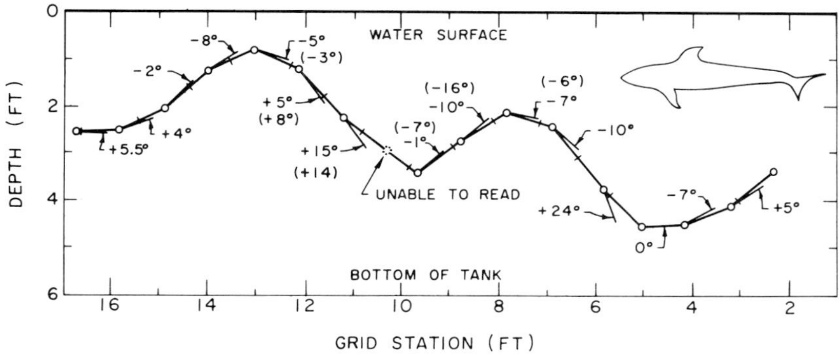

FIG. 16. Tail movements of a porpoise in cruising. The angles with arrows are the incidence angles of the tail relative to the path of the tail base measured by Lang and Daybell (1963); the angles in parenthesis are the theoretical prediction at the corresponding position. (Experimental data—Courtesy of Dr. T. G. Lang.)

dimensional characteristics (provided by the theory discussed in the previous section) for each strip, so as to give a qualitative comparison between the theory and the experimental results. For this purpose a representative run (Run No. 15-22) was selected from the experiments of Lang and Daybell; the path traversed by the base of the porpoise's tailfluke and the tail "feathering angle" (i.e., the angle between the tail and its own trajectory) are reproduced in Fig. 16. The test data were used by Wu (1971b) to obtain the following estimates:

$$C_{T_0} = 0.11, \qquad \xi_0 = 8.3, \qquad \bar{C}_{T_0} = C_{T_0}/\xi_0{}^2 = 1.6 \times 10^{-3}. \tag{4.67}$$

The corresponding σ_c [see Eq. (4.57)] and the reduced frequency σ (based on the estimated half-chord of tail) are

$$\sigma_c \simeq 0.01, \qquad \sigma \simeq 0.23 \tag{4.68}$$

Suppose that this tail movement was performed at the optimum condition, we can determine the following values for the efficiency, the amplitude ratio Z_p, and the phase advance α_p of the pitching mode from Figs. 14 and 15 for the above $\bar{C}_{T_0}$ and σ:

$$\eta = 0.99, \qquad Z_p = 0.104, \qquad \alpha_p = 263°. \tag{4.69}$$

The feathering angle of the tail, α_{tail}, can further be deduced as

$$\alpha_{tail} \simeq 0.55 \sin \omega t - 0.862 \sin(\omega t - 7°). \tag{4.70}$$

This predicted α_{tail} is shown in Fig. 16 within parentheses directly below the experimental data of Lang and Daybell. This comparison, however, should be properly qualified since the application of the two-dimensional theory tends to overestimate the efficiency; also the determination of the effective mean chord is crude, and the accuracy of the measured α_{tail} was originally claimed to be somewhat uncertain. These rather obscure circumstances notwithstanding, it is still of significance to note that the general trend of the theoretical prediction is in fair agreement with the experimental measurements.

In terms of Lighthill's form (4.62) of the lateral motion, the pitch axis position b and the feathering parameter θ corresponding to Z_p and α_p given by (4.69) follow from (4.63) and (4.66) to give

$$b/l \simeq 0.585, \qquad \theta = 0.92. \tag{4.71}$$

The above value of b/l locates the pitch axis at about the 0.8-chord point from the leading edge, which is well in the favorable range predicted by Lighthill (1970). Furthermore, the θ value is quite close to the state of complete feathering. Finally, an interpolation check with Figs. 13–15 shows that the observed reduced frequency $\sigma = 0.23$ is somewhat greater than the σ_m ($\simeq$0.14 for the $\bar{C}_{T_0}$ at hand); the leading-edge suction at this σ is nevertheless still reasonably small ($C_S/C_T \simeq 0.4$).

In conclusion, we summarize the main features of the tail movement as just analyzed. (i) The estimated reduced frequency $\sigma = 0.23$ is large compared with the critical frequency $\sigma_c = 0.01$, but lies well in the range in which the leading-edge suction is not large. (ii) The loading parameter $\bar{C}_{T_0}$ ($=1.6 \times 10^{-3}$ as estimated) is very small, mainly owing to the large amplitude of heaving. (iii) The phase difference $\alpha_p = 263°$ between the pitching (about mid-chord) and heaving modes falls in the range of σ where α_p is nearly stationary. (iv) With pitching kept at a rather small amplitude (pitching–heaving amplitude ratio $Z_p = 0.11$ in this case) but with the correct phase α_p, an impressively high efficiency ($\eta \simeq 0.99$) can be achieved. (v) When the pitching is referred to the heaving by a 90° phase advance, the pitch axis is at about the 0.8-chord point, and the feathering ($\theta = 0.92$) is nearly complete. It seems quite conclusive that (i) is the primary condition for selecting the frequency σ in practice.

E. Movements of Bird's Wing in Flapping Flight

The foregoing two-dimensional theory can also be used to discuss qualitatively the optimal movement of a bird's wing in flapping flight, as most species of migrating birds have wings of large aspect ratio and there must be a considerable saving of energy when the wing movement is optimum. The strip theory has been adopted by Wu (1971b) to provide a first-order estimate, leaving the effect of finite span as a further refinement. A somewhat superficial difference between fish propulsion and bird flight arises from the need in the latter case of adding on to the oscillatory motion the constant angle of attack required for supporting the body weight in air. But this steady component can be accounted for separately; it does not correlate with the oscillatory component in the balance of mean energy.

The general picture is roughly as follows: As the wing flaps up and down, the pitching amplitude increases with the distance outward from the body, reaching a nearly horizontal position at the top and bottom of each stroke. Such a wing movement, according to this simple strip theory argument, is the most efficient, and leaves behind the least possible vorticity in overcoming a given frictional drag. This crude picture may be further refined by employing a more accurate lifting-line or lifting-surface theory and by including physiological considerations about limitations of physical structure, muscular power, metabolic rate, and other factors.

V. Concluding Remarks

The hydromechanics of the swimming of fishes and cetaceans has been developed so far on the basis of small amplitude theory. Under natural conditions, however, the amplitude of the undulatory transverse waves of the body motion is usually moderate or large, even during the optimum performance at high hydromechanical efficiencies. While the simplified slender-body theory and the two-dimensional theory, if applied properly as demonstrated in the foregoing, can provide useful information and preliminary physical insight that would be valuable for deeper understanding, it is nevertheless very difficult to give here an estimate of their degree of accuracy. Such estimations must come from a more complete and improved theory and from a set of well-planned and precisely executed experiments with live specimens or models under conditions in close simulation to nature. Further developments in such theoretical and experimental approaches are certainly important and desirable; both approaches appear challenging and the major steps in these directions seem to call for much ingenuity and new concept. The main lines of future research may include (i) an extension of slender-body theory to large amplitude motion, (ii) the development of a three-dimensional, large amplitude, lifting-surface or lifting-line theory which will be sufficiently accurate to provide reliable pressure forces near the leading edge; (iii) the calculation and measurements of unsteady boundary layers around a body in lateral motions of finite amplitude. Some of these new thoughts have been expounded and explored (in regard to their feasibility) in the great studies by Lighthill (1969, 1970), which the readers may also find as rich resources of other interesting ideas in this developing field.

Aside from the undulatory mode of transverse waves, which has formed the central theme of the present discussion, there are other varieties of body motions, such as (a) traveling waves primarily along fringe belts, or along a fin only, as used by some flatfishes like knifefish (see, e.g., Hertel, 1963) and halibut; (b) bottom swimmers relying on some ground effect as in the swimming of flatfishes and rays (the latter possessing greatly developed pectoral fins whose motion resembles the flapping wings of birds); (c) actual ejecting of liquid as employed by squids; (d) propulsion by bending of abdominal limbs of lobsters, shrimps, prawns, and by waving of a large number of dense tassels underneath a star fish. Some of these problems have already received attention, some others would require only a minor extension of the present theories, and still others may call for major efforts. The item (d) represents perhaps the low efficiency end of the spectrum of aquatic animals not microscopically small, emerging into a common ground with (e) flagellated propulsion of protozoa and bacteria, and (f) ciliated propulsion of numerous microorganisms by waving movements of dense cilia attached to the body surface. These latter cases of microorganism locomotion, a subject as broad and interesting as the larger aquatic animals discussed here, are considered to be out of the original scope of the present survey; they will not be pursued further here, but hopefully will be treated in a future study.

ACKNOWLEDGMENT

I wish to acknowledge with deep appreciation and gratitude the extremely valuable discussions with Professor M. James Lighthill, leading to a much enlightened understanding of this interesting subject. The first version of this review article had been completed shortly before I had the pleasure and privilege of learning from Professor Lighthill his most recent, not yet then published, great study (1970), in the light of which this article has been subsequently extended (the revision would have been more extensive had the time permitted). I am deeply indebted to Professor Lighthill for letting me quote and refer to his important contributions in over a decade. I am grateful to Dr. Tom Lang and Professors Charles DePrima and Duen-pao Wang for interesting and valuable discussions. Expressions of appreciation and thanks are also due Sir James Gray and Sir G. I. Taylor for their stimulating and pioneering contributions that have never failed to attract from many of us a strong interest in this subject. Also, I would like to acknowledge the courtesy of the *Journal of Fluid Mechanics* for letting me cite various articles and contributions as well as my own articles. Finally, I treasure with sincere thanks the untiring encouragement from Professor Chia-Shun Yih.

Preparation of this article was partially sponsored by the National Science Foundation, under Grant GK 10216, and by the Office of Naval Research, under Contract N00014-67-A0094-0012.

REFERENCES

Ashley, H., Windall, S., and Landahl, M. T. (1965). New directions in lifting surface theory. *AIAA J.* **3**, 3–16.

Bainbridge, R. (1960). Speed and stamina in three fish. *J. Exp. Biol.* **37**, 129–153.

Bainbridge, R. (1961). Problems of fish locomotion. *Symp. Zool. Soc. London* **5**, 13–31.

Bainbridge, R. (1963). Caudal fin and body movement in the propulsion of some fish. *J. Exp. Biol.* **40**, 23–56.

Benjamin, T. B. (1960). Effects of a flexible boundary on hydrodynamic stability. *J. Fluid Mech.* **9**, 513–532.

Betchov, R. (1959). Rep. ES-29174. Douglas Aircraft Co.

Breder, C. M. (1926). Locomotion of fishes. *Zoologica* **4**, 159–297.

Chwang, A. T., and Wu, T. Y. (1971). A note on the helical movement of micro-organisms. *Proc. Roy. Soc., Ser. B* (in press).

Fabula, A. G., Hoyt, J. W., and Crawford, H. R. (1963). *Amer. Phys. Soc. Meet., Buffalo, New York.*

Fung, Y. C. (1967). Elasticity of soft tissues in simple elongation. *Amer. J. Physiol.* **213**, 1532–1544.

Gadd, G. E. (1952). Some hydrodynamical aspects of the swimming of snakes and eels. *Phil. Mag.* [7] **43**, 663–670.

Gadd, G. E. (1963). Some hydrodynamical aspects of swimming. Ship Rep. No. 45. Natl. Phys. Lab., London.

Gray, J. (1948). Aspects of the locomotion of whales. *Nature (London)* **161**, 199–200.

Gray, J. (1949), Aquatic locomotion. *Nature (London)* **164**, 1073–1075.

Gray, J. (1968). "Animal Locomotion." Weidenfeld & Nicolson, London.

Gray, J., and Hancock, G. J. (1955). The propulsion of sea-urchin spermatozoa. *J. Exp. Biol.* **32**, 802–814.

Greidanus, J. H., van de Vooren, A. I., and Bergh, H. (1952). Experimental determination of the aerodynamic coefficients of an oscillating wing in incompressible, two-dimensional flow. Reps. F-101-F-104. Natl. Luchtvaart-Lab., Amsterdam.

Hertel, H. (1963). "Structure-Form-Movement." Otto Krausskopf Verlag, Mainz (English translation, Reinhold, New York, 1966).

Hoyt, J. W. (1970). Private communication.

Johannessen, C. L., and Harder, J. A. (1960). Sustained swimming speeds of dolphins. *Science* **132**, 1550–1551.

Jones, D. S. (1957). The unsteady motion of a thin aerofoil in an incompressible fluid. *Commun. Pure Appl. Math.* **10**, 1–21.

Kelly, H. R. (1960). Fish propulsion, a supplement to the theory of Smith and Stone. Tech. Note 40606-3. Naval Ordnance Test Station, China Lake, California.

Kelly, H. R. (1961). Fish propulsion hydrodynamics. *In* "Developments in Mechanics" (J. E. Lay and L. E. Malvern, eds.), Vol. I, pp. 442–450. Plenum Press, New York.

Kramer, M. O. (1960). Boundary layer stabilization by distributed damping. *J. Amer. Soc. Nav. Eng.* **72**, 25–33.

Küssner, H. G., and Schwarz, L. (1940). Der schwingende Flügel mit aerodynamisch ausgeglichenem Ruder. *Luftfahrt-Forsch.* **17**, 337–354.

Landahl, M. T. (1961). *J. Fluid Mech.* **13**, 609–632.

Lang, T. G. (1962). Analysis of the predicted and observed speeds of porpoises, whales, and fish. Tech. Note P5015-21. Naval Ordnance Test Station, Pasadena, California.

Lang, T. G. (1966). Hydrodynamic analysis of cetacean performance. "Whales, Dolphins and Porpoises" (K. S. Norris, ed.). Univ. of California Press, Berkeley, California.

Lang, T. G., and Daybell, D. A. (1963). Porpoise performance tests in a sea-water tank. NAVWEPS Rep. 8060, NOTS TP 3063. Naval Ordnance Test Station, China Lake, California.

Lang, T. G. and Norris, K. S. (1966). *Science* **151**, 588.

Lang, T. G. and Pryor, K. (1966). *Science* **152**, 531.

Lighthill, M. J. (1960a). Note on the swimming of slender fish. *J. Fluid Mech.* **9**, 305–317.

Lighthill, M. J. (1960b). Mathematics and aeronautics. *J. Roy. Aeronaut. Soc.* **64**, 375–394.

Lighthill, M. J. (1969). Hydrodynamics of aquatic animal propulsion. *Annu. Rev. Fluid Mech.* **1**, 413–445.

Lighthill, M. J. (1970). Aquatic animal propulsion of high hydromechanical efficiency. *J. Fluid Mech.* **44**, 265-301.

Lighthill, M. J. (1971). Private communication.

Muskhelishvili, N. I. (1953). "Singular Integral Equations." P. Noordhoff Ltd., Groningen, Holland.

Osborne, M. F. M. (1960). The hydrodynamical performance of migratory salmon. *J. Exp. Biol.* **38**, 365–390.

Pao, S. K., and Siekmann, J. (1964). Note on the Smith-Stone theory of fish propulsion. *Proc. Roy. Soc., Ser. A* **280**, 398–408.

Ransleben, G. E., Jr., and Abramson, H. N. (1962). Experimental determination of oscillatory lift and moment distributions on fully submerged flexible hydrofoils. Rep. No. 2. Southwest Res. Inst.

Saffman, P. G. (1967). The self-propulsion of a deformable body in a perfect fluid. *J. Fluid Mech.* **28**, 385–389.

Schwarz, L. (1940). Berechnung der Druckverteilung einer harmonisch sich verformenden Tragfläche in ebener Stromung. *Luftfahrt-Forsch.* **17**, 379–386.

Siekmann, J. (1962). Theoretical studies of sea animal locomotion. Part 1. *Ing.-Arch.* **31**, 214–228.

Siekmann, J. (1963). On a pulsating jet from the end of a tube, with application to the propulsion of certain aquatic animals. *J. Fluid Mech.* **15**, 399–418.

Smith, E. H., and Stone, D. E. (1961). Perfect fluid forces in fish propulsion: The solution of the problem in an elliptic cylinder coordinate system. *Proc. Roy. Soc., Ser. A* **261**, 316–328.

Taylor, G. I. (1952). The action of waving cylindrical tails in propelling microscopic organisms. *Proc. Roy. Soc., Ser. A* **211**, 225–239.

Theodorsen, T. (1934). General theory of aerodynamic instability and the mechanism of flutter. *Nat. Adv. Comm. Aeronaut., Tech. Rep.* **496**.

Uldrick, J. P., and Siekmann, J. (1964). On the swimming of a flexible plate of arbitrary finite thickness. *J. Fluid Mech.* **20**, 1–33.

Van Driest, E. R., and Blumer, C. B. (1963). Rep. SID 63-390. North Am. Aviation, Inc.

Veltkamp, G. W. (1958). The drag on a vibrating aerofoil in incompressible flow. *Indag. Math.* **20**, 278–297.

von Kármán, T., and Burgers, J. M. (1943). General aerodynamic theory—perfect fluids. *In* "Aerodynamic Theory" (W. F. Durand, ed.), Vol. II, Div. E, pp. 304–310. Springer, Berlin.

von Kármán, T., and Sears, W. R. (1938). Airfoil theory for nonuniform motion. *J. Aeronaut. Sci.* **5**, 379–390.

Walters, V., and Fiersteine, H. L. (1964). *Nature (London)* **202**, 208–209.

Wang, P. K. C. (1966). Optimum propulsion of an oscillating hydrofoil. *IEEE Trans. Auto. Control* **11**, 645–651.

Wu, T. Y. (1961). Swimming of a waving plate. *J. Fluid Mech.* **10**, 321–344.

Wu, T. Y. (1962). Accelerated swimming of a waving plate. *Proc. 4th Symp. Nav. Hydrodyn.*, 1962 ONR/ACR-92, pp. 457–473. U.S. Govt. Printing Office, Washington, D.C.

Wu, T. Y. (1966). The mechanics of swimming. *In* "Biomechanics" (Y. C. Fung, ed.), p. 112. Am. Soc. Mech. Eng., New York.

Wu, T. Y. (1968). Fluid mechanics of swimming propulsion. *Proc. 8th Symp. Nav. Hydrodyn.*, (in press).

Wu, T. Y. (1971a). Hydromechanics of swimming propulsion. Part 1. Swimming of a two-dimensional flexible plate at variable forward speeds in an inviscid fluid. *J. Fluid Mech.* **46**, 337–355.

Wu, T. Y. (1971b). Hydromechanics of swimming propulsion. Part 2. Some optimum shape problems. *J. Fluid Mech.* **46**, 521–544.

Wu, T. Y. (1971c). Hydromechanics of swimming propulsion. Part 3. Swimming of slender fish with side fins and its optimum movements. *J. Fluid Mech.* **46**, 545–568.

Biomechanics: A Survey of the Blood Flow Problem

Y. C. FUNG

Department of Engineering Sciences
(including Bioengineering)
University of California at San Diego, La Jolla, California

I. Introduction

What kind of problem in biology and medicine leads to a significant investigation in applied mechanics? What can applied mechanics contribute to the biomedical field? Answers to these questions are what one wishes to know before devoting his time to the study of biomechanics.

Perhaps the best way to answer these questions is to survey briefly the history of biomechanics, and to illustrate one or two cases in some depth. In this article, we shall consider the problems of blood flow. Our attention will be focused on the question of proper mathematical formulation of the problems. Therefore, we shall discuss the constitutive equations of the various tissues involved, the geometrical configurations and dimensions of the system, and the possible boundary conditions. The mathematical framework of the problems of wave propagation in arteries and peristalsis in small vessels will be discussed. The scope of the existing literature on physiologically important problems will be presented.

II. Historical Remarks

Since science and scientists have always been concerned with the world of the living, it is natural that biomechanics has roots which reach back to the dawn of civilization. Of the ancient philosophers, Aristotle (384–322 B. C.) was an eloquent speaker on the connection between "physics"—which for him was a general description of the universe—and the study of living things. However, biomechanics in the form we understand it now probably began with Galileo and Harvey. William Harvey (1578–1658) discovered blood circulation in 1615 although, having no microscope, he never saw the capillary blood vessels. This should make us appreciate his conviction in logical reasoning even more deeply today, because without the ability to see the passage from the arteries to the veins, the discovery of circulation must be regarded as "theoretical." The actual discovery of capillaries was made by Marcello Malpighi (1628–1694) in 1661, 46 years after Harvey made the capillaries a logical necessity.

A contemporary of William Harvey was Galileo (1564–1642) who was a student of medicine before he became famous as a physicist. He discovered the constancy of the period of a pendulum, and used the pendulum to measure the pulse rate of people, expressing the results quantitatively in terms of the length of a pendulum synchronous with the beat. He invented the thermoscope, and made the first microscope in the modern sense in 1609, although rudimentary microscopes had been made earlier by J. Jansen and his son Zacharias in 1590.

Young Galileo's fame was so great and his lectures at Padua so popular that his influence on biomechanics went far beyond his personal contributions mentioned above. Harvey studied at Padua (1598–1601) while Galileo was active there. The essential part of Harvey's demonstration of circulation is the result not of mere observation but of the application of Galileo's principle of measurement. Having shown that the blood can only leave the ventricle of the heart in one direction, he turned to measure the capacity of the heart. He found it to be two ounces. The heart beats 72 times a minute so that in 1 hour it throws into the system $2 \times 72 \times 60$ ounces = 8640 ounces = 540 pounds of blood! Where can all this blood come from? Where can it all go? He deduced the existence of circulation as a necessary consequence of the function of the heart.

Another colleague of Galileo, Santorio Santorio (1561–1636), a professor of medicine at Padua, used Galileo's method of measurement and philosophy to compare the weight of the human body at different times and in different circumstances. He found that the body loses weight by mere exposure, a process which he assigned to "insensible perspiration." His experiments laid the foundation of the modern study of "metabolism."

The physical discoveries of Galileo and the demonstrations of Santorio and Harvey gave a great impetus to the attempt to explain vital processes in terms of mechanics. Galileo showed that mathematics was the essential key to science without which nature could not be properly understood. This idea was expounded further by the great philosopher, René Descartes, who was about 30 years younger than Galileo. Descartes developed the thesis that the universe can be, and can only be, understood by mathematical thinking. He applied this principle to biology as well as to cosmology. In simpler problems such as the optics of the eye glasses, and of the eyeball itself, he met with great success. However, this approach sometimes fails. In a work published posthumously (1662 and 1664), he proposed a complete human physiology solely upon theoretical grounds. He set forth a very complicated model of the animal structure, including the system of nerves. But subsequent investigations failed to confirm many of his findings. These errors of fact caused a loss of confidence in Descartes' approach—a lesson which should be borne in mind by all theoreticians.

Other attempts a little less ambitious than Descartes' were more successful. Giovanni Alfonso Borelli (1608–1679) was an eminent Italian geometer and astronomer. His *On Motion of Animals* (1680) is a classic. He was successful in clarifying the muscular movement and body dynamics. He treated the flight of birds and the swimming of fishes, the movements of the heart and of the intestines.

Robert Boyle (1627–1691) studied the lung, and discussed the function of air in water with respect to fish respiration. Robert Hooke (1635–1703) gave us Hooke's law in mechanics, and the word "cell" to biology to designate the elementary entities of life. (His famous book *Micrographia*

(1664) was reprinted recently by Dover Publications.) Leonhard Euler (1707–1783) wrote a definitive paper in 1775 on the blood flow in the arteries. Thomas Young (1773–1829), who gave us the Young's modulus in elasticity and demonstrated the wave theory of light, was a physician in London. He worked on astigmatism in lenses and on color vision. Jean Poiseuille (1799–1869) invented the mercury manometer to measure the blood pressure in the dog's aorta (1828) while he was a medical student, and discovered his law of viscous flow (1840).

To Herrmann von Helmholtz (1821–1894), Fig. 1, might go the title

FIG. 1. Portrait of Herrmann von Helmholtz. From the frontispiece to Wissenschaftliche Abhandlungen von Helmholtz. Leipzig, Johann Ambrosius Barth, 1895. Photo by Giacomo Brogi in 1891.

"Father of Biomechanics." He was professor of physiology and pathology at Königsberg, professor of anatomy and physiology at Bonn, professor of physiology at Heidelberg, and finally professor of physics in Berlin

(1871). He wrote his paper on the conservation of energy while he was in military service, immediately after graduation from medical school. His contributions ranged over mathematics, hydrodynamics, elasticity, vibration theory, optics, acoustics, thermodynamics, electrodynamics, physiology, and medicine. He discovered the focusing mechanism of the eye and, following Young, formulated the three-color theory of color vision. He invented the phakoscope to study the changes in the lens, the ophthalmoscope to view the retina, the ophthalmometer for the measurement of eye dimensions, and the stereoscope with interpupillary distance adjustments for stereo vision. He studied the mechanism of hearing and invented the Helmholtz resonator. His theory of the permanence of vortex lines lies at the very foundation of modern fluid mechanics. His book *Sensations of Tone* is popular even today. He was the first to determine the velocity of the nerve pulse, giving the rate 30 meter/sec, and to show that the heat released by muscular contraction is an important source of animal heat.

There are a number of other names in biomechanics which are equally familiar to engineers. The physiologist Adolf Fick (1829–1901) was the author of Fick's law in mass transfer. The hydrodynamicists Diederik Johannes Korteweg (1848–1941) and Horace Lamb (1849–1934) wrote beautiful papers on wave propagation in blood vessels. Otto Frank (1865–1944) worked out a hydrodynamic theory of the heart. Balthasar van der Pol (1889–1963) wrote in 1926 about the modeling of the heart with the nonlinear (van der Pol) oscillators, and was able to simulate the heart with four van der Pol oscillators to produce a realistic-looking electrocardiograph (van der Pol and van der Mark, 1929).

With such a rich background we must say, however, that biological fluid mechanics fell outside the main stream of development of fliud mechanics for a long time. The serious study of blood flow is a fairly recent endeavor. Many bio-fluid-mechanical problems remain to be discovered and solved.

III. Basic Information Required for Formulating Blood Flow Problems

For the blood flow problem we must consider the blood, the blood vessels, the tissues surrounding the blood vessels, the geometry of the vascular system, and the driving forces. Let us discuss these elements one by one.

A. Blood Rheology

Blood is a marvellous fluid which nurtures life, contains many enzymes and hormones, knows when to flow and when to clot, and transports oxygen and carbon dioxide between the lungs and the cells of the tissues. We can leave most of these important functions of blood to physiologists, biochemists, and pathological chemists, however; for fluid mechanics the most important information we need is the constitutive equation.

Blood is a suspension of cells in an aqueous solution of electrolytes and nonelectrolytes. By centrifugation, the blood is separated into *plasma* and *cells*. The plasma is about 90% water by weight, 7% plasma protein, 1% inorganic substances, and 1% organic substances. The cellular contents are essentially all *erythrocytes*, or *red cells*, with *white cells* of various categories making up less than 1/600th of the total cellular volume, and *platelets* less than 1/800th of the cellular volume. Normally the red cells occupy about 50% of the bood volume (the rest is plasma). They are small, and number about 5 million/mm^3. The normal white cell count is considered to be from 5000 to 8000/mm^3, and platelets from 250,000 to 300,000/mm^3. Human red cells are disk shaped, with a diameter of 8.5×10^{-4} cm and thickness 2.4×10^{-4} cm. White cells are more rounded and have many types. Platelets are much smaller, and have a diameter of about 2.5×10^{-4} cm.

If the blood is allowed to clot, a straw-colored fluid called *serum* is expressed into the plasma when the clot spontaneously contracts. Serum is similar to plasma in composition, but with one important colloidal protein, *fibrinogen*, removed while forming the clot. Most of the platelets are enmeshed in the clot.

The specific gravity of red cells is about 1.10, that of plasma is 1.03. When plasma was tested in a viscometer, it was found to behave like a Newtonian viscous fluid (Merrill *et al.* 1965, Gabe and Zazt, 1968), with a coefficient of viscosity about 1.2 cP (Gregersen, 1967; Chien *et al.*, 1966, 1971). When whole blood was tested in a viscometer, its non-Newtonian character was revealed. Figure 2 shows the variation of the viscosity of blood with the strain rate when the blood is tested in a Couette-flow viscometer, the gap width of which is much larger than the diameter of the individual red cells. The viscosity of blood varies with the *hematocrit*, i.e., the percentage of the total volume of blood occupied by the cells. It varies also with temperature (see Fig. 3), and with disease state, if any.

There is a question about what happens to the blood viscosity when the strain rate is reduced to zero. Cokelet *et al.* (1963) deduced the existence of a finite yield stress. They say that at a vanishing shear rate

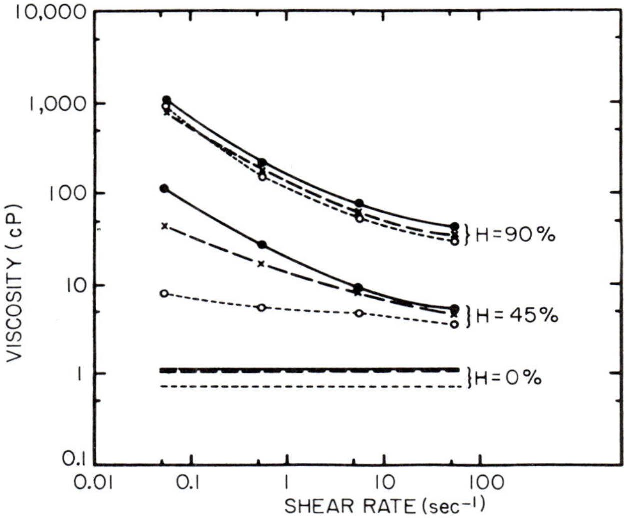

FIG. 2. The viscosity–shear rate relations in whole blood (●——●), defibrinated blood (×– –×), and washed cells in Ringer (○– – –○) at 45 and 90% red cell concentrations. (From Chien *et al.*, 1966.)

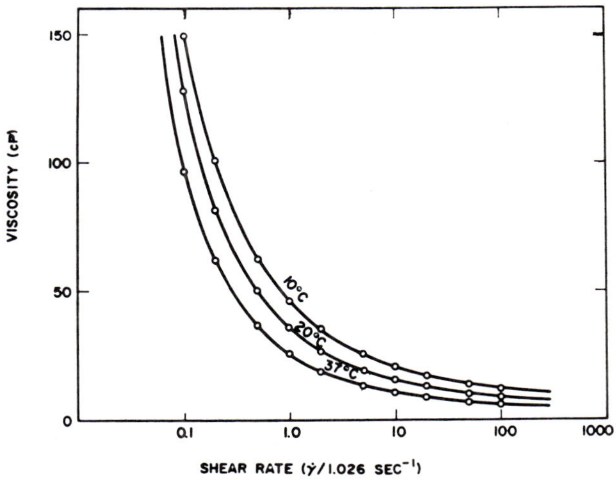

FIG. 3. The variation of the viscosity of human blood ($\eta = 100\ \tau/\dot{\gamma}$) with the shear rate $\dot{\gamma}$ and the temperature for a male donor, containing acid-citrate-dextrose, reconstituted from plasma and red cells to the original hematocrit of 44.8. From Merrill *et al.* (1963a, p. 257).

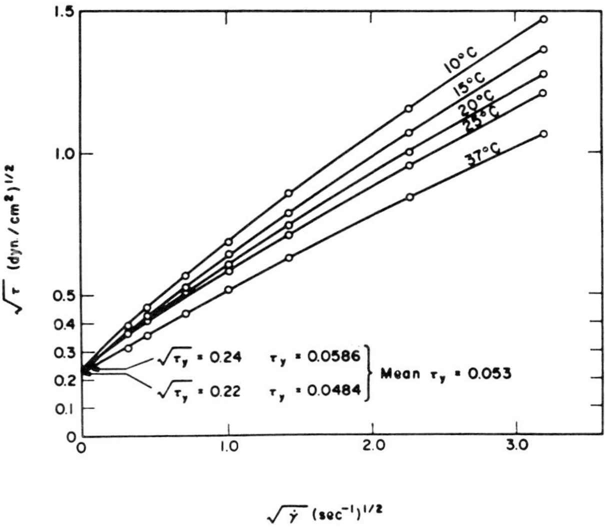

FIG. 4. The variation of the square root of shear stress $(\tau)^{1/2}$ with the square root of shear rate $(\dot{\gamma})^{1/2}$ and the temperature. Human blood. According to Merrill *et al.* (1963a, p. 257).

the blood behaves like an elastic solid (see Fig. 4). They deduced this conclusion on the basis that their torque measuring device had a rapid response, and could be used to measure transient effects. They studied the time history of the torque after the rotating bob had been stopped suddenly, and compared the transient response for blood with that for a clay suspension which is known to have a finite yield stress. They showed that the values deduced from this experiment agreed within a few percent with those obtained by extrapolation of the Casson plot as shown in Fig. 4. Merrill *et al.* (1965) also used a capillary viscometer to see if blood in a capillary could maintain a pressure difference across the tube ends without any detectable fluid flow. Such a pressure difference was detected and found to agree with the yield stress determined by Cokelet's method.

It must be understood that by "existence" of a yield stress is meant that no sensible flow can be detected in a fluid under a shearing stress in a finite interval of time (say, 15 min). The difficulty of determining the yield stress of blood as $\dot{\gamma} \to 0$ is compounded by the fact that an experiment for very small shear rate is necessarily a transient one if that experiment is to be executed in a finite interval of time. For blood the analysis is further complicated by the migration of red cells away from the walls of the viscometer when $\dot{\gamma}$ is smaller than about 1 sec^{-1}. Cokelet's analysis takes these factors into account, and requires considerable manipulation of the raw data (see Cokelet, 1971). If the maximum transient shear stress reached after the start of an experiment at constant shear rate is plotted directly with respect to the nominal shear rate, the result would appear as shown in Fig. 5 by Chien *et al.* (1966). The differences between the plots of Figs. 4 and 5 are caused mainly by the data analysis procedures, and partly reflect the dynamics of the instrument as well as of the blood state.

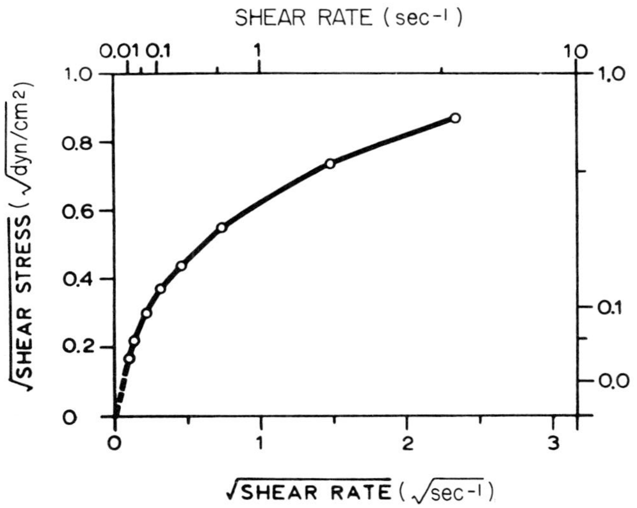

FIG. 5. Casson plot for whole blood at a hematocrit of 51.7 (temp. = 37°C) according to Chien *et al.* (1966).

The data of Cokelet *et al.* for small shear rate, say $\dot{\gamma} < 10\ \text{sec}^{-1}$, and for hematocrit less than 40%, can be described approximately by Casson's equation (1959)

$$(\tau)^{1/2} = (\tau_y)^{1/2} + b(\dot{\gamma})^{1/2}, \tag{3.1}$$

where τ is the shear stress, $\dot{\gamma}$ is the shear strain rate, τ_y is the yield stress in shear, and b is a constant. Note that the yield stress τ_y given by Cokelet *et al.* is very small: of the order of 0.05 dyn/cm^2, and is almost independent of the temperature in the range 10°–37°C. They state that τ_y is markedly influenced by the macromolecular composition of the suspending fluid. Suspension of red cells in saline plus albumin has zero yield stress; suspension of red cells in plasma containing fibrinogen has a finite yield stress.

B. The Red Blood Cell

The human red blood cell is small and extremely deformable, and flows through capillary blood vessels whose diameter is about the same as that of the red cell. In a static condition under a uniform external pressure the red cell appears as a biconcave disk (see Fig. 6). When it is

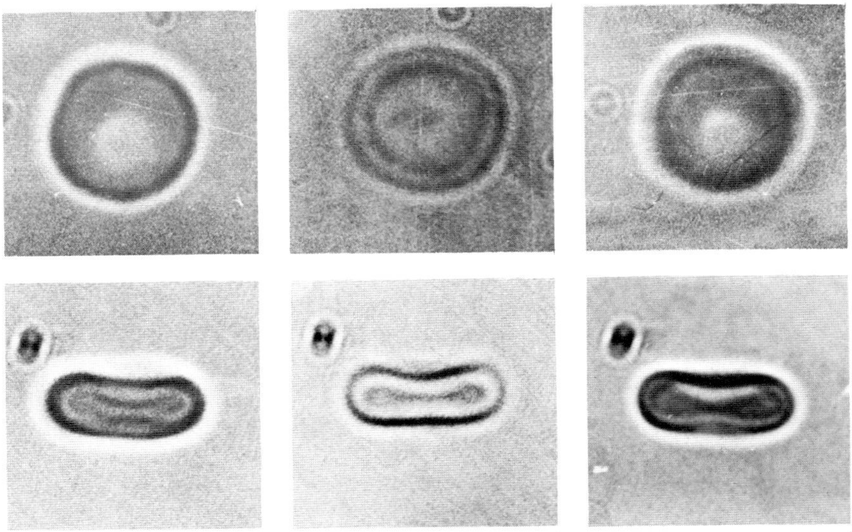

Fig. 6. Red cell suspended in Eagle solution. The side view appears as a disk; while the top view is circular. The three photographs were taken from the same red cell with slight changes in the focus of the microscope. This illustrates the fact that one must be careful in interpreting microscopic photographs. One must not attribute undue significance to the details which merely reflect the diffraction of light through a body.

forced to flow through a small capillary blood vessel, the red cell is stressed and deformed as shown in Fig. 7. The mechanical interaction between the red cell and the endothelial cells of the capillary blood vessel is an important and fascinating problem. The exchange of water and solutes depends on how tightly they are pressed together; the pressure–flow relationship depends on the magnitude of the shear stress imposed by the blood vessel wall on the red cells—any possible damage to the red cell, hemolysis, depends on how much the cell membrane is stressed. Hemolysis, however, is strongly influenced by interaction of blood with solid surface, and is a very complex phenomenon (see Blackshear, 1971). For red cells in flowing blood, see also Bloch (1962), and Branemark and Lindström (1963).

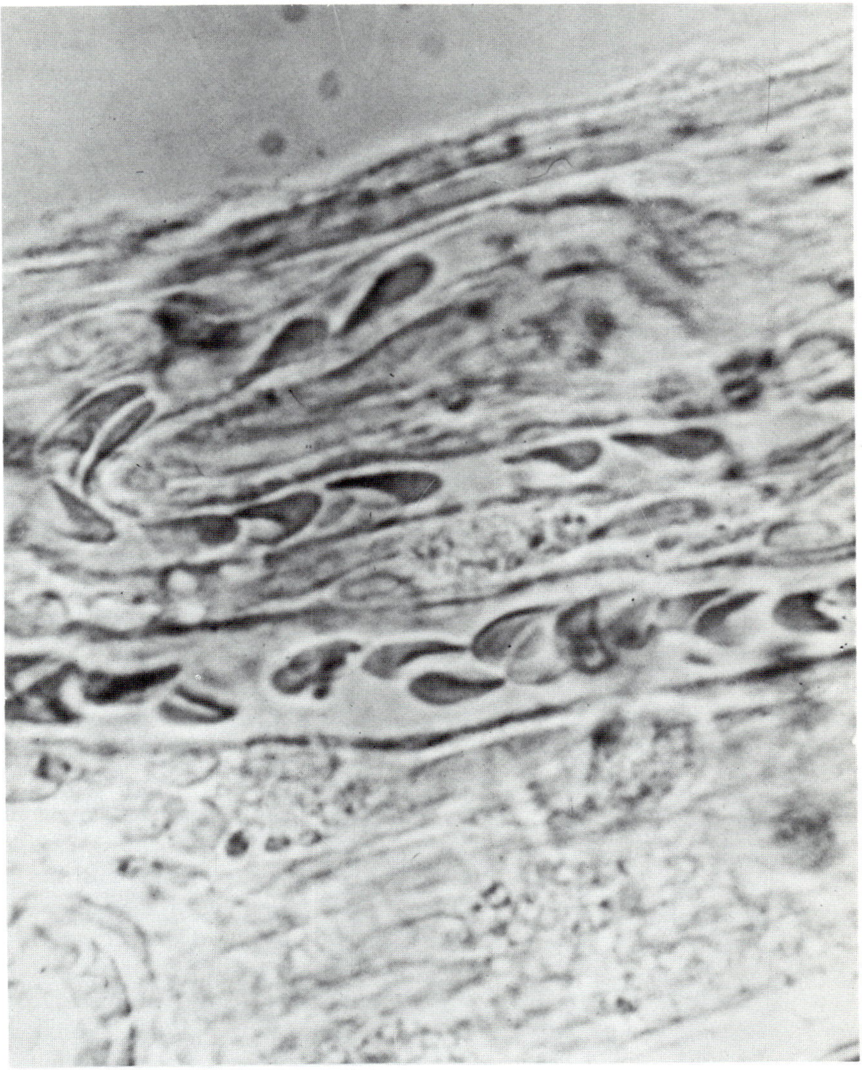

FIG. 7. Blood flow in the capillary blood vessels in the mesentery of a dog, showing the deformation of the red cells. Courtesy of Dr. Ted Bond. See Guest *et al.* (1963).

A number of books are available which summarize the physiology, chemistry, and pathology of the red cells. See Ponder (1948), Bessis (1956), Macfarlane and Robb-Smith (1961), and Harris (1965).

A theoretical analysis of the mechanics of the red cell based solely on the hypothesis that the cell contents behave like a liquid was presented by Fung (1966b). The consequences of the fact that a red cell can be transformed into a sphere when it is placed in a hypotonic solution are analyzed by Fung and Tong (1968), who deduced a necessary condition relating the distribution of the extensional rigidity (product of the Young's modulus and the wall thickness) to the surface tension and the pressure difference across the cell membrane.

The strongest indication of the correctness of the hypothesis that the interior of the red cells is a viscous liquid is the agreement of theoretical predictions with experimentally observed features. Among these the most important is the flexibility of the red cells. As is shown in Fig. 7, a red cell can be so deformed in a capillary blood vessel that its trailing edge looks like a cusp. A thin-walled, liquid-filled shell can be folded locally into a cusp without difficulty, but a homogeneous solid body cannot be deformed into a cusp without a local intensification of the external pressure towards the cusp. This is illustrated in Fig. 8. For a red

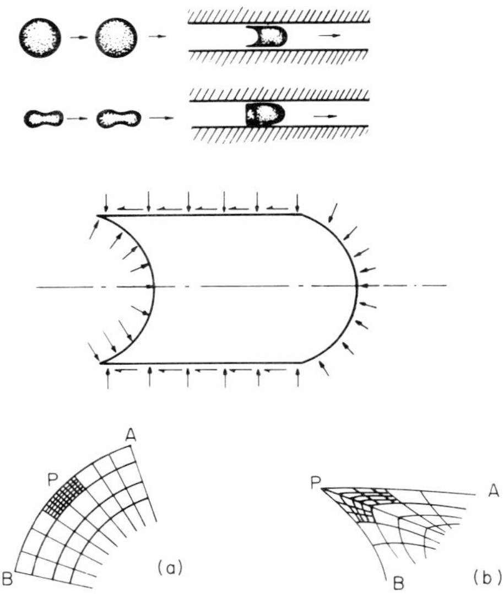

FIG. 8. The formation of a cusp at a point on the edge of a homogeneous solid through deformation calls for a localized concentration of surface stresses. In a capillary blood vessel such a stress concentration is absent. Hence the solid-interior hypothesis cannot be supported.

cell moving in a capillary blood vessel, the only forces acting on it are the shear stress and the pressure, which are more-or-less uniform. Hence the cusp cannot be explained by any simple, solid-interior model of the red cell.

The viscosity of the contents of the red cells has been measured by Cokelet and Meiselman (1968) and Schmidt–Nielsen and Taylor (1968). They packed red cells by centrifugation, hemolyzed the cells either by ultrasonic vibration or by chemical agents, then measured the viscosity after the ghosts (the cell membranes) were removed. At a hemoglobin concentration about 33% by weight which is found in red cells, a viscosity of about 6.0 cP was found for the cell contents. About the same time, Dintenfass (1968) obtained the same figure from theoretical calculations of the internal viscosity of red cells on the basis of viscometric flow of a suspension of liquid droplets.

Perhaps the most astonishing demonstration of the flexibility of the red blood cell is the experiment of Gregersen *et al.* (1967; Gregersen, 1967) on squeezing red cells through small cylindrical tunnels. They used sheets of polycarbonate paper which was irradiated and etched, thus producing randomly distributed circular cylindrical tunnels of rather uniform diameter. By varying the intensity of the radiation and the process of etching, the diameters of the cylindrical tunnels can be controlled. When blood is placed on the upper side of the paper, the red cells can be drained through the holes under the influence of gravity or a pressure difference. It was found that a red blood cell whose diameter is 8 μm can flow through tunnels of diameter as small as 2.4 μm without rupture. Forced flow through tunnels of smaller diameter will cause hemolysis. Hypotonically sphered red cells cannot pass through such small holes. Neither can artificially hardened red cells. Teitel (1965) has studied such squeezing *in vivo*, and measured the degree of hemolysis as the red cells pass through the small capillary blood vessels in the spleen. He suggests that it is nature's way of eliminating undesired spherocytes in the blood.

C. The Dependence of Apparent Blood Viscosity on the Size of the Measuring Instruments

Since blood is a suspension of red cells which must maintain their integrity, it is to be expected that when the dimensions of the tube through which blood flows are comparable with the red cell diameter, the blood cannot be regarded as a homogeneous fluid. The plasma can still be treated as a viscous fluid, but the red cells must be treated as flexible bodies well packed in plasma.

If a cylindrical tube is used as a viscometer, the viscosity measured is independent of the tube diameter as long as it is more than 1 or 2 mm.* When, however, tubes of smaller diameter are used, the apparent viscosity is found to be smaller. In Fig. 9a is shown the data obtained

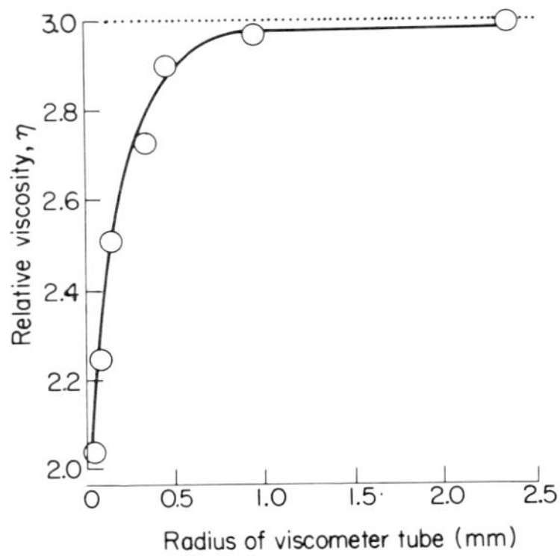

FIG. 9a. The decrease of apparent relative viscosity of blood (η) when flowing through circular cylindrical tubes with very small diameters. From Haynes (1960), with permission. For tubes of more than 1 mm radius, there is no further change in relative viscosity.

by Kumin and analyzed by Haynes (1960). The viscosity of the blood is expressed as relative viscosity: its ratio to that of water at the same temperature. Since the absolute viscosity of water is the same however small the diameter of the tube, the apparent viscosity is seen to decrease with decreasing tube size. This tendency is continued to very small tubes. In Fig. 9b is shown the data by Braasch and Jenett (1968) on pig blood. It is seen that the relative viscosity is decreased even when the tube diameter is only 6 μm. To interpret this data we should remember that the diameter of the erythrocytes of pig blood is about 6 μm. In a flow in the capillary tube the red cell is elongated as a whole, bulged out at the front end, and buckled at the rear end; so that the cross-sectional dimension can be considerably smaller than the diameter of a resting cell. In other words, even though the smallest tube used by Braasch and Jenett has a diameter about equal to the largest dimension of the red cell in the resting state, there will still be considerable clearance

* Assume that the flow rate is so high that the pressure drop is proportional to the flow rate.

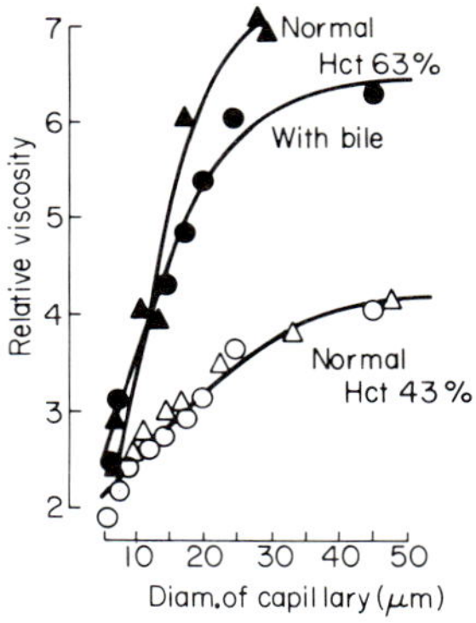

FIG. 9b. The dependence of the relative viscosity of blood on the diameter of the capillary tubes used in the measurement. All measurements were made at room temperature. The curve "with bile" refers to a solution with 50 ml serum + 50 ml erythrocytes + 1.5 ml bile, at a hematocrit of 60 %. As a control, 0.9 % NaCl was used in place of bile, at hematocrit 43 %. The curve for 63 % hematocrit shows the effect of hemoconcentration in blood which had a relative viscosity of 10.2 when measured in a 500 μm tube. From Braasch and Jenett (1968).

between the red cell and the tube wall in the condition of flow under which the measurements were made.

This has long been known as the Fahraeus–Lindqvist effect. It is undoubtedly a revelation of the individuality of the red cells—that the blood can no longer be treated as a homogeneous viscous fluid when the tube diameter is smaller than 0.1 cm. What is surprising is that the individuality of the red cells (of diamter 8×10^{-4} cm) should appear in such large tubes (with tube diameters up to 100 cell diameters).

Fahraeus and Lindqvist wrote about this variation of apparent viscosity with tube diameter in 1931. Similar effects were found in couette flow and cone-plate viscometers. There are at least two basic reasons for this phenomenon. The principal reason was pointed out by Fahraeus (1929) and detailed by others (see Cokelet, 1971), that when blood flows from a reservoir into a cylindrical tube, the volume concentration of red cells (hematocrit) in the tube is smaller than that in the reservoir. The smaller the tube the larger is this effect. The other reason is the migration of red cells away from the wall—a complex problem which has been reviewed exhaustively by Goldsmith and Mason (1967). It is interesting to note that the nonlinear convective inertia terms in the Navier–Stokes equation play an important role in this problem, even though the particle Reynolds number is still very small (say, in the range 10^{-3} to 5×10^{-2}.) Eirich (1937) first observed such migration of neutrally buoyant spheres in Poiseuille flow. Segré and Silberberg (1963) showed that in dilute suspension of spheres in a Poiseuille flow (ratio of sphere radius to tube radius = 0.16) the spheres reached a stable equilibrium position at a distance equal to 0.63 times the tube radius from the axis, independent of the initial positions of release. Thus, particles introduced near the wall migrated inward, and those near the axis migrated outward, during flow, leading to an accumulation of spheres in an intermediate annulus. This phenomenon was termed the "tubular pinch effect." A number of subsequent investigations confirmed the result, and showed that it applied to rigid rods and disks, and also to rectangular ducts, but not to deformable spheres and flexible fibers, which always migrated to the tube axis, as in the creeping flow (or Stokes) regions. Buoyant spheres, however, migrate to the wall or to the tube axis depending on whether they were lighter or denser than the fluid and whether Poiseuille flow was up or down the tube. The two-way migration of neutrally buoyant rigid spheres, rods, and disks has been observed also in oscillatory and pulsatile flow (see Goldsmith and Mason, 1967, p. 216). A complete discussion of the mathematical problem is given by Brenner (1965).

In extremely small tubes it is expected that the red blood cells will encounter increasing resistance because of the tight fit between the red cells and the tube wall. Therefore the trend of decreasing relative viscosity with decreasing tube diameter must reach an end and then a reversed trend will begin. This has been demonstrated by Dintenfass (1967).

The smallest blood vessels have a diameter equal to or smaller than that of the red blood cells. The individual red cells are severely deformed in passing through these vessels. In dealing with problems of microcirculation, therefore, the blood cannot be treated as a homogeneous fluid. It must be regarded as a suspension of red cells in plasma, which itself is a Newtonian viscous fluid.

From the fluid-mechanical point of view the account of red blood cells is incomplete unless the mechanical properties of the red cell membrane are recorded. We would like to know the cell membrane's elasticity and elastic moduli, its thickness, the surface tension between the cell membrane and the plasma outside and the hemoglobin inside, and the electrostatic forces existing in the cell and on its membrane. It is most likely that there are proteins surrounding or attached to the cell membrane so that the mechanical properties of the cell membrane are effectively modified. Evidence suggesting this is the effectiveness of the lysing and sphering and antisphering agents in changing the red cell shape, and the seemingly great strength of the red cell membrane in hypotonic swelling. We should like to know the interaction between the red cell membrane and the endothelial cell membrane when they rub against each other in the capillary blood vessels. Also, we should like to know the permeability of the cell membrane with respect to ions and molecules of different kinds. But we do not have the data. Much work involving the mechanical properties of red cells is speculative. It is hoped, however, as we have remarked before, that a collection of solutions to boundary-value problems based on such hypothetical mechanical properties and meticulous comparison with experimental results will eventually solve the entire problem.

D. The Large Blood Vessels

For the purpose of blood flow analysis, we need to know the elasticity and viscoelasticity of the blood vessels, as well as the interfacial conditions between the blood and the blood vessel.

Since blood vessels have been studied so extensively for so long, it would be natural to assume that we have all the data about the mechanical properties of blood vessels. But this is not so. In fact we do not have the stress–strain–history law for any vessel, large or small.

The general characteristics of the blood vessel elasticity are known. When an artery of the dog was subjected to a longitudinal stretch while its cylindrical geometry and diameter were kept constant by varying internal pressure, a tension–stretch relationship as shown in Fig. 10 was obtained. If a strain cycle was imposed, a hysteresis loop was obtained. The hysteresis changes with the number of strain reversals, as is shown in Fig. 11. Similarly, when the longitudinal stretch was kept constant while the circumferential stretch was varied by inflation, the stress-stretch relation as shown in Fig. 12 was obtained.

These results show that the stress–strain–history relationship for a blood vessel is nonlinear over the physiological range. However, if one is concerned only with a small range of strain and stress, it is always possible to linearize. A number of reports on the elasticity and viscoelasticity of arteries in small ranges of strain are available, the most extensive being those of Hardung (1952, 1953, 1962), Peterson *et al.* (1960), Peterson (1962), Bergel (1961a,b), Apter (1964), Apter *et al.* (1966, 1967), Apter and Marquez (1968), Patel *et al.* (1964) and Patel *et al.* (1969). In a small range of strain the linearized elasticity may assume the form of Hooke's law. Then it is possible to speak of the Young's modulus, and the Poisson's ratio. That the material of the arteries may be considered incompressible was verified by Patel *et al.*

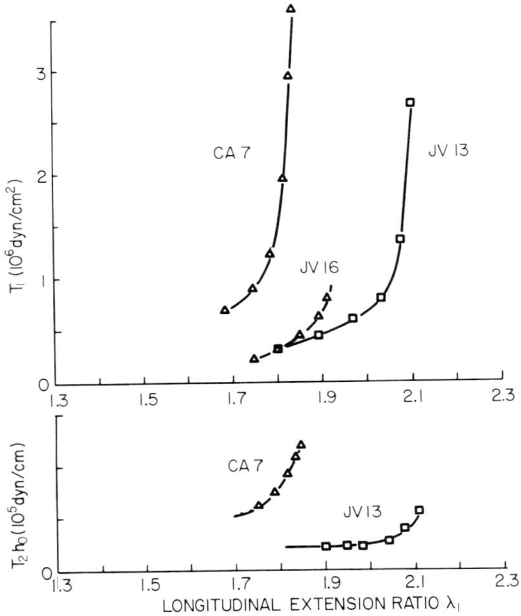

FIG. 10. The Langrangian stresses acting on blood vessels vs. the longitudinal extension ratio. The longitudinal stretch λ_1 was increased at a low strain rate while the lateral extension ratio λ_2 was kept at 1. Note that many of these curves can be made to coincide roughly if curves are shifted horizontally. $T_1 \sim \lambda_1$. From Lee *et al.* (1967).

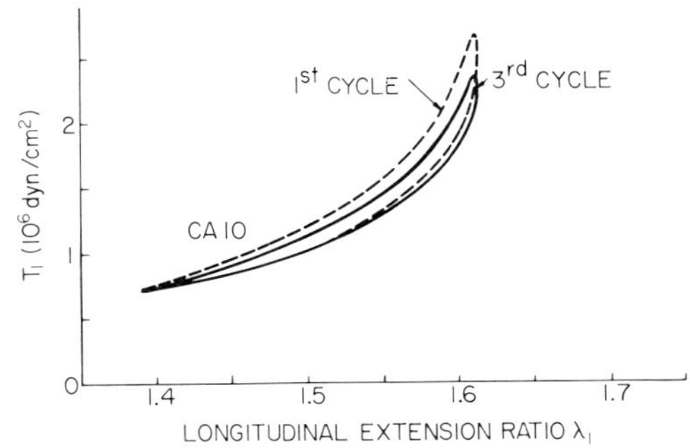

FIG. 11. The hysteresis curves of a carotid artery stretched longitudinally without pressurization. Stretchings were performed within physiological length variation of the arteries. There was a decrease in stress response as cycling (0.21 cycle/min) proceeded. T_1 = longitudinal stress, T_2 = circumferential stress. From Lee *et al.* (1967).

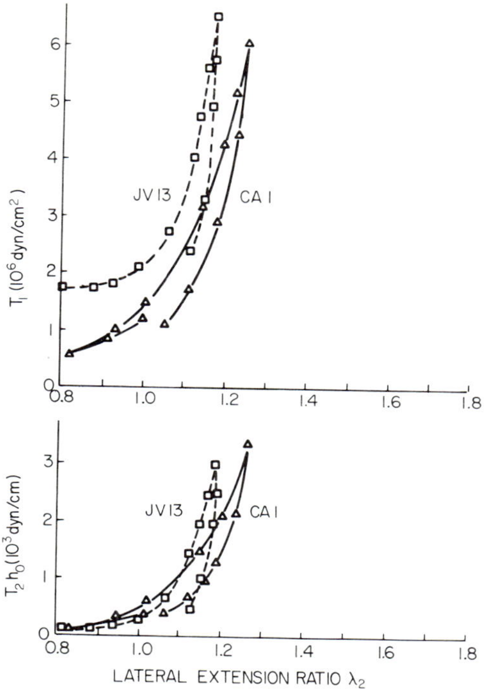

FIG. 12. The Lagrangian stresses acting on the blood vessels vs. the lateral extension ratio. The inflation was first decreased from the physiological state by steps, then increased to about 20% above, and finally decreased back to the physiological state. The longitudinal extension was kept constant. T_1 = longitudinal stress, T_2 = circumferential stress. From Lee *et al.* (1967).

(1969), who found that the bulk modulus was several orders of magnitude larger than the Young's modulus.

It is essential for any linearization of a nonlinear material to state in what region the linearization is carried out. For the blood vessels it is obviously necessary to specify the values of the stress and strain about which the linearization was made. But this is not always feasible. Blood vessels in a living animal are not in a relaxed state, and generally it is impossible to know what are the stress and strain in them. Most published data strive to meet the conditions of being "physiological," even though it is difficult to define that word quantitatively. For example, the measured Young's modulus of the carotid artery of a dog would depend on whether the dog's head was held up or down!

With this understanding we quote in Table 1 the static incremental modulus of elasticity of the aorta and arteries from Bergel (1961a).

Further collection of data and detailed discussions can be found in Bergel (1971). Discussion on the formulation of the viscoelasticity of these vessels is given by Fung (1971c).

TABLE 1

MEAN VALUES FOR STATIC INCREMENTAL MODULES OF ELASTICITY[a,b]

Pressure (torr)	Thoracic aorta	Abdominal aorta	Femoral artery	Carotid artery
40	1.2 ± 0.1(6)	1.6 ± 0.4(4)	1.2 ± 0.2(6)	1.0 ± 0.2(7)
100	4.3 ± 0.4(12)	8.9 ± 3.5(8)	6.9 ± 1.0(9)	6.4 ± 1.0(12)
160	9.9 ± 0.5(6)	12.4 ± 2.2(4)	12.1 ± 2.4(6)	12.2 ± 2.7(7)
220	18.1 ± 2.8(5)	18.0 ± 5.5(3)	20.4 ± 4.4(6)	12.2 ± 1.5(7)

[a] The number of measurements is shown in parentheses. Some additional specimens were studied at 100 torr before making dynamic measurements, and these have been included. From Bergel (1961a, p. 449, with permission).

[b] In dyn/cm^2 × 10^6 ± S.D. of mean.

E. THE CAPILLARY BLOOD VESSELS

Whereas large blood vessels may be regarded as isolated tubes, small blood vessels are so integrated with the surrounding media that the latter contribute to the rigidity of the blood vessels. It is known that the smaller arteries are more rigid. Some of this rigidity must be caused by the surrounding tissues.

The smallest blood vessels, the capillaries, are found to be very rigid. Baez *et al.* (1960) found no change in capillary lumen over a wide range of static pressure in rat mesenteries. It is difficult to explain this rigidity on the basis of the constituents of the capillary blood vessel wall. In 1965 I suggested that some of the rigidity must be contributed by the surrounding tissue. To evaluate this contribution, Zweifach, Intaglietta, and I measured the elasticity of the tissue and then calculated the distensibility of the capillary blood vessels under the hypothesis that the blood vessel is in direct contact with the surrounding tissue. The experiment was made on the avascular portion of the mesentery of the rabbit (see Fung, 1966a,b; Fung *et al.*, 1966).

Microscopic examinations show that in the rabbit mesentery the capillaries are imbedded in the tissue matrix. The basement membrane blends almost imperceptibly with the general tissue matrix. The avascular regions contain no large blood vessels, and they have very few capillaries, but in other respects appear to be the same as the rest of the mesentery. For the rabbit they are membranes of a thickness varying between 30 and 65 μm. A torsion test was selected. From the torque–rotation relationship the nonlinear elastic modulus can be determined.

The experimental setup consits of a circular membrane clamped on two concentric rigid circular disks. The outer disk was fixed in space, while the inner disk was subjected to a known torque. The rotation of the inner disk was recorded. The test showed that the material was approximately elastic: it had a hysteresis loop similar to those of the large arteries shown in Figs. 10–12 (see Fig. 13). Taking the branch corre-

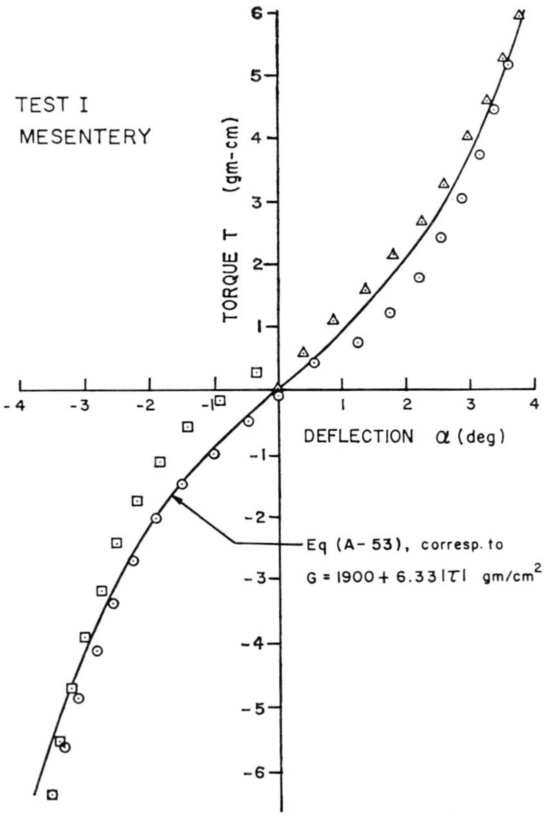

FIG. 13. A typical "stress–strain" relationship of a mesentery membrane subjected to shearing stress. From Fung *et al.* (1966). Triangles: increasing strain; circles: decreasing strain; squares: return stroke.

sponding to the increasing strain, we obtained a nonlinear stress–strain relationship. The results show that the shear modulus of elasticity G is not a constant, but depends on the magnitude of the stresses in the mesentery. It was found that the experimental data fit reasonably well the expression

$$G = \mu + C_2 \mid \tau \mid, \tag{3.2}$$

where μ depends on the initial tensions of the membrane, $\mid \tau \mid$ is the absolute value of the shear stress due to the torque load, and C_2 is a dimensionless constant. Since the tension in the mesentery membrane was difficult to measure and was not controlled, the value of μ varied from one specimen to another. For the rabbit mesentery and diaphram, μ ranged over 0.22 to 1.9×10^6 dyn/cm^2, whereas C_2 ranged over 5.60 to 11.0 (average 8.23, S.D. 2.1).

The shear modulus so determined was that of the entire mesentery in the plane state of stress. Since the mesothelial covers were no more than 10% of the entire thickness, the modulus was assumed to be applicable to the gel in the mesentery.

To deal with three-dimensional problems, a generalization of Eq. (3.2) is necessary. We should like to express it in a form which is independent of the orientation of the frame of reference. Based on the invariance arguments, it is natural to replace the shear $\mid \tau \mid$ in Eq. (3.2) by the *octahedral shear stress* τ_0, which is invariant with respect to the rotation of coordinates, and which is reduced to $(\frac{2}{3})^{1/2} \mid \tau \mid$ in the case of simple shear. Expressed in the cylindrical polar coordinates, τ_0 is

$$\tau_0 = (\tfrac{2}{3})^{1/2} \{\tfrac{1}{6}[(\sigma_{rr} - \sigma_{\theta\theta})^2 + (\sigma_{\theta\theta} - \sigma_{zz})^2 + (\sigma_{zz} - \sigma_{rr})^2] + \tau_{r\theta}^2 + \tau_{\theta z}^2 + \tau_{zr}^2\}^{1/2}. \tag{3.3}$$

A stress tensor has three invariants. Besides τ_0, the other two invariants are the *mean stress*

$$\sigma_0 = \tfrac{1}{3}(\sigma_{rr} + \sigma_{\theta\theta} + \sigma_{zz}) \tag{3.4}$$

and the product of the three principal stress deviators. Kauderer (1958, pp. 19–21) has shown that the deviatoric strain energy function is independent of the third invariant. (Kauderer's proof can be generalized to our case in which G is a function of both σ_0 and τ_0.) Hence we proposed the following basic form:

$$G = \mu(\sigma_0) + C\tau_0, \tag{3.5}$$

where μ is a function of σ_0. Equation (3.5) reduces to (3.2) in the case of simple shear in plane stress.

The corresponding invariants of the strain tensor are the mean strain e_0 and the octahedral shear strain ψ_0:

$$e_0 = \tfrac{1}{3}(e_{rr} + e_{\theta\theta} + e_{zz}),$$
$$\psi_0 = (\tfrac{2}{3})^{1/2}\{\tfrac{1}{6}[(e_{rr} - e_{\theta\theta})^2 + (e_{\theta\theta} - e_{zz})^2 + (e_{zz} - e_{rr})^2] + e_{r\theta}^2 + e_{\theta z}^2 + e_{zr}^2\}^{1/2}. \tag{3.6}$$

If the material is assumed to be isotropic, the stress–strain law is best posed in the form

$$\sigma_0 = 3Ke_0\,, \tag{3.7}$$

$$\sigma'_{ij} = 2Ge'_{ij} \qquad (i, j = 1, 2, 3, \text{ or } r, \theta, z), \tag{3.8}$$

where σ'_{ij} and e'_{ij} are the stress deviators and strain deviators, respectively:

$$\sigma'_{ij} = \sigma_{ij} - \sigma_0\delta_{ij}\,, \qquad e'_{ij} = e_{ij} - e_0\delta_{ij}\,. \tag{3.9}$$

K is the bulk modulus which may depend on σ_0 or e_0. For biological tissues K is so much larger than G that the assumption of incompressibility is usually valid.

From (3.5) and (3.8) we obtain

$$\psi_0 = \tau_0/2(\mu + C\tau_0), \tag{3.10}$$

$$\tau_0 = 2\mu\psi_0/(1 - 2C\psi_0), \tag{3.11}$$

$$\sigma'_{ij} = 2\mu[1/(1 - 2C\psi_0)]\, e'_{ij}\,, \tag{3.12}$$

$$e'_{ij} = (1/2\mu)\{1/[1 + (C\tau_0/\mu)]\}\, \sigma'_{ij}\,. \tag{3.13}$$

With this stress–strain relationship, we analyzed the distensibility of a tube buried in a sheet, under the boundary conditions that the tube is subjected to an internal pressure while the surfaces of the sheet are free (Fung, 1966a,b). From the final result the theoretical distensibility curves are obtained for $\mu = 1 \times 10^6$ dyn/cm^2, and $C = 8$, and with geometrical relations shown in Fig. 14a. Figure 14b shows that the smallest blood vessels are more rigid than the larger ones. If we compute the contribution of the surrounding tissue to the total rigidity of the blood vessels, we obtain the following result: capillary, 99.7%; centrally located venule, 61.3%; arteriole, 45.2%, eccentrically located venule, 41.7%; terminal artery, 11.8%. Thus the elastic rigidity of the capillary in the mesentery is derived almost entirely from the surrounding tissue. The contribution of surrounding tissue becomes less important for larger vessels.

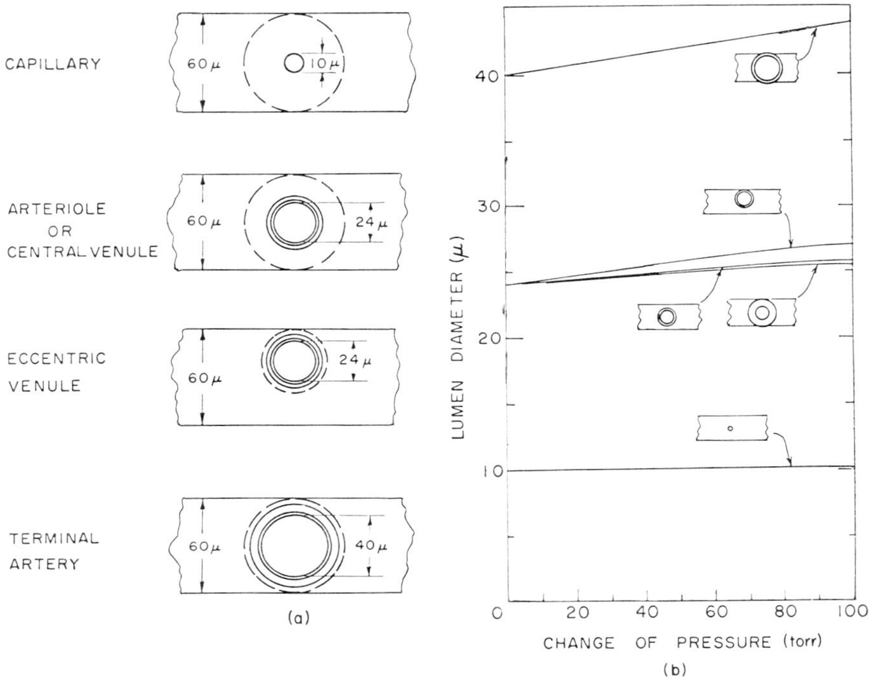

FIG. 14. (a) Typical dimensions of microscopic blood vessels in a mesentery membrane 60 μ thick. (b) The pressure–lumen-diameter relationship for four types of small blood vessels. From Fung (1966b).

We conclude, therefore, that if the surrounding tissue is well integrated with the vessel, then the larger the tissue relative to the vessel, the more significant is its contribution to the elasticity of the vessel. In the case of the smallest capillary blood vessels, the surrounding tissue is very much larger than the vessels, and the rigidity of the capillary blood vessel is derived almost entirely from the surrounding tissue. Hence, as far as the mechanical properties are concerned, a capillary blood vessel is more aptly described as *a tunnel in a gel.*

F. THE KINEMATIC AND DYNAMIC BOUNDARY CONDITIONS

All blood vessels should not be regarded as long circular cylindrical tubes. In various organs vascular beds assume various geometries. For example, in the alveoli of the lung, the capillary blood vessels are so short that their lengths are about the same or even shorter than their diameters.

A network of such short tubes is better analyzed without thinking of each segment as a tube.

An illustration is given in Fig. 15 which shows a plan view of a cat's alveolus in the lung. The multiply-connected area is the blood vessel network. This network is essentially two-dimensional, in the form of a

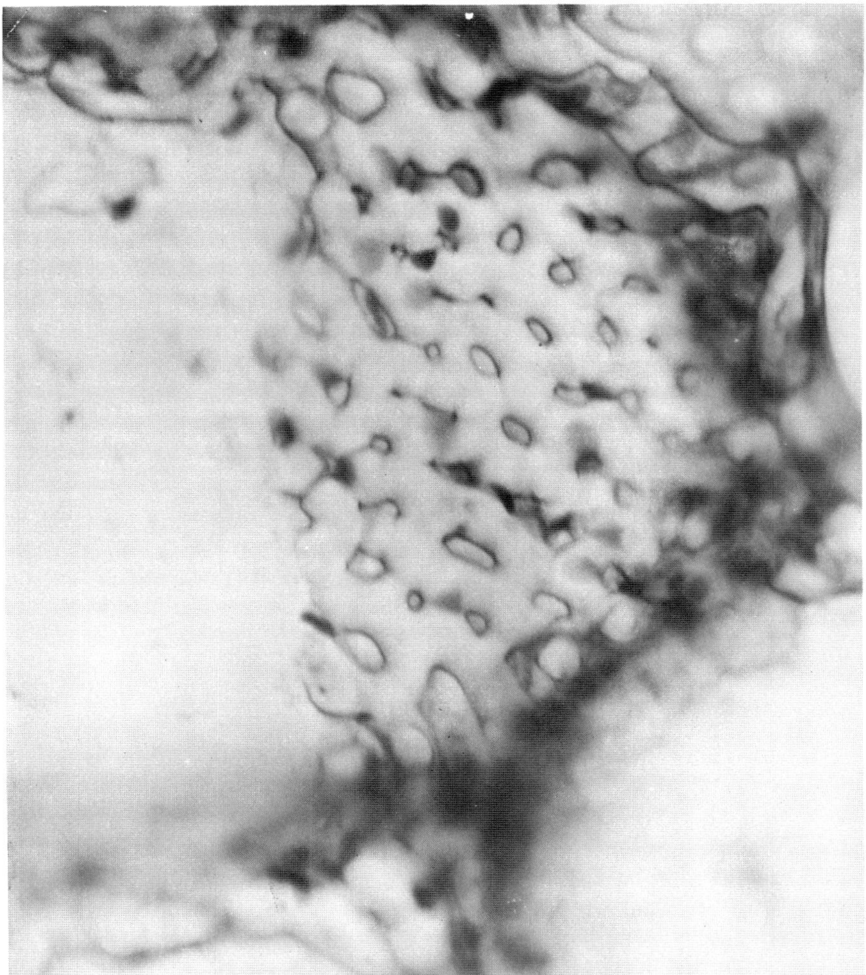

FIG. 15. A plan view of a cat's alveolus in the lung ($\times$ 1000). From Fung and Sobin (1969).

sheet. In cross section, the sheet appears as in Fig. 16. The thickness of the sheet is about 8 μm and the linear dimension of each sheet in an alveolus is of the order of 200 μm. In the plan view, the area occupied by the blood is about 90% of the total area. Thus it is more accurate to represent the alveolar capillary network as a sheet with dispersed

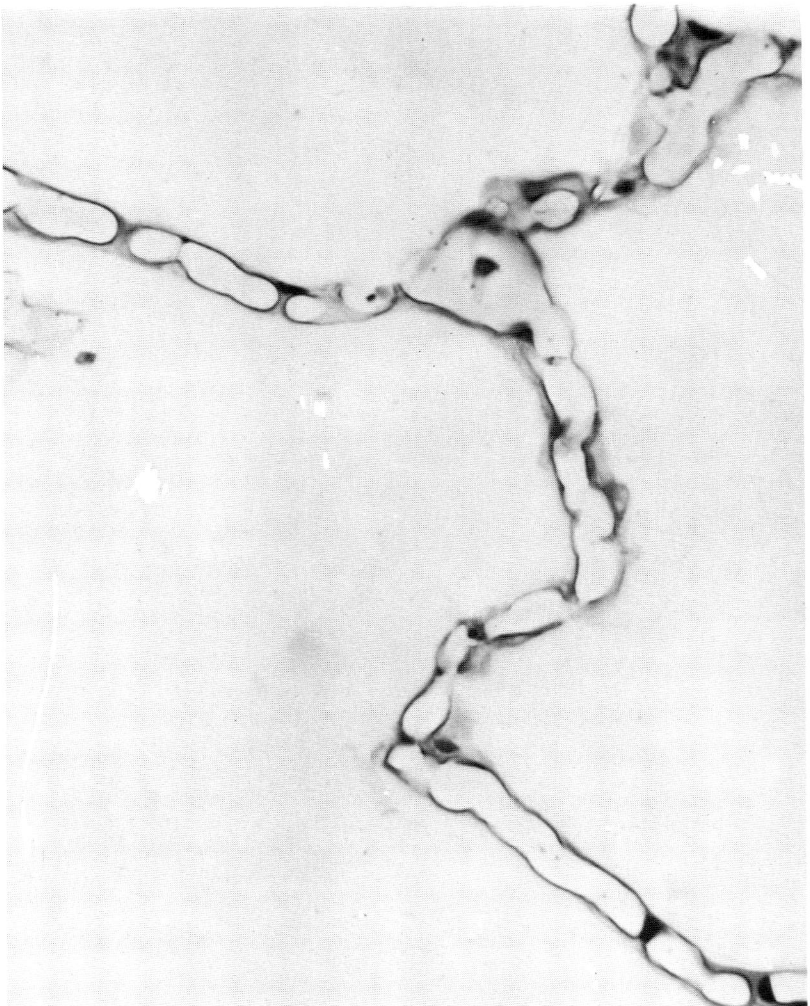

FIG. 16. Cross section of an alveolar sheet of the cat's lung ($\times$ 1000). From Fung and Sobin (1969).

obstructions in the form of "posts" whose average diameter is about 3 μm. A careful anatomical documentation of the pulmonary alveolar sheet is given by Sobin *et al.* (1970). Other organs have other special vascular beds.

The force driving the circulation is well known—the pressure generated by the heart. Other forces that must be considered in microcirculation are the osmotic forces, the surface tension, and the gravitation. The cell membranes of the endothelial cells lining the blood vessels are semipermeable membranes through which water and electrolytes and molecules of molecular weight less than about 1000 pass freely. Substances with a molecular weight greater than 1000 encounter an increasing resistance to their passage through the capillary wall, until at molecular weights of about 10,000 to 20,000 relatively little penetration is achieved. The semipermeable membrane creates a molecular imbalance on the two sides of the membrane which results in an osmotic pressure acting on the membrane.

Therefore, we cannot always treat the boundary wall as impermeable. Generally speaking, mass transfer occurs across living cells at all times, and blood vessels are no exception. For large blood vessels which serve mainly as conduits the structure is such that the fluid transfer across the vessel wall is negligible compared with the fluid transported in the vessel. For capillary blood vessels, whose main function is to nourish the surrounding tissue and to remove the waste products, the mass transfer across blood vessel wall is significant.

It was in 1896 that E. H. Starling put forward a hypothesis concerning the transport of fluid from blood plasma to the surrounding tissue accross the walls of the capillary blood vessels. Starling's hypothesis may be expressed in the formula

$$\dot{m} = K(p_{\rm i} - p_{\rm o} - \pi_{\rm i} + \pi_{\rm o}), \tag{3.14}$$

where $\dot{m}$ stands for the rate of fluid mass movement across the wall, expressed in mass/area/sec, p_i is the hydrostatic pressure of the plasma in the capillary, p_o is the hydrostatic pressure in the interstitial fluid outside the capillary, π_i is the colloid osmotic pressure of the plasma inside the capillary, and π_o is the colloid osmotic pressure of the interstitial fluid outside the capillary. With the pressures expressed in dyn/cm^2, the constant K has the physical dimension $[T/L]$ and hence can be expressed in the units of sec/cm. The constant K is often called the *filtration constant*. For the capillaries of the mesentery of the rabbit, K lies in the range 3×10^{-10} to 25×10^{-10} sec/cm, according to Zweifach and Intaglietta (1968).

Starling's hypothesis has been verified in laboratory experiments for a number of artificial membranes such as collodion (Mauro, 1957; Meschia and Setnikar, 1958) and, accordingly, is often referred to as Starling's law. For single capillary blood vessels, acceptance of Starling's hypothesis is based on the celebrated work of Landis which was published in 1927. Landis introduced micropipettes into the lumens of capillaries and determined the pressure within them by estimating the height of a column of colored fluid (connected to the pipette) necessary to allow the colored fluid to flow into the capillary. He studied the movement of fluid out of the capillaries, first by observing whether dyes injected into the vessel left the capillaries; and second, by occluding the capillary so that fluid movement could only be due to escape through the walls, and estimated the rate of fluid movement by the rate at which a red cell moved towards or away from the blocked end. By these means Landis was able to obtain evidence confirming the Starling hypothesis. Later, the studies of the effects of variation of the colloid osmotic pressure by

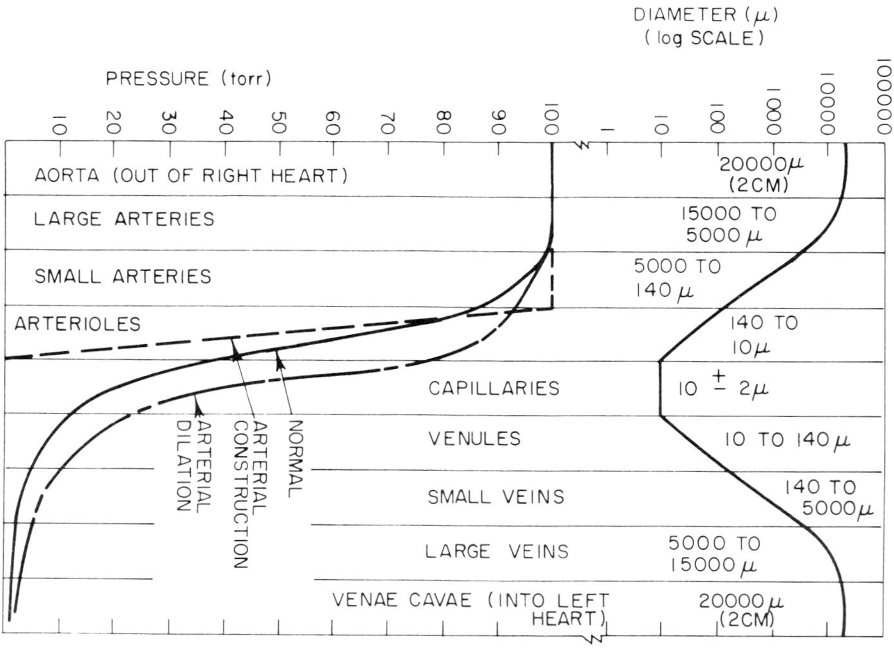

FIG. 17. Pressure as a function of the type and the diameter of the blood vessels of the circulatory system. From Guyton (1961, p. 376), as redrawn by Merrill and Wells (1961, p. 665).

Hyman (1944; Hyman *et al.*, 1952) and Danielli (1940, 1941) in the frog, and by Pappenheimer and Soto-Rivera (1948) in mammals, gave further support to the validity of the Starling hypothesis in an over-all manner. Recently, Intaglietta and Zweifach (1966; see also, Zweifach and Intaglietta, 1968) introduced a number of improvements to the Landis-type experiments, and found that the filtration constant K varies from vessel to vessel in the capillary bed.

As a ummary of the vascular system in which blood flows, Fig. 17 presents the pressure as a function of the type of the blood vessels and the diameters of the blood vessels (Guyton, 1961). Table 2 shows the range of Reynolds number and other quantities in the blood vessels as summarized by Attinger (1964, p. 3).

TABLE 2

Approximate Dimensions and Blood Velocities in Various Segments of the Cardiovascular System for a 13 kg Dog[a]

Segment	Number	Diameter (mm)	Cross section (cm^2)	Length (cm)	Volume (ml)	Blood velocity (cm/sec)	Reynolds number
Left atrium					25		
Left ventricle					25		
Aorta	1	10	0.8	40	30	50	2500
Large arteries	40	3	3.0	20	60	13.4	201
Main arterial branches	600	1	5	10	50	8	40
Terminal arteries	1800	0.6	7.0	1.0	5	6	9
Arterioles	40×10^6	0.02	125	0.2	25	0.32	0.03
Capillaries	12×10^8	0.008	600	0.1	60	0.07	0.003
Venules	80×10^6	0.03	570	0.2	114	0.07	0.01
Small veins	1800	1.5	30	1.0	30	1.3	9.8
Main veins	600	2.4	27	10	270	1.48	18
Large veins	40	6.0	11	20	220	3.6	108
Venae cavae	1	12.5	1.2	40	50	33.4	2090
Right atrium					25		
Right ventricle					25		
Main pulmonary artery	1	12[b]	1.1	2.4	24 (both rows)	36.4	2090
Lobar pulmonary artery branches	9	4[b]	1.19	17.9		33.6	670
Smaller arteries and arterioles					18		
Pulmonary capillaries	6×10^8	0.008	300	0.05	16	0.14	0.006
Pulmonary veins					52 (both rows)		
Large pulmonary veins	4						

[a] Assumed cardiac output: 2.4 liters/min. From Attinger (1964, p. 3).

[b] Mean of major and minor semi-axes of the elleptic cross section.

IV. Boundary-Value Problems

In the following we shall review briefly those problems in circulation physiology that have received some attention from the point of view of continuum mechanics.

The problem of pulsatile blood flow in arteries is perhaps the most studied problem in biomechanics. Analytical work was initiated by Euler (1775), and continuing contributions came from Thomas Young, E. & W. Weber, Moens, Korteweg, Lamb, and others. Rudinger (1966) and Skalak (1966) have given reviews on the historical development of the subject. Recent work in this field includes the books of McDonald (1960), Attinger (1964), and Wetterer and Kenner (1968).

Widespread interest in this field is attested by the large number of publications that appeared in recent years. In the following we shall summarize the literature under several headings: harmonic traveling waves in circular cylindrical tubes, nonlinear effects, geometric perturbations such as entry, branching, etc., and the arterial system as a whole.

A. Harmonic Traveling Waves in a Circular Cylindrical Tube

The mathematical problem of propagation of pulse wave in the artery is the same as the problem of water hammer in civil engineering. As will be seen presently, a large number of papers pertain to this subject. In earlier papers the blood was treated as an incompressible and nonviscous fluid. The tube was treated as a membrane which can resist stretching but not bending. In later papers these assumptions were gradually relaxed.

Arteries and veins with diameters of 0.1 cm up to 1 cm or more in humans are so large compared with the red blood cells (diameter about 8×10^{-4} cm) that it is legitimate to treat the blood as a homogeneous fluid. In dealing with the problems of pulse wave propagation, the individuality of the red blood cells, platelets, and other cells can be disregarded. Rheological investigations of blood in an apparatus whose dimension is comparable to that of arteries show that at a sufficiently high rate of shear, blood is approximately Newtonian.

Arteries and veins are more or less alike in size (although larger veins are more numerous) but fluid mechanically they are different. Many veins have valves, but the arteries have not. The pressure and pressure gradient in the veins are low; those in the arteries are high. Pulsation is a predominant feature in the arteries, it is not in the veins. These differences make it expedient to treat arteries and veins separately. The literature is concerned almost exclusively with arteries. For veins, see Brecher (1956).

To illustrate the mathematical problem, we may consider a simplified case with the following assumptions:

(a) The fluid is homogeneous, viscous, and Newtonian.
(b) The wall material is isotropic and linearly viscoelastic.
(c) The fluid motion is laminar.
(d) The motion is so small that squares and higher order products of displacements and velocities and their derivatives are negligible.

Then the field equations are the linearized Navier–Stokes equations for the blood, the linearized Navier's equation for the wall, and the equation of continuity. The boundary conditions are the continuity of stresses and velocities at the fluid–solid interface, and appropriate conditions on the external surface of the tube and at the ends of the tube. Methods of handling these equations are well-known in the theories of hydrodynamics and elasticity. For example, for an *axisymmetric* traveling wave in a tube of *incompressible* viscoelastic material we have, for the fluid,

$$\frac{\partial v_r}{\partial t} = -\frac{1}{\rho}\frac{\partial p}{\partial r} + \nu\left(\frac{\partial^2 v_r}{\partial r^2} + \frac{1}{r}\frac{\partial v_r}{\partial r} - \frac{v_r}{r^2} + \frac{\partial^2 v_r}{\partial x^2}\right), \tag{4.1}$$

$$\frac{\partial v_x}{\partial t} = -\frac{1}{\rho}\frac{\partial p}{\partial x} + \nu\left(\frac{\partial^2 v_x}{\partial r^2} + \frac{1}{r}\frac{\partial v_x}{\partial r} + \frac{\partial^2 v_x}{\partial x^2}\right), \tag{4.2}$$

$$\frac{\partial v_x}{\partial x} + \frac{\partial v_r}{\partial r} + \frac{v_r}{r} = 0, \tag{4.3}$$

and, for the tube wall,

$$\frac{\rho_{\mathrm{w}}}{\mu^*}\frac{\partial^2 u_r}{\partial t^2} = \frac{\partial^2 u_r}{\partial r^2} + \frac{1}{r}\frac{\partial u_r}{\partial r} - \frac{u_r}{r^2} + \frac{\partial^2 u_r}{\partial x^2} - \frac{1}{\mu^*}\frac{\partial \Omega}{\partial r}, \tag{4.4}$$

$$\frac{\rho_{\mathrm{w}}}{\mu^*}\frac{\partial^2 u_x}{\partial t^2} = \frac{\partial^2 u_x}{\partial r^2} + \frac{1}{r}\frac{\partial u_x}{\partial r} + \frac{\partial^2 u_x}{\partial x^2} - \frac{1}{\mu^*}\frac{\partial \Omega}{\partial x}, \tag{4.5}$$

$$\frac{\partial u_x}{\partial x} + \frac{\partial u_r}{\partial r} + \frac{u_r}{r} = 0, \tag{4.6}$$

where v_x, v_r, v_θ are the velocity components of the fluid, u_x, u_r, and u_θ are the displacement components of the wall, μ^* is the dynamic modulus of rigidity of the wall, and Ω is a finite pressure, which must be introduced since we have assumed the material to be incompressible. ν is the kinematic viscosity of the blood, μ is the coefficient of viscosity, ρ is the density of the blood, ρ_w is that of the wall material.

The boundary conditions are, if the external surface of the tube is stress-free, and if the inner and outer radii of the tube are a and b, respectively,

$$v_r = 0 \qquad \text{at} \quad r = 0, \tag{4.7}$$

$$\partial v_x/\partial r = 0 \qquad \text{at} \quad r = 0, \tag{4.8}$$

$$v_r = \partial u_r/\partial t \qquad \text{at} \quad r = a, \tag{4.9}$$

$$v_x = \partial u_x/\partial t \qquad \text{at} \quad r = a, \tag{4.10}$$

$$\mu(\partial v_r/\partial x + \partial v_x/\partial r) = \mu^*(\partial u_r/\partial x + \partial u_x/\partial r) \qquad \text{at} \quad r = a, \tag{4.11}$$

$$-p + 2\mu(\partial v_r/\partial r) = -\Omega + 2\mu^*(\partial u_r/\partial r) \qquad \text{at} \quad r = a, \tag{4.12}$$

$$\mu^*(\partial u_r/\partial x + \partial u_x/\partial r) = 0 \qquad \text{at} \quad r = b, \tag{4.13a}$$

$$-\Omega + 2\mu^*(\partial u_r/\partial r) = 0 \qquad \text{at} \quad r = b. \tag{4.13b}$$

The solution that satisfies the boundary conditions (4.7) and (4.8) may be posed in the following form:*

$$v_x = -\sum_{n=0}^{N} i\{A_1\gamma_n J_0(i\gamma_n r) + A_2\varkappa_n J_0(i\varkappa_n r)\} \exp i(n\omega t - \gamma_n x), \tag{4.14}$$

$$v_r = -\sum_{n=0}^{N} i\gamma_n\{A_1 J_1(i\gamma_n r) + A_2 J_1(i\varkappa_n r)\} \exp i(n\omega t - \gamma_n x), \tag{4.15}$$

$$p = \sum_{n=0}^{N} A_3 J_0(i\gamma_n r) \exp i(n\omega t - \gamma_n x), \tag{4.16}$$

$$u_r = \sum_{n=0}^{N} - i\gamma_n\{A_4 J_1(k_n r) + B_4 Y_1(k_n r) + A_5 J_1(i\gamma_n r) + B_5 Y_1(i\gamma_n r)\} \cdot \exp i(n\omega t - \gamma_n x), \tag{4.17}$$

$$u_x = \sum_{n=0}^{N} - \{k_n A_4 J_0(k_n r) + k_n B_4 Y_0(k_n r) + i\gamma_n A_5 J_0(i\gamma_n r) + i\gamma_n B_5 Y_0(i\gamma_n r)\} \cdot \exp i(n\omega t - \gamma_n x), \tag{4.18}$$

$$\Omega = \sum_{n=0}^{N} \{A_6 J_0(i\gamma_n r) + B_6 Y_0(i\gamma_n r)\} \exp i(n\omega t - \gamma_n x), \tag{4.19}$$

where ω is the angular frequency, n is the harmonic number, γ_n is the propagation constant of the nth harmonic, N is a constant, the A's

* In a strict notation the A's and B's should have a subscript n because their values may be different for each n.

and B's are complex constants, J_0 and J_1 are Bessel functions of the first kind, Y_0 and Y_1 are Bessel functions of the second kind, and

$$\varkappa_n{}^2 = (in\omega/\nu) + \gamma_n{}^2, \qquad k_n{}^2 = n^2\omega^2\rho_W/\mu^* - \gamma_n{}^2. \tag{4.20}$$

$$A_1 = (i/n\omega\rho)\, A_3\,, \qquad A_6 = n^2\omega^2\rho_W A_5\,, \qquad B_6 = n^2\omega^2\rho_W B_6\,. \tag{4.21}$$

When these solutions are substituted into the boundary conditions (4.9)–(4.14) six linear, homogeneous, and simultaneous equations in six unknown coefficients A_1, A_2,..., B_5 are obtained. For a nontrivial solution the determinant of the coefficients of A_1, A_2,..., B_5 must vanish. This determinantal equation

$$\Delta(\gamma_n\,, \varkappa_n\,, k_n\,, \mu, \mu^*, a, b) = 0 \tag{4.22}$$

is the frequency equation for the pulse wave. If $\gamma_n = \beta + i\alpha$ is solved with other parameters assigned, then β is the *wave number*, α is the *attenuation coefficient*, and $C_p = \omega/\beta$ is the *phase velocity*.

It is evident that extensive numerical calculations are necessary to obtain detailed information. A great deal has been published. The reader is referred to the original papers summarized in Table 3.

Concerning waves in blood vessels the following remarks may be useful:

(1) As to the effect of wall thickness, Klip (1962) points out that when experimental results on longitudinal waves were compared with the theory, the predictions of the theory were found to be far more accurate for thick-walled tubes than for thin-walled ones. When the ratio of the wall thickness to the inner diameter of the cylinder is less than 1/20 the agreement between theory and experiment proved bad. Several reasons may be suggested for this: (a) the effect of initial stress was neglected, (b) the small displacement required for the linearization of the differential equations is hard to be reconciled with the requirements of measurability in experiments; (c) inhomogeneity in material and imperfections in geometry are more serious for thin-walled cylinders.

(2) Jones *et al.* (1968) showed that the faster waves are more sensitive to variations in the elastic properties of the medium surrounding the blood vessels. At high Reynolds numbers the attenuation due to fluid viscosity over a fixed length was found to be substantially greater for the fast waves than for the slow waves. At very low Reynolds numbers the slow waves are more strongly damped. A comparison with the *in vivo* data shows that fluid viscosity alone cannot account for the observed attenuation and does not have the proper frequency dependence. For physiologically meaningful parameter values the damping due to blood

TABLE 3

SUMMARY OF CONTRIBUTIONS TO HARMONIC WAVE PROPAGATION IN ARTERIES

	Fluid				Wall					Waves			Experiments		
Reference[a]	Inviscid	Viscous	Compressible	Incompressible	Membrane	Finite thickness	Elastic	Rigid	Viscoelastic	Axisymmetric	Asymmetric	Number of modes	Rigid tube	Flexible tube	Animal
(1)[b]	×			×			×								
(2)[c]	×			×	×		×			×		1			
(3)[d]	×			×	×		×								
(4)[e]	×			×	×		×								
(5)[f]													×	×	
(6)[g]	×		×		×		×								
(7)[h]	×		×		×		×					2			
(8)[i]	×		×		×			×					×	×	
(9)		×		×	×		×	×							
(10)[j]															×
(11)[k]	×	×		×											
(12)[l]	×			×	×		×		×	×		1			
(13)[m]	×	×	×	×	×		×								
(14)[n]														×	
(15)[o]		×		×	×		×			×		1			
(16)[p]		×		×	×		×		×	×		1			
(17)[q]	×	×		×	×		×			×		1			
(18)[r]		×		×			×	×							
(19)[s]															×
(20)[t]									×					×	
(21)[u]												2		×	

(22)[v]													×
(23)[w]		×	×	×	×	×	×	×	×	×	3		
(24)[x]												×	
(25)[y]		×		×	×								×
(26)[z]		×		×	×		×		×		2		
(27)[aa]		×	×		×	×	×		×				
(28)[bb]	×				×	×	×		×			×	
(29)[cc]	×		×			×	×		×	×			
(30)[dd]	×			×		×		×	×				
(31)[ee]													×
(32)[ff]		×		×		×	×	×	×	×			

[a] Numbers refer to references: (1) Euler (1775); (2) Young (1808, 1809); (3) E. Weber and W. Weber (1825); (4) Resal (1876); (5) Moens (1878); (6) Korteweg (1878); (7) Lamb (1897–1898); (8) Joukowsky (1900); (9) Witzig (1914); (10) Hamilton, Remington, and Dow (1939); (11) Branson (1945); (12) King (1947b); (13) Lambossy (1950, 1951); (14) Müller (1935, 1951, 1959); (15) Jacobs (1953); (16) Morgan and Kiely (1954); (17) Morgan and Ferrante (1955); (18) Womersley (1955a, b 1957); (19) Landowne (1958); (20) Taylor (1959); (21) Van Citters (1960); (22) McDonald (1960, 1965); (23) Klip (1962); (24) Hardung (1962); (25) Fry *et al.* (1964); (26) Atabek and Lew (1966); (27) Rubinow and Keller (1968); (28) Anliker and Raman (1966); (29) Anliker and Maxwell (1966); (30) Cox (1968); (31) Cox (1970); (32) Jones *et al.* (1968).

[b] Pressure p related to cross-sectional area s by $s = s_0 p(c + p)^{-1}$ or $s = s_0(1 - e^{p/c})$. Gave details for $c = \infty$. Euler's eqs. were solved by Lambert (1956).

[c] Derived the wave speed formula $c_0 = (hE/2a\rho)^{1/2}$, h = thickness of tube, a = radius, ρ = density of fluid. Young's c_0 is known as Moens–Korteweg formula.

[d] Rederived Young's formula.

[e] Rederived Young's formula.

[f] Gave wave speed $c = \text{const.} \times c_0$; const. = 0.8, 0.9, 1.036 depending on various conditions.

[g] Obtained $c_1 = (2a\rho/hE + \rho/K)^{-1/2}$, ρ = density of fluid, K = bulk modulus of fluid, c_1 is known to engineers in water-hammer, and was derived independently by Joukowsky (1900).

(*continued*)

Footnotes to Table 3 continued

[h] Derived phase velocity of long waves, c_0 and $c_2 = [E/(1 - \sigma^2)\rho_W]^{1/2}$; σ = Poisson's ratio, ρ_W = density of wall.
[i] Exhaustive treatment of water-hammer. Riemann's method. Detailed wave form analysis.
[j] Arterial waves related to stroke volume and cardiac ejection.
[k] Results for viscous flow incorrect because he used an equation for conservation of mass which is valid for inviscid flow only.
[l] Hookean tube.
[m] Historical review.
[n] Cf. Klip (1962).
[o] Nonlinear elasticity. Perturbation of mean flow.
[p] Long wave. Hooke's law.
[q] Long wave. Perturbation on steady flow.
[r] Compared with experiments. Considered tethering. Branching. Variation of crosssection. Extensive theory.
[s] Induced waves in human brachial and radial arteries.
[t] Impedance method electric analog.
[u] Found 2 longitudinal modes.
[v] Thorough examination of theories and experiments from the physiological point of view.
[w] Discussed experiments, extensive theory.
[x] Input impedance and reflection at ends and branches.
[y] Evaluation of parameters.
[z] Effect of prestress demonstrated.
[aa] External tissues represented by 4-parameter impedance, showed significant effect. Found frequency cutoff in inviscid membrane case.
[bb] Korotkoff sound.
[cc] Dispersion. Initial strain. Shows axisymmetric waves are mildly dispersive, asymmetric waves are highly dispersive, and exhibit cutoff phenomenon.
[dd] Detailed numerical results.
[ee] Detailed experimental results.
[ff] Extensive parametric study. Include surrounding tissues.

viscosity is much less than that due to the viscoelasticity of the wall material for both families of waves.

(3) A particular problem of interest is the generation of Korotkow sound in the arteries. This is the sound used by physicians when they measure the patient's blood pressure by inflating a cuff on the arm. If the stethoscope bell is placed over a peripheral artery (e.g., the brachial near the elbow or the radial at the wrist), nothing can be heard at all in normal circumstances. When, however, a wide cuff containing an inflatable bag is wrapped around the upper arm and the bag is inflated to pressures above the systolic pressure and the pressure is allowed to fall slowly, characteristic sounds are heard from the brachial artery (the Korotkow sound). It is the interpretation of this sound that is of interest. According to Burton (1965), the first, sharp, tapping sound may be interpreted as the systolic pressure. As the bag pressure falls the sound becomes louder and more extended in time, then reaches a maximum intensity and begins to diminish, eventually to "disappear." At a pressure just below that where the sound begins to diminish, there is a change in the character of the sound, known as "muffling." The sound loses its ringing, staccato quality and becomes a thumping. Burton (1965) recommends taking the muffling as the indication for diastolic pressure, while many clinicians take the "disappearance" of the sound as a criterion. Burton says "the more one inquires into the basis of the Riva–Rocci indirect method of measuring arterial pressure (by inflating a bag), the less confidence one has in its accuracy, both for systolic and diastolic pressures. Yet it remains an invaluable aid to diagnostic medicine, in no way replaced in practice by the availability of methods involving direct arterial puncture. No one has, as yet, invented any better indirect method."

It is natural that the problem of Korotkow sound should attract the attention of workers in applied mechanics. The nature of the sound depends on whether the flow is turbulent or not. The generation of the sound depends on elastic vibration of the blood vessel wall; its propagation depends on the hydroelastic stability of the blood vessel under the compressive load. Depending on the stability, the waves in the wall are amplified or attenuated in the direction of propagation. In many aspects it is similar to the panel-flutter problem in aeronautics.

Anliker and Raman (1966) presented an analysis of the Korotkow sounds at diastole as a phenomenon of dynamic instability of fluid-filled shells. McCutcheon and Rushmer (1967) offered an experimental critique in which the nonlinearity of the blood vessel elasticity is emphasized. Many other authors contributed to the analysis of this problem.

(4) Experimental study of the elastic wave propagation was pursued in recent years with the principal objective of deriving the elastic and viscoelastic constants of the blood vessel wall. Introduction of artificial, high-frequency, symmetric, and antisymmetric excitations with short wavelengths by Anliker *et al.* (1968a, b) added a new dimension to the study of pulse waves. Use of new probes (Ling *et al.*, 1968) helps further development.

(5) What are the use of these investigations to the clinician? Let us quote Willem Klip, whose theoretical and experimental investigation on this subject is one of the most thorough. In his 1962 monograph, Klip gave the following evaluation of his work on wave propagation in a rubber tube of great length:

> Can the results obtained in these investigations be of any use to the clinician?
>
> The answer is difficult to give, it lies somewhere between "Yes" and "No."
>
> If "use" is meant to be "direct use," we must say "No." What the clinician wants is a method that will strip the "pulse wave velocity" he measures from the influences of all the perturbing phenomena like reflection against branches and "periphery," like respiration, digestion, changes in blood pressure, frequency composition, and other kinds, a method that will give him the elastic parameters of the vessel walls, so that he can use these numbers as a measure of the degree of arteriosclerosis. We cannot give him this.
>
> However, if the clinician ever wants to achieve his aims it is useful for him to have an insight into the influence on the phase velocity and the damping, of the few parameters that were discussed here.

B. Effects of Nonlinearity

The two most often used methods of taking nonlinear terms into account are characterized by (1) the use of approximate one-dimensional equations which are then integrated numerically, utilizing the method of characteristics, and (2) the use of perturbation methods.

The one-dimensional approach is based on

(a) the equation of continuity

$$v(\partial s/\partial z) + s(\partial v/\partial z) + \partial s/\partial t + \Psi = 0, \tag{4.23}$$

(b) the equation of motion

$$\partial v/\partial t + v(\partial v/\partial z) = -(1/\rho)\, \partial p/\partial z + f, \tag{4.24}$$

(c) the equation relating the cross-sectional area s to the pressure p

$$s = s(p, z, t). \tag{4.25}$$

As was pointed out by Skalak (1966), these equations are precisely of the form proposed by Euler (1775), except for the frictional term f and the seepage per unit length through the wall, Ψ. The seepage Ψ is introduced to allow for the effect of side branches and division of blood vessels. Streeter *et al.* (1963, 1964) considered Ψ pressure-dependent; Iberall (1964) and Skalak and Stathis (1966) considered Ψ flow-dependent. For f, Streeter *et al.* (1963) used $f = kv^n/a$ where k and n were adjusted to fit experimental data.

On eliminating s, the slopes of the characteristics are found to be

$$\frac{dz}{dt} = v \pm \left[\frac{s}{\rho(\partial s/\partial p)_{z,t}}\right]^{1/2} = v \pm c. \tag{4.26}$$

The wave velocity c relative to the fluid is the local value of Young's velocity. Along the characteristics the differential equations become

$$\frac{dv}{dt} + \frac{1}{\rho c}\frac{dp}{dt} = f \pm \frac{vc}{s}\left(\frac{\partial s}{\partial z}\right)_{p,t} \mp \frac{c}{s}\left(\frac{\partial s}{\partial t}\right)_{p,z} \mp \frac{\Psi c}{s}, \tag{4.27}$$

where the $\pm$ signs follow those in (4.26). These equations can be integrated numerically from any intial conditions and appropriate boundary conditions. Such calculations were made first by Lambert (1956) for blood flow. Streeter *et al.* (1963) showed that the method is practical.

With converging waves there is the possibility of shock wave formation. Lambert (1956, 1958) derived the shock condition. Mollo–Christensen (1966) showed that self-preserving waves are possible only for certain amplitudes. Rudinger (1970) analyzed the shock formation in some detail and showed that with relevant physiological parameters such waves should be detectable in humans; he obtained an experimental confirmation by scaled models in a simplified setup.

When details of the three-dimensional flow field is desired, the *method of perturbation*, or expansion of the differential equations and boundary conditions in powers of certain small parameters, is often used. In this category are the papers by Jacobs (1953) Morgan and Ferrante (1955), and Uchida (1956), who assumed that blood flow consists of a small oscillatory perturbation superposed on a steady flow: an assumption which is not valid for human circulation. Womersley (1957) used perturbation methods to estimate the effect of the variation in tube cross section during a pulse, and to correct for quadratic terms in the equation of motion of the fluid.

Evans (1962a, b) studied the effects of variation of tube radius and physical properties with distance along the artery. Skalak *et al.* (1966) analyzed the convective acceleration at increased cardiac output. Lee (1966) developed a long-wave theory which includes geometric and constitutive nonlinearities.

C. Cardiovascular System As a Whole

The simplest theory of the cardiovascular system is the *Windkessel* theory proposed by Otto Frank in 1899. The idea was stated earlier by Hales (1733) and Weber (1850). In this theory, the heart and the vascular system are represented by an elastic chamber and a rigid vessel of constant resistance. See Fig. 18. Let i be the inflow (cm^3/sec) into this

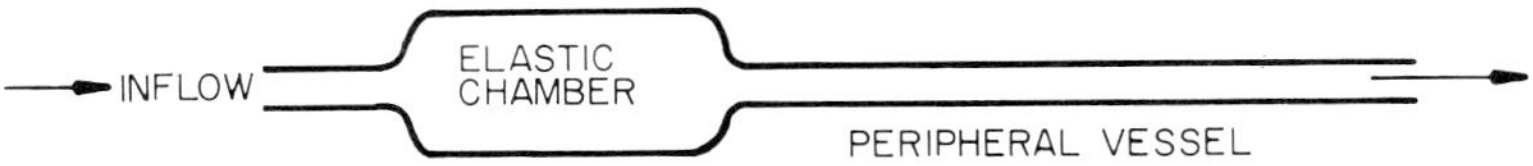

FIG. 18. A conceptual model (Windkessel) of circulation system.

system. Part of this inflow is sent through the peripheral blood vessels, and part of it is used to distend the elastic chamber. If p is the blood pressure (pressure in the left ventricle of the heart), the flow in the peripheral vessel is equal to p/R, where R is called *peripheral resistance*. For the elastic chamber, its change of volume is assumed to be proportional to the pressure. The rate of change of the volume of the elastic chamber is therefore proportional to dp/dt. Let the constant of proportionality be written as K_0. Then, on equating the inflow to the sum of the rate of change of volume of the elastic chamber and the outflow p/R, we obtain

$$i = K_0(dp/dt) + p/R. \tag{4.28}$$

It is found that this equation works remarkably well in correlating experimental data on the total blood flow i with the blood pressure p, particularly during diastole. (See Aperia, 1940; Hamilton *et al.*, 1939.) Hence in spite of the simplicity of the underlying assumptions it is quite useful.

To obtain greater detail, one must represent the cardiovascular system more realistically. Frank himself later (1926, 1930) turned to the elastic tube model discussed in the previous section. Further work can be done either analytically or numerically. In an analytical study one may replace the real system by a simpler one. For example, an artery with many branches may be approximated by a porous tube, etc. On the other hand, if a numerical approach is taken one could represent the the cardiovascular system by an analog model with many elements and many parameters. Many people are developing this "brute-force" approach. Table 4 presents a summary of earlier attempts in these directions. More recent work generally uses digital computers. The scope of a truly modern large scale computation applied to biomechanics, patient care, and rehabilitation can be seen in Attinger (1971) and Noordergraaf *et al.* (1963, 1964).

TABLE 4

CARDIOVASCULAR SYSTEM ANALYSES

Author(s)	Organ	Theory	Remarks
(I) Idealization			
Womersley (1957)	Artery	Tapered tube	
Evans (1962b)	Artery	Tapered tube	
Streeter *et al.* (1963)	Artery	Tapered tube	
Gessner (1964)	Artery	Graphic method	
Iberall (1964)	Artery	Tapered or porous tube	
Wiener (1964)	Lung	Tethered tubes	
Bergel and Milnor (1965)	Artery	Nonuniform system	Experiments show remarkably flat input impedance vs. frequency curve compared with that of any single tube
Morkin *et al.* (1965)	Lung		Experiment
M. G. Taylor (1965)	Artery	Transmission line analog	
McDonald and Gessner (1966)	Artery	Increase of elasticity with distance from the heart	Experiments
M. G. Taylor, (1966a)	Artery	Random branding	
Wiener *et al.* (1966)	Lung	42 generations, random length	
Wylie (1966)	Artery	Tapered tube, nonlinear wall	
McDonald (1968)			
(II) Systems Approach			
Beneken (1965)		Analog computer	
Noordergraaf *et al.* (1963)		Analog computers	
Gabe (1965)		Analog computer	

D. DETAILED GEOMETRICAL EFFECTS

The flow conditions at the ends, branches, curved sections, and other geometric features are of great interest because the cardiovascular system is composed of many blood vessels and, therefore, has many junctions. The flow separation, cavitation, or turbulence generated at the geometric irregularities in large blood vessels are often suspected to be related to arterio-atherosclerosis.

The "end effects" have been studied by many authors. If a tube ends in a big reservoir, the velocity distribution at the junction (entrance section) of the tube and the reservoir may usually be assumed to be uniform. Under this assumption Atabek and Chang (1961) studied periodic flows in the entrance region of a circular tube. Later, Atabek (1964) generalized this study to entry flow with a nonuniform velocity distribution at the entrance on the basis of boundary layer approximation. The end effects in side-branching tubes and nonNewtonian effects are also discussed. The analysis is valid for sufficiently large Reynolds numbers. Kuchar and Ostrach (1966) analyzed the end effects on a similar basis for steady flow. For smaller Reynolds numbers the boundary layer approximation is not valid; the radial velocity component may become as large as 20 to 30 % of the maximum axial velocity. Lew and Fung (1969c, 1970a) treated the case steady entry flow with uniform velocity distribution at the entry section, for an arbitrary Reynolds number. In the limiting case of zero Reynolds number the "entry length" (beyond which the flow deviates by less than 1 % from the Poiseuille profile) is found to be 1.3 times the radius of the tube. For larger Reynolds numbers the entry length depends on the Reynolds number. Beyond a Reynolds number (based on the tube radius and mean flow velocity) of 50 the entry length is given by the well-known formula

$$L/a = 0.16\,\mathbf{R} = 0.16\,aU/\nu, \tag{4.29}$$

where $\mathbf{R}$ is the Reynolds number, a is the tube radius, U is the mean flow speed, ν is the kinematic viscosity of fluid, and L is the entry length.

Another group of investigations was concerned with the flow patterns due to curvature and branching. Womersley (1958) and Bugliarello (1966) presented analyses and experimental results. Attinger (1964) and Attinger *et al.* (1966) showed flow patterns revealed by streaming birefringence photography for pulsatile flow in round and elliptical tubes with Y junctions. Stehbens (1959), Krovetz (1965) and others used models to demonstrate the eixtence of flow separation in pulsatile flow; while Freis and Heath (1964) showed the presence of the same in a dog's aorta.

Extensive experiments on laminar-turbulence transition were performed by Sarpkaya (1966) and Yellin (1966b). Sarpkaya presented a theoretical analysis of the points of inflection in the velocity profile as an indicator for the stability of the flow. He concluded that for the same maximum pressure gradient, pulsatile flow is more stable than the corresponding steady Poiseuille flow.

Many pathological studies suggested the existence of a link between artherosclerosis and hydrodynamics. Holman (1954) demonstrated *in vivo* that hydrodynamic factors such as turbulence and pressure oscillations lead to progressive structural weakening of vascular walls. Rodbard (1959) discussed the role of hydrodynamics in embryology and pathology of the vascular lining. Scharfstein *et al.* (1963) and Gutstein *et al.* (1963) believed that endothelial injury is produced by high shear stress. Roach (1963) made a thorough study on the production and time course of "post-stenotic dilatation" in the femoral and carotid arteries of adult dogs and concluded that the post-stenotic dilatation always develops if "turbulence" (indicated by a certain murmur) is present, and never develops if such a murmur is absent.

A trail-blazing paper was published by Fry (1968) who determined the critical shear stress in relation to changes in the endothelial cells of the thoracic aorta. This furnishes a foundation for the "high shear stress" theory of artherosclerosis. Further work demonstrated the correlation between local changes of vessel wall structure with the shear stresses. On the other hand, Caro, Fitz-Gerald, and Schroeter (1969, 1971) showed that the distribution of *early* atheroma in man is coincident with those regions in which arterial wall shear rate is relatively low, while the development of lesions is retarded in regions where wall shear rate is relatively high. They suggested that the atherogenesis is associated with shear dependent mass transport phenomena.

On the mathematical side, Lee and Fung (1970) presented an analysis of flow in locally constricted tubes at low Reynolds numbers. A circular cylindrical tube with a localized disturbance on the wall is transformed conformally into a circular cylinder; the Navier–Stokes equation is then solved by the method of iteration. Forrester and Young (1970) presented further results.

V. Microcirculation

The boundary-value problems discussed so far are characterized by dimensions so large compared with the red blood cells that the blood may be regarded as a homogeneous fluid. If we go down the scale of blood vessels, we soon come to vessels whose diameters are so small as to be comparable with those of the red cells. Then we can no longer ignore the individuality of the red cells. Problems in which we must pay attention to the red cells and the plasma separately are called problems of microcirculation.

A fortunate feature of microcirculation is that the scale is so small and the flow is so slow that the inertia force due to convective acceleration may be neglected in comparison with the surface forces due to viscosity.

The equations of motion are therefore greatly simplified. The motion of the plasma (a Newtonian fluid) is described by Stokes' equation. The motion of the fluid inside the red cell probably can also be described by Stokes' equation.

However, other complications occur. Small blood vessels are more intimately embedded in the surrounding tissue. Therefore in general the system to be considered must involve more than just the blood vessel. At a fluid–solid interface mass transfer usually occurs, so that permeability must be considered and the boundary conditions for the fluid are different from the ordinary "no slip" conditions (vanishing velocity). The nature of the boundary conditions often leads to very complex nonlinear boundary-value problems.

There is no doubt that the subject of microcirculation is important. Every cell in an animal must live near a capillary blood vessel to obtain the needed oxygen, water, and other materials, and to remove unwanted substances. On the average a cell cannot live if it is farther away from a blood capillary than about one-thousandth of an inch. Thus it is estimated that on the average there are from 1000 to 60,000 miles of capillaries in

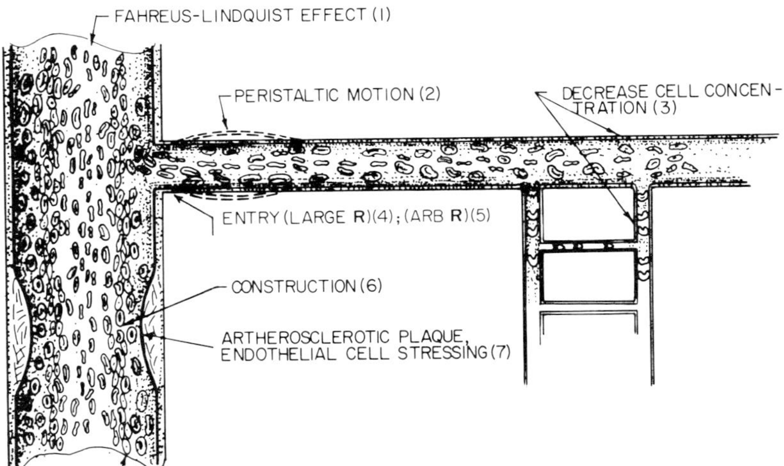

FIG. 19. A listing of some problems on blood flow in small blood vessels investigated recently. Numbers indicated references: (1) Haynes (1960); Burton (1965); Whitmore (1963). (2) Fung and Yih (1967, 1968); Yin and Fung (1969, 1971); Burns and Parkes (1967); Shapiro *et al.* (1969). (3) Seshadri and Sutera (1968, 1970); Svanes and Zweifach (1968). (4) Atabek and Chang (1961); (5) Lew and Fung (1969c, 1970a). (6) Lee and Fung (1970). (7) Fry (1968). (8) Segré and Silberberg (1962).

our bodies. To understand the function of our bodies we must study the details of microcirculation.

Serious studies of the mechanics of microcirculation are of recent origin. Some problems that have been investigated are summarized in Figs. 19 and 20. In the small arteries, we have the problem of reduced viscosity of the blood, the "plasma skimming" of the red cells, and the nonuniform distribution of the red blood cells in the tube. For arterioles there is the problem of peristaltic motion. For capillaries there are the problems of decreased cell concentration, on–off flow in a network, the red cell–capillary interaction, the flow of plasma between red cells, the deformation of the red cells and the endothelial cells, the permeation across the blood vessel wall, etc.

Several recent reviews have dealt with the mechanics of microcirculation. Fung (1969a) discussed the concerted methods of approach: experiments *in vivo*, macroscopic modeling, and mathematical analysis.

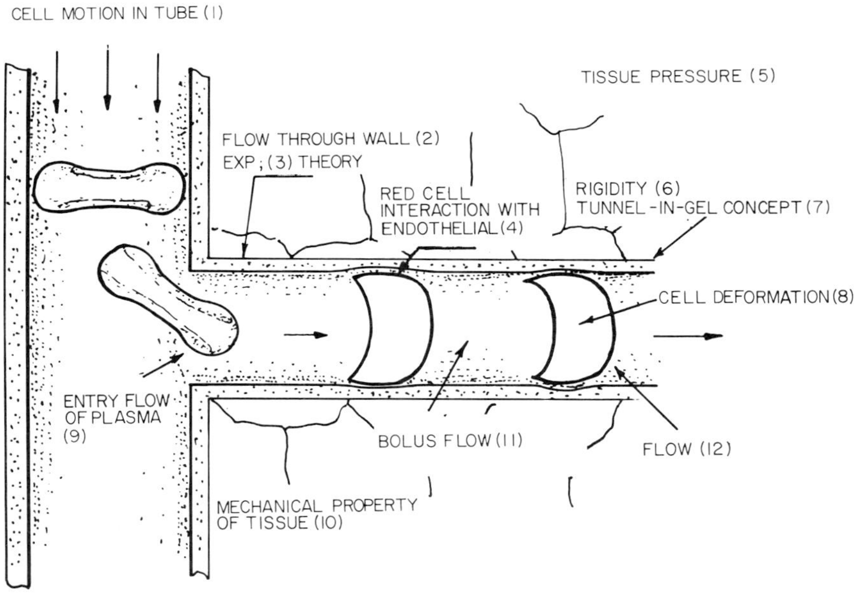

FIG. 20. A listing of some problems on blood flow in the capillary blood vessels investigated recently. Numbers indicate references: (1) Sutera and Hochmuth (1968); Lee and Fung (1969b). (2) Zweifach and Intaglietta (1968). (3) Lew and Fung (1969a). (4) Lighthill (1968). (5) Guyton (1963, 1965; *et al.*, 1966a, b); Scholander *et al.* (1968). (6) Baez *et al.* (1960). (7) Fung *et al.* (1966). (8) Lee and Fung (1969b). (9) Lew and Fung (1969c). (10) Fung (1967); Blatz *et al.* (1969). (11) Lew and Fung (1969b); Prothero and Burton (1961, 1962a,b); Aroesty and Gross (1970). (12) Barnard *et al.* (1968).

Skalak (1971) gave an excellent review on the problems of red cell deformation, particulate flow, red cell and blood vessel interaction, and overall flow pattern. Fung and Zweifach (1971) presented a concise review of the anatomy and physiology of capillary beds, and experimental and analytical results on entry flow, bolus flow, plug flow, and cell–vessel interaction. See also Bugliarello (1969).

Two aspects not discussed in these reviews are the flow in specialized beds and the flow in tubes with permeable walls. Of the specialized capillary beds, the pulmonary alveoli as shown in Figs. 15 and 16 are unique. Sobin *et al.* (1970) measured the alveolar sheet geometry. The flow of plasma in such a sheet has been analyzed by Lee and Fung (1968, 1969a), and Lee (1969). The flow in an elastic alveolar sheet has been analyzed by Fung and Sobin (1969) and a summary review is given by Fung (1969b).

Solutions of flow in tubes with permeable walls have been given by Lew and Fung (1969a) and Oka and Murata (1970). Lack of knowledge about the tissue space (the P_i, π_i terms in Eq. (3.14) is a principal weakness at the present time.

Therefore, in the following we shall discuss only one problem which was not considered in the reviews named above, namely, the peristaltic pumping.

Peristalsis

In the wing of a bat the arterioles, venules, and lymph nodes can be seen to vary their diameters periodically as if in peristaltic pumping (see Nicoll and Webb, 1946; Nicoll, (1954). They suggest that peristalsis plays a role in blood circulation. An idealized problem of peristalsis, as a sinusoidal traveling wave moving on the walls of a two-dimensional channel or a circular cylindrical tube, has been studied in considerable detail. Shapiro and his associates (1969) examined the propulsion and particle paths under the assumption of large wavelength and zero Reynolds number. This amounts to the zeroth order theory in an expansion in terms of the wavelength parameter α (the tube radius divided by the wavelength). Higher order theory in the α expansion was advanced by Zien and Ostrach (1971) and Li (1971). On the other hand, Fung and Yih (1968) solved the problem on the basis of an expansion in a power series in terms of the amplitude parameter ϵ (the wave amplitude divided by the tube radius) for finite Reynolds number and finite wavelength. Yin and Fung (1969) extended the procedure to the case of a circular cylindrical tube. Other solutions have been given by

Burns and Parkes (1967) and Hanin (1968). A comparison between theory and experiment was presented by Yin and Fung (1971).

These studies show that peristalsis of the kind considered—a harmonic traveling wave along a uniform channel or tube—provides little power in pumping as long as the tube (or channel) is not practically occluded. Therefore, in practical applications, peristaltic pumps (as used in heart-lung machines, etc.) should use fully occluded tubes, or tubes with valves such as those provided by nature in the lymph vessels, venules, and veins.

On the other hand, it is evident that a full understanding of biological peristalsis must include the action of the muscles in the tube wall. It is the stimulated muscle that provides the motive power to push the fluid along. A theoretical analysis that takes the muscle action into account was given by Fung (1971a, b). His physiological model is that of the ureter whose smooth muscle cells are assumed to be uniformly distributed and oriented spirally in the ureter wall. The slug of fluid is assumed to be so slender that its length is much greater than the diameter of the ureter. The Reynolds number of the flow is of the order of 1. Under these conditions the geometric profile of the slug of fluid, the pressure distributions in the fluid and in the ureteral wall, and the tensile stress in the ureteral smooth muscle can be determined. The results show that for geometric parameters appropriate to the human ureter the additional tension in the ureteral wall caused by the opening of the ureteral lumen due to the passage of a bolus of urine is very small—in the range of 10 to 20 dyn/cm if the bolus is less than 10 cm long. The corresponding rise in static pressure is also small: the pressure at the rear end of the bolus is only slightly higher than that at the front end. For a bolus of 5 cm length a pressure rise of less than 1 torr would be sufficient. The muscle contractile element begins exerting its tension at the section where the diameter of the bolus is the maximum. As soon as the muscle contraction begins, the ureteral lumen closes rapidly: full closure is obtained in a distance of no more than a few millimeters. Muscle contraction continues beyond this point of closure which marks the end of the bolus. Since the materials are incompressible the muscle contraction beyond the end of the bolus must be isometric. In such an isometric contraction the tension in the muscle continues to rise until a peak is reached after some time, then the tension gradually decays. The time delay for the tension to reach the maximum in an isometric contraction of the ureteral muscle must mean that there is a time delay in the pressure rise inside the ureter. This theoretical prediction was indeed found to be true. Barry *et al.* (1971) have found, from inserted catheter-tip pressure gauges, that the pressure in the ureter of the dog rises only slightly when the bolus of fluid passes the pressure gauges, but the pressure continues to rise after the bolus has gone past the gauges. Eventually a peak of 20 or 30 torr is reached before the pressure slowly subsides and a second bolus comes along. (This high value of peak pressure, however, might be an artifact caused by the presence of the catheter.) Physiologically, one could argue that the reserve power of the muscle to impose a pressure much higher than what is needed to move the fluid is useful in cleansing the ureter, stopping reflux, sending fluid through the ureterovesicular junction (valve) against a possibly higher pressure in the bladder, collecting fluid from the pelvis of the kidney, and effecting fluid transport even when only a small segment of the entire ureter can function normally, as in the case of an artificial silastic ureter implanted by surgery.

It will be of interest to note the mathematical structure of the problem because it involves properties of the muscle which are not familiar in conventional mechanics. Let us take a set of polar coordinates (r, θ, x) with the x axis coinciding with the axis of the ureter (see Fig. 21). We

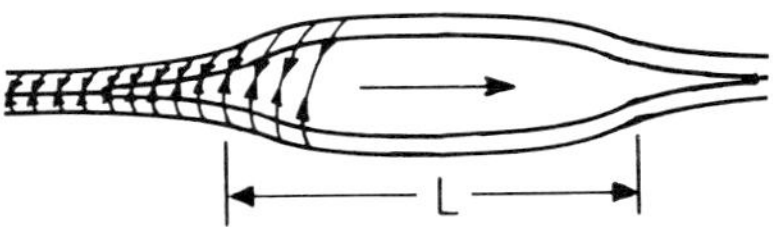

FIG. 21. Schematic drawing of a solitary peristaltic wave in the ureter, showing passive response of ureteral wall to the moving bolus of fluid in front, and active muscle contraction in the rear.

denote the velocity components by v_r, v_θ, v_x, the normal stresses by σ_r, σ_θ, σ_x, and the shear stresses by $\sigma_{r\theta}$, σ_{rx}, $\sigma_{x\theta}$. The inner wall of ureter is located at $r = r_i(x, t)$; the outer at $r = r_o(x, t)$, where t denotes time. The motion is assumed to be axisymmetric. Then the equations of equilibrium of the tube wall yield the relation

$$p_i r_i - p_o r_o = (r_o - r_i) \cdot \langle T \rangle \equiv \int_{r_i}^{r_o} \sigma_\theta \, dr, \qquad (5.1)$$

where p_i, p_o are the pressures in the lumen and outside the ureter, respectively, and $\langle T \rangle$ is the average tensile stress in the wall. For the fluid, the Stokes equations

$$-\frac{\partial p}{\partial x} + \mu \frac{1}{r} \frac{\partial}{\partial r}\left(r \frac{\partial v_x}{\partial r}\right) = 0, \qquad \frac{\partial p}{\partial r} = 0, \qquad (5.2a,b)$$

and the equation of continuity for an incompressible fluid,

$$\frac{1}{r}\frac{\partial}{\partial r}(rv_r) + \frac{1}{r}\frac{\partial v_\theta}{\partial \theta} + \frac{\partial v_x}{\partial x} = 0, \tag{5.3}$$

are subjected to the boundary conditions

$$v_x = v_r = 0 \quad \text{at} \quad r = r_i, \qquad v_r = \partial v_x/\partial r = 0 \quad \text{at} \quad r = 0. \tag{5.4}$$

Noting from (5.2b) that p is independent of r, we obtain from (5.2a) and (5.4) that

$$v_x(r, x, t) = U(x, t)[1 - (r^2/r_i^2)]. \tag{5.5}$$

On the other hand, (5.3) and (5.4) yield for a symmetric flow the relation

$$v_r(r, x, t) = -\frac{1}{r}\int_0^r r\frac{\partial v_x(r, x, t)}{\partial x}\,dr. \tag{5.6}$$

On the inner wall of the tube, the radial velocity $v_r(r_i, x, t)$ is precisely the velocity of the wall $\partial r_i(x, t)/\partial t$. On substituting (5.5) into (5.6), integrating, then setting $r = r_i$, and equating it with $\partial r_i(x, t)/\partial t$, we obtain

$$\partial r_i(x, t)/\partial t = -\tfrac{1}{4}r_i\,\partial U(x, t)/\partial x. \tag{5.7}$$

In a steady peristaltic motion for which the whole pattern moves to the right at a constant velocity c, r_i and U are functions of the single variable $x - ct \equiv \zeta$, and Eq. (5.7) can be integrated to give

$$\log[r_i(\zeta)/A] = (1/4c)\,U(\zeta), \tag{5.8}$$

with an integration constant A. Thus the velocity U is positive (agreeing with the direction of propagation of the peristaltic wave) when $r_i > A$; it is negative when $r_i < A$. Backward flow (U negative) occurs if the tube is open with radius $<A$, as a consequence of the conservation of mass. This backward flow can be stopped if the ureter is closed off both at the front and the rear when $r_i = A$ as shown in Fig. 22, as indeed it does in reality.

The action of the muscle requires some explanation. When a muscle fiber is in the resting state, it is soft and behaves like other soft tissues. When it is stimulated (in the heart, by electric pacemaker; in the ureter, by pacemaker or by stretching; in the skeletal muscle, by nerves), it shortens and generates tension. A heart muscle or a ureteral smooth muscle, when stimulated by a single electric pulse of sufficient magnitude, becomes "active;" and each cycle of activity has to run its course:

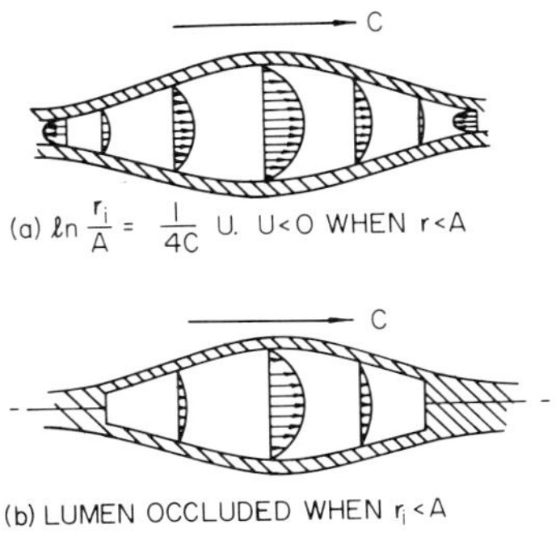

FIG. 22. Fluid velocity distribution in a long wave as demanded by the equation of continuity, and Stokes equation. (a) Condition given by Eq. (5.8). (b) Back flow is prevented by closing off the front and the rear end at the section when the lumen is equal to A.

repeated stimulation at a frequency higher than that dictated by the activity cycle will not generate a state of constant tension. On the other hand, skeletal muscle can be kept in a state of constant tension (tetanized) when stimulated at a sufficiently high frequency. The mechanical behavior of these muscles has been a subject of intensive study for at least half of a century, but an unequivocal description still does not exist. The least understood are the smooth muscules. In the following, we shall use a formulation orginally proposed for the heart muscule (Fung, 1970); it contains an element of tentativeness in need of further validation.

Referring for details to the original paper, we list the conclusion as follows: The total tensile stress T in a muscle fiber is the sum of two terms P and S, called the stresses in the *parallel* and *series* elements, respectively:

$$T = P + S. \tag{5.9}$$

Stresses will be defined in the Lagrangian sense and referred to a fixed reference cross-sectional area. P is a function of muscle length, whereas

$$\begin{aligned} S &= 0 \quad \text{for a resting muscle,} \\ &\neq 0 \quad \text{for an active muscle.} \end{aligned} \tag{5.10}$$

The length of a muscle fiber L can be resolved into the length of myosin-containing fibers M, the length of actin-containing fibers C, the *insertion* of actin and myosin Δ, and the *extension* η of the series elastic element:

$$L = (M + 2C) - \Delta + \eta. \tag{5.11}$$

The sum $M + 2C - \Delta$ is called the length of the *contractile element.* M and C are constant, Δ is variable, and η depends on the tension S.

The elasticity of the parallel and series elements is represented by the empirical equations

$$P = (P^* + \hat{\beta})\, e^{\hat{\alpha}(L-L^*)} - \hat{\beta}, \tag{5.12}$$

$$S = (S^* + \beta)\, e^{\alpha(\eta-\eta^*)} - \beta \tag{5.13}$$

in which $\hat{\alpha}$, $\hat{\beta}$, P^*, L^*, and α, β, S^*, and η^* are physiological constants.

To determine S, we must know η or Δ and L. The rate of insertion $d\Delta/dt$ is called the *contractile element velocity* or the velocity of shortening; it is a function of L, Δ, S, and t and is represented by the equation

$$\frac{d\Delta}{dt} = \frac{b(L) \operatorname{sgn} |\, S_0 f(t) - S\,|^n}{a(L) + S}, \tag{5.14}$$

in which $a(L)$, $b(L)$ are functions of L, S_0 is the peak tensile stress arrived in an isometric contraction at length L, n is an exponent whose value lies between 0 and 1, and $f(t)$ is a function of the time after stimulation, t. We propose

$$f(t) = \lambda \sin \pi/2[(t + t_0)/t_m], \tag{5.15}$$

with

$$1/\lambda = \sin \pi/2[(t_{\mathrm{ip}} + t_0)/t_m], \tag{5.16}$$

where t_0 and t_m are constants, and t_{ip} is the time at which the maximum tension is reached in an isometric contraction (L = const.) of the muscle. The symbol sgn stands for $\pm$ sign of the quantity $S_0 f(t) - S$. If $S_0 f(t) > S$, $d\Delta/dt$ is positive (muscle shortens); if $S_0 f(t) < S$, the sign of $d\Delta/dt$ reverses and the muscle lengthens.

The analysis of the ureter is complicated because of the variable distribution of tensile stress σ_θ. To simplify the analysis, note that according to Eq. (5.1) we need only the mean stress $\langle T \rangle$. Let a *neutral surface* with radius r_{N} be defined so that

$$\langle T \rangle = \sigma_\theta(r_{\mathrm{N}}). \tag{5.17}$$

Let the neutral surface be located at $r = R_{\mathrm{N}}$ when the ureter is unstressed,*

* A definition of the "natural state" or "unstressed state" of the ureter is required. It is natural to consider a ureter as unstressed as it is grown in the body under a constant peritoneal (approximately atmospheric) pressure. Then the muscle tension $T = P + S$ is the stress above the peritoneal pressure p_0. With the internal pressure p_i measured as the gauge pressure with respect to p_0, then we should set $p_0 = 0$ in our equations.

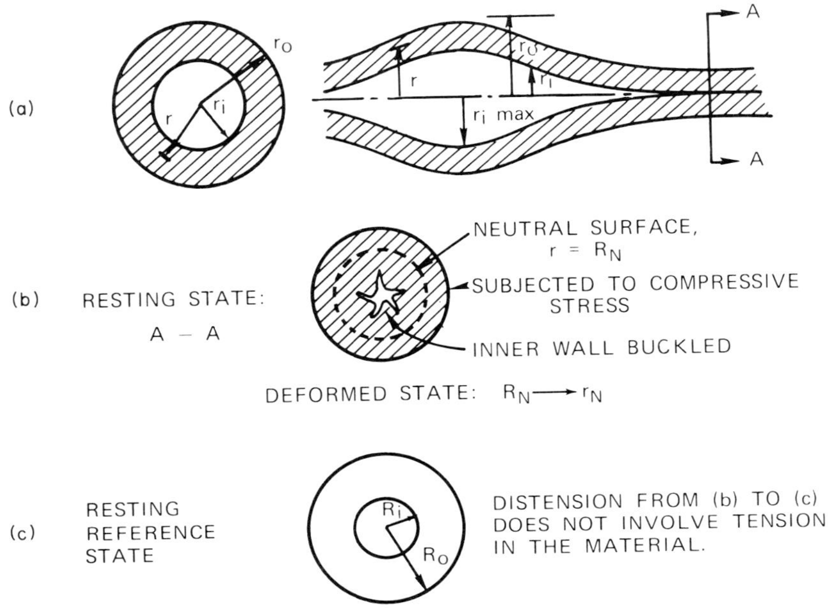

FIG. 23. Notations. Subscript "i" refers to inner wall, "o" refers to outer wall, $r_{i\,max}$ is the maximum radius of the bolus. The "neutral surface" is indicated by a subscript "N."

and at $r = r_N$ when the ureter is distended (see Fig. 23). Then we can take R_N as the unit of length for the muscle and write r_N/R_N and $r_N{}^*/R_N$ in place of L and L^* in Eq. (5.12), and η/R_N and η^*/R_N in place of η and η^* in Eq. (5.13). Then, for the ureteral wall we have, from Eqs. (5.9)–(5.14),

$$\frac{dT}{dt} = \left[\frac{\hat{\alpha}}{R_N}(P + \hat{\beta}) + \frac{\alpha}{R_N}(S + \beta)\right]\frac{dr_N}{dt}$$

$$+ \left[\frac{\alpha}{R_N}(S + \beta) + \frac{\partial S}{\partial \Delta}\right]\frac{b(L)\,\mathrm{sgn}\,|\,S_0 f(t) - S\,|^n}{a(L) + S}, \qquad (5.18)$$

where the parameters $\hat{\alpha}$, $\hat{\beta}$, α, β, etc. are values appropriate for the ureter with respect to circumferential distension, and T stands for $\langle T \rangle$.

In a persitaltic wave both r_N and T as functions of space and time are functions of $\zeta = x - ct$. Hence, $d/dt = -c d/d\zeta$. Substituting this and $S = T - P$ into (5.18), we obtain the first basic equation:

$$-c \frac{dT}{d\zeta} = -c[\alpha T - (\alpha - \hat{\alpha})P + \alpha\beta + \hat{\alpha}\hat{\beta}] \frac{1}{R_N} \frac{dr_N}{d\zeta}$$
$$+ \left[\alpha(T - P) + \alpha\beta + \frac{\partial S}{\partial \Delta} R_N\right] \frac{1}{R_N} \frac{b \operatorname{sgn} | S_0 f(t) - T + P |^n}{a + T - P}. \tag{5.19}$$

A second equation is obtained by combining (5.8) and (5.2a), introducing ζ, and using (5.5):

$$-\frac{dp_i}{d\zeta} = \frac{4\mu}{r_i^2} U = \frac{16c\mu}{r_i^2} \log \frac{r_i}{A}. \tag{5.20}$$

The left-hand side can be expressed in terms of T through Eq. (5.1), which, by differentiating with respect to ζ, and setting $p_0 = 0$, yields

$$(r_o - r_i) \frac{dT}{d\zeta} + \left(\frac{dr_o}{d\zeta} - \frac{dr_i}{d\zeta}\right) T = r_i \frac{dp_i}{d\zeta} + p_i \frac{dr_i}{d\zeta}. \tag{5.21}$$

Eliminating $dp_i/d\zeta$ from (5.20) and (5.21), and expressing the radii r_i and r_o in terms of r_N by the condition of incompressibility of the ureteral wall, we obtain

$$\frac{dT}{d\zeta} = K \frac{dr_N}{d\zeta} + N, \tag{5.22}$$

where

$$K = T \frac{r_N}{r_i} \left(\frac{1}{r_o} + \frac{1}{r_i}\right),$$

$$N = -\frac{16c\mu}{r_i(r_o - r_i)} \log \frac{r_i}{A}.$$

Equations (5.19) and (5.22) govern the segment of the ureter in which the muscle is active. The corresponding equations in the *passive segment* can be obtained by letting $S = T - P = 0$. These segments are joined together at a section which may be designated $\zeta = 0$, where

$$T |_{\zeta=0} = P |_{\zeta=0}. \tag{5.23}$$

The peristaltic features discussed above are obtained from these equations. Numerical results are presented by Fung (1971b).

The method presented above can be extended to the analysis of the heart, the geometry of which is more complicated. Other organs such as the precapillary sphinctors, lymph ducts, intestines, and the uterus, with their functions dependent on the muscle, can be analyzed by a scheme similar to or as an extension of the above.

VI. Conclusion

There is no doubt that mechanics is a necessary part of life science. It has not been used extensively in biology and medicine because traditionally it is not part of a biologist's or physician's standard training. However, since life science and physical science are one and the same, as von Helmholtz, Du Bois-Reymond, and others had maintained long ago, the renewed interest in biomechanics is bound to come.

Yet the biological world is a very complex one. In this article we have tried to show that great care must be exercized to assemble the pieces of information to formulate a meaningful boundary-values problem. Rash idealization and oversimplification is probably the single most important cause for most biologists not to take biomechanical publications seriously.

Blood circulation is currently the most popular subject in biomechanics, as the sizable literature attests. The value of mechanics has been shown very clearly in the development of heart-assist devices. The development of artificial heart valves demonstrated again the importance of fluid mechanics. Undoubtedly the programs on artificial heart, lung, kidney, etc. would rely heavily on biomechanics as well as biomaterials.

In microcircualtion research mechanics offers a powerful tool. It offers such details about the blood flow that are difficult to obtain by conventional means at the disposal of physiologists.

Many other aspects of biomechanics are not touched upon in the present article. The determination of the mechanical properties of living tissues, the use of such information in the study of pathogenesis, the application of biomechanics to diagnosis, surgery, and patient care, the treatment of head injury, rupture, hernia, joints, and many problems in orthopedics, are subjects of widespread interest.

Space and ocean exploration present many new and unfamiliar problems to biomechanics. But a review of biomechanics in these fields, as well as of the problems of swimming, flying, walking, etc., would have to be done by other authors more competent in these fields and more familiar with these problems than the present writer.

REFERENCES

ANLIKER, M., and MAXWELL, J. A. (1966). The dispersion of waves in blood vessels. *In* "Biomechanics" (Y. C. Fung, ed.), pp. 47–67. Am. Soc. Mech. Eng., New York.

ANLIKER, M., and RAMAN, K. R. (1966). Korotkoff sounds at diastole—a phenomenon of dynamic instability of fluid-filled shells. *Int. J. Solids Struct.* **2**, 467–492.

ANLIKER, M., MORITZ, W. E., and OGDEN, E. (1968a). Transmission characteristics of axial waves in blood vessels. *J. Biomech.* **1**, 235–246.

ANLIKER, M., HISTAND, M. B., and OGDEN, E. (1968b). Dispersion and attenuation of small artificial pressure waves in the aorta. *Circ. Res.* **13**, 539–551.

APERIA, A. (1940). Hemodynamical studies. *Skand. Arch. Physiol.* **83**, Suppl. 16.

APTER, J. T. (1964). Mathematical development of a physical model of some visco-elastic properties of the aorta. *Bull. Math. Biophys.* **26**, 367–388.

APTER, J. T., and MARQUEZ, E. (1968). V. Effects of sinusoidal forcing at low and at resonance frequencies. *Circ. Res.* **22**, 393–404.

APTER, J. T., RABINOWITZ, M., and CUMMINGS, D. H. (1966). Correlation of visco-elastic properties of large arteries with microscopic structure. I. Methods used and their justification (Apter). II. Collagen, elastin and muscle determined chemically, histologically and physiologically: III. Circumferential viscous and elastin constants measured *in vitro* (Apter). *Circ. Res.* **19**, 104–121.

APTER, J. T., RABINOWITZ, M., and CUMMINGS, D. H. (1967). IV. Thermal responses of elastin, collagen, smooth muscle and intact arteries. *Circ. Res.* **21**, 901–918.

AROESTY, J., and GROSS, J. F. (1970). Convection and diffusion in the microcirculation. *Microvascular Res.* **2**, 247–267.

ATABEK, H. B. (1964). End effects. *In* "Pulsatile Blood Flow" (E. O. Attinger, ed.), Chapter 10, pp. 201–220. McGraw-Hill, New York.

ATABEK, H. B., and CHANG, C. C. (1961). Oscillatory flow near the entry of a circular tube. *Z. Angew. Math. Phys.* **12**, 185–201.

ATABEK, H. B., and LEW, H. S. (1966). Wave propagation through a viscous incompressible fluid contained in an initially stressed elastic tube. *Biophys. J.* **6**, 481–503.

ATTINGER, E. O. (1964). Flow patterns and vascular geometry. *In* "Pulsatile Blood Flow" (E. O. Attinger, ed.), Chapter 9, pp. 179–200. McGraw-Hill, New York.

ATTINGER, E. O. (1971). Biomechanics, patient care and rehabilitation. *In* "Biomechanics, Its Foundations and Objectives" (Y. C. Fung *et al.*, eds.). Prentice-Hall, Englewood Cliffs, New Jersey.

ATTINGER, E. O., SUGAWARA, H., NAVARRO, A., and ANNÉ, A. (1966). Pulsatile flow patterns in distensible tubes. *Circ. Res.* **18**, 447–456.

BAEZ, S., LAMPORT, H., and BAEZ, A. (1960). Pressure effects in living microscopic vessels. *In* "Flow Properties of Blood and Other Biological System" (A. L. Copley and G. Stainsby, eds.), pp. 122–136. Pergamon Press, Oxford.

BARNARD, A. C. L., LOPEZ, L., AND HELLUMS, J. D. (1968). Basic theory of blood flow in capillaries. *Microvascular Res.* **1**, 23–34.

BARRY, W. F., JR., ABSHER, R. G., and BOYARSKY, S. (1971). Correlation of cineradiographic image and pressure tracing of ureteral activity. *Workshop Hydrodyn. Upper Urinary Tract*, 1969. *In* "Urodynamics: Hydrodynamics of the Ureter and Renal Pelvis" (S. Boyarsky, ed.). Academic Press. New York.

BENEKEN, J. E. W. (1965). "A Mathematical Approach to Cardio-Vascular Function; The Uncontrolled Human System." Inst. Med. Phys., Utrecht, Holland.

BERGEL, D. H. (1961a). The static elastic properties of the arterial wall. *J. Physiol. (London)* **156**, 445–457.

BERGEL, D. H. (1961b). The dynamic elastic properties of the arterial wall. *J. Physiol. (London)* **156**, 458–469.

BERGEL, D. H. (1971). The properties of blood vessels, or the spring of the arteries. *In* "Biomechanics, Its Foundations and Objectives" (Y. C. Fung *et al.*, eds.). Prentice-Hall, Englewood Cliffs, New Jersey.

BERGEL, D. H., and MILNOR, W. R. (1965). Pulmonary vascular impedance in the dog. *Circ. Res.* **16**, 401–415.

BESSIS, M. (1956). "Cytology of the Blood and Blood-forming Organs." Grune & Stratton, New York.

BLACKSHEAR, P. L. (1971). Mechanical hemolysis in flowing blood. *In* "Biomechanics, Its Foundations and Objectives" (Y. C. Fung *et al.*, eds.). Prentice-Hall, Englewood Cliffs, New Jersey.

BLATZ, P. J., CHU, B. M., AND WAYLAND, H. (1969). On the mechanical behavior of elastic animal tissue. *Trans. Soc. Rheol.* **13**, 83–102.

BLOCH, E. H. (1962). A quantitative study of the hemodynamics in the living microvascular system. *Amer. J. Anat.* **110**, 125–153.

BORELLI, G. A. (1680). "De Motu Animalium."

BOYLE, R. (1684). "Memoirs for the Natural History of Human Blood."

BRAASCH, D., and JENETT, W. (1968). Erythrocytenflexihilität, Hämokonzentration und Reibungswiderstand in Glascapillaren mit Durchmessern zwischen 6 his 50 μ. *Pfluegers Arch. Gesamte Physiol. Menschen Tiere* **302**, 245–254.

BRANEMARK, P.-I., and LINDSTRÖM, J. (1963). Shape of circulating blood corpuscles *Biorheology* **1**, 139–142.

BRANSON, H. (1945). The flaw of a viscous fluid in an elastic tube: A model of the femoral artery. *Bull. Math. Biophys.* **7**, 181–188.

BRECHER, G. A. (1956). "Venous Return." Grune & Stratton, New York.

BRENNER, H. (1965). Hydrodynamics resistance of particles at small Reynolds numbers. *Advan. Chem. Eng.* **6**, 287–438.

BUGLIARELLO, G. (1966). Microcirculation hemodynamics and other biological flow problems. *In* "Biomedical Fluid Mechanics Symposium," pp. 192–208. Am. Soc. Mech. Eng., New York.

BUGLIARELLO, G. (1969). Some aspects of the biomechanics of the microcirculation, anno 1969. *In* "Development in Mechanics." *Proc. Midwest Mech. Conf.* **5**, 921–962.

BURNS, J. C., and PARKES, T. (1967). Peristaltic motion. *J. Fluid Mech.* **29**, 731–743.

BURTON, A. C. (1962). Physical principles of circulatory phenomena: The physical equilibria of the heart and blood vessels. *In* "Handbook of Physiology" (Am. Physiol. Soc.), Sect. 2, Circulation (W. F. Hamilton and P. Dow, eds), Vol. I, Chapter 6, pp. 85–106. Williams & Wilkins, Baltimore, Maryland.

BURTON, A. C. (1965). "Physiology and Biophysics of the Circulation." Year Book Publ., Chicago, Illinois.

CAREW, T. E., VAISHNAV, R. N., and PATEL, D. J. (1968). Compressibility of the arterial wall. *Circ. Res.* **23**, 61.

CARO, C. G., and MCDONALD, D. A. (1961). The relation of pressure and flow in the pulmonary vascular bed. *J. Physiol. (London)* **157**, 426–453.

CARO, C. G., FITZ-GERALD, J. M., AND SCHROTER, R. C. (1969). Arterial wall shear and distribution of early atheroma in man. *Nature (London)* **223**, 1159–1161.

CARO, C. G., FITZ-GERALD, J. M., AND SCHROTER, R. C. (1971). Atheroma and arterial wall shear: Observation, correlation and proposal of a shear dependent mass transfer mechanism for atherogenesis. *Proc. Roy. Soc. Ser. B.* **177**, 109–159.

CASSON, N. (1959). A flow equation for pigment–oil suspensions of the printing-ink type. *In* "Rheology of Disperse Systems" (C. C. Mill, ed.). Pergamon Press, Oxford.

CHIEN, S., USAMI, S., TAYLOR, M. LUNDBERG, J. L., and GREGERSEN, M. I. (1966). Effects of hematocrit and plasma proteins of human blood rheology at low shear rates. *J. Appl. Physiol.* **21**, 81–87.

CHIEN, S., USAMI, S., DELLENBACK, R. T., and GREGERSEN, M. I. (1967a). Blood viscosity: Influence of erythrocyte deformation. *Science* **157**, 827–829.

CHIEN, S., USAMI, S., DELLENBACK, R. T., and GREGERSEN, M. I. (1967b). Blood viscosity: Influence of erythrocyte aggregation. *Science* **157**, 829–831.

CHIEN, S., BRYANT, C. A., and LUSE, S. A. (1971). Mechanical hemolysis of erythrocytes during passage through micropores. *Microvascular Res.* (in press).

COKELET, G. R. (1971). The rheology of human blood. *In* "Biomechanics, Its Foundations and Objectives" (Y. C. Fung *et al.*, eds.). Prentice-Hall, Englewood Cliffs, New Jersey.

COKELET, G. R., and MEISELMAN, H. J. (1968). Rheological comparison of hemoglobin solutions and erythrocyte suspensions. *Science* **162**, 275–277.

COKELET, G. R., MERRILL, E. W., GILLILAND, E. R., SHIN, H., BRITTEN, A., and WELLS, R. E. (1963). The rheology of human blood-measurement near and at a zero shear rate. *Trans. Soc. Rheol.* **7**, 303–317.

COPLEY, A. L., and STAINSBY, G., eds. (1960). "Flow Properties of Blood and Other Biological Systems." Pergamon Press, Oxford.

COX, R. (1968). Wave propagation through a Newtonian fluid contained within a thick-walled, viscoelastic tube. *Biophys. J.* **8**, 691–709.

COX, R. H. (1970). Blood flow and pressure propagation in the canine femoral artery. *J. Biomech.* **2**, 131–150.

DANIELLI, J. F. (1940). Capillary permeability and oedema in the perfused frog. *J. Physiol. (London)* **98**, 109–129.

DANIELLI, J. F. (1941). A method for estimating the fraction of the volume of a muscle contained in the vascular system. *J. Physiol. (London)* **100**, 239–245.

DESCARTES, RENÉ (1664). "De l'Homme."

DINTENFASS, L. (1967). Inversion of the Fahraeus-Lindquist phenomenon in blood flow through capillaries of diminishing radius. *Nature (London)* **215**, 1099–1100.

DINTENFASS, L. (1968). Internal viscosity of the red cell and a blood viscosity equation. *Nature (London)* **219**, 956–958.

EIRICH, F., AND GOLDSCHMID, O. (1937). Untersuchungen über die Viskosität von Suspensionen und Lösungen. *Kolloid Z.* **81**, 7–18.

EULER, L. (1775). Principia pro motu sanguins per arterias determinado. Opera posthuma mathematica et physica anno 1844 detecta, ediderunt P. H. Fuss et N. Fuss. Petropoli, Apud Eggers et socios, Vol. 2, pp. 814–823.

EVANS, R. L. (1962a). A unifying approach to blood flow theory. *J. Theor. Biophys.* **3**, 392–411.

EVANS, R. L. (1962b). Pulsatile flow in vessels whose distensibility and size vary with site. *Phys. Med. & Biol.* **7**, 105–116.

FAHRAEUS, R. (1929). The suspension stability of the blood. *Physiol. Rev.* **9**, 241–274.

FAHRAEUS, R., AND LINDQVIST, T. (1931). The viscosity of the blood in narrow capillary tubes. *Amer. J. Physiol.* **96**, 562–568.

FORRESTER, J. H., AND YOUNG, D. F. (1970). Flow through a converging-diverging tube and its implications in occlusive vascular disease. *J. Biomech.* **3**, 297–316.

FOX, E. A., and SAIBEL, E. (1963b). Attempts in the mathematical analysis of blood flow. *Trans. Soc. Rheol.* **7**, 25–31.

FRANK, O. (1899). Die Grundform des arteriellen Pulses. *Z. Biol.* **37**, 483–526.

FRANK, O. (1926). Die Theorie der Pulswellen. *Z. Biol.* **85**, 91–130.

FRANK, O. (1930). Schatzung des Schlagvolumens des menschlichen Herzens auf Grund der Wellen- und Windkessel theorie. *Z. Biol.* **90**, 405–409.

FRASHER, W. G. (1966). What is known about the physiology of large blood vessels. *In* "Biomechanics" (Y. C. Fung, ed.), pp. 1–19. Am. Soc. Mech. Eng., New York.

FREIS, E. D., and HEATH, W. C. (1964). Hydrodynamics of aortic blood flow. *Circ. Res.* **14**, 105–116.

FREY-WYSSLING, A. (1952). "Deformation and Flow in Biological Systems." Wiley (Interscience), New York.

FRY, D. L. (1968). Acute endothelial changes associated with increased blood velocity gradients. *Circ. Res.* **22**, 165–197.

FRY, D. L., NOBLE, F. W., and MALLOS, A. J. (1957). An electric device for instantaneous computation of aortic blood velocity. *Circ. Res.* **5**, 75–78.

FRY, D. L., GRIGGS, D. M., JR., and GREENFIELD, J. C., JR. (1964). *In vivo* studies of pulsatile blood flow: The relationship of the pressure gradient to the blood velocity. *In* "Pulsatile Blood Flow" (E. O. Attinger, ed.), Chapter 5, pp. 101–114. McGraw-Hill, New York.

FUNG, Y. C. (1966a). Microscopic blood vessels in the mesentery. *In* "Biomechanics" (Y. C. Fung, ed.), pp. 151–166. Am. Soc. Mech. Eng., New York.

FUNG, Y. C. (1966b). Theoretical considerations of the elasticity of red cells and small blood vessels. *Fed. Proc., Fed. Amer. Soc. Exp. Biol.* **25**, 1761–1772.

FUNG, Y. C. (1967). The elasticity of soft tissues in simple elongation. *Amer. J. Physiol.* **213**, 1532–1544.

FUNG, Y. C. (1968). Biomechanics, its scope, history and some problems of continuum mechanics in physiology. *Appl. Mech. Rev.* **21**, 1–20.

FUNG, Y. C. (1969a). Blood flow in the capillary bed. *J. Biomech.* **2**, 353–373.

FUNG, Y. C. (1969b). Studies on the blood flow in the lung. *Proc. Can. Congr. Appl. Mech., 2nd*, 1969 pp. 433–454.

FUNG, Y. C. (1970). Mathematical representation of the mechanical properties of the heart muscle. *J. Biomech.* **3**, 381–404.

FUNG, Y. C. (1971a). Peristaltic pumping: A bioengineering model. *In* "Urodynamics: Hydrodynamics of the Ureter and Renal Pelvis" (S. Boyarsky, ed.). Academic Press, New York.

FUNG, Y. C. (1971b). Muscle controlled flow. *Proc. 12th Midwestern Mech. Conf.* Univ. of Notre Dame, South Bend, Indiana.

FUNG, Y. C. (1971c). Stress-strain-history relations of soft tissues in simple elongation. *In* "Biomechanics, Its Foundations and Objectives" (Y. C. Fung *et al.*, eds.). Prentice-Hall, Englewood Cliffs, New Jersey.

FUNG, Y. C., and SOBIN, S. S. (1969). Theory of sheet flow in lung alveoli. *J. Appl. Physiol.* **26**, 472–488.

FUNG, Y. C., and TONG, P. (1968). Theory of the sphering of red blood cells. *J. Biophys.* **8**, 175–198.

FUNG, Y. C., and YIH, C. S. (1968). Peristaltic transport. *J. Appl. Mech.* **35**, 669–675.

FUNG, Y. C., and ZWEIFACH, B. W. (1971). Microcirculation: Blood flow in capillary blood vessels. *Ann. Rev. Fluid Mech.* **3**, 189–210.

FUNG, Y. C., ZWEIFACH, B. W., and INTAGLIETTA, M. (1966). Elastic environment of the capillary bed. *Circ. Res.* **19**, 441–461.

GABE, I. T. (1965). An analogue computer deriving oscillatory arterial blood flow from the pressure gradient. *Phys. Med. & Biol.* **10**, 407–415.

GABE, I. T., AND ZAZT, L. (1968). Studies of plasma viscosity under conditions of oscillatory flow *Biorheology* **5**, 86 (Abstract).

GESSNER, U. (1964). A graphic method of computing pulsatile blood flow parameters. *In* "Pulsatile Blood Flow" (E. O. Attinger, ed.), Chapter 23, pp. 389–405. McGraw-Hill, New York.

GOLDSMITH, H. L., AND MASON, S. G. (1967). The microrheology of dispersions. *In* "Rheology, Theory and Applications" (F. R. Eirich, ed.), Vol. 4, Chapter 2, pp. 85–250. Academic Press, New York.

GOMEZ, D. M. (1964). "Hemodynamique et angiocinétique; étude rationnelle des lois régissant les phénomènes cardio-vaculaires." Hermann, Paris.

GREGERSEN, M. I. (1967). Factors regulating blood viscosity: Relation to problems of the microcirculation. *In* "Colloque International, Les Concepts de Claude Bernard sur le Milieu," pp. 231–244. Masson, Paris.

GREGERSEN, M. I., BRYANT, C. A., HAMMERLE, W. E., USAMI, S., and CHIEN, S. (1967). Flow characteristics of human erythrocytes through polycarbonate sieves. *Science* **157**, 825–827.

GUEST, M. M., BOND, T. P., COOPER, R. G., and DERRICK, J. R. (1963). Red blood cells: Change in shape in capillaries. *Science* **142**, 1319–1321.

GUTSTEIN, W. H., LAZZARINI-ROBERTSON, A., JR., and LATAILLADE, J. N. (1963). The role of local arterial irritability in the development of arterio-atherosclerosis. *Amer. J. Pathol.* **42**, 61–71.

GUYTON, A. C. (1961). "Medical Physiology." Saunders, Philadelphia, Pennsylvania.

GUYTON, A. C. (1963). A concept of negative interstitial pressure based on pressures in implanted perforated capsules. *Circ. Res.* **12**, 399–414.

GUYTON, A. C. (1965). II. Pressure-volume curves of interstitial space. *Circ. Res.* **16**, 452–460.

GUYTON, A. C., SCHEEL, K., and MURPHREE, D. (1966a). III. Its effect on resistance to tissue fluid mobility. *Circ. Res.* **19**, 412–419.

GUYTON, A. C., PRATHER, J., SCHEEL, K., AND MCGEHEE, J. (1966b). IV. Interstitial fluid pressure: Its effect on fluid movement through the capillary wall. *Circ. Res.* **19**, 1022–1030.

HALES, S. (1733). "Statical Essays," Vol. II. Innays & Manby, London (reprinted by Hafner, New York, 1964).

HAMBURGER, W. W. (1939). The earliest known reference to the heart and circulation. The Edwin Smith Surgical Papyrus, circa 3000 B.C. *Amer. Heart J.* **17**, 259–274.

HAMILTON, W. F., REMINGTON, J. W., and DOW, P. (1939). An experimental study of the standing wave in the pulse propagated through the aorta. *Amer. J. Physiol.* **125**, 48–59.

HANIN, M. (1968). The flow through a channel due to transversely oscillating walls. *Isr. J. Technol.* **6**, 67–71.

HARDUNG, V. (1952). Ueber eine Methode zur Messung der Dynamischen Elastizität und Viskosität Kautschukähnlicher Kärper, insbesondere von Blutgefäszen und anderen Elastischen Grewebeteilen. *Helv. Physiol. Pharmacol. Acta* **10**, 482–498.

HARDUNG, V. (1953). Vergleichende Messungen der dynansischen Elastizität und Viskosität von Blutgefässen, Kautschak, and synthetischen Elastomeren. *Helv. Physiol. Pharmacol. Acta* **11**, 194–211.

HARDUNG, V. (1962). Propagation of pulse waves in visco-elastic tubings. *In* "Handbook of Physiology" (Am. Physiol. Soc., J. Field, ed.), Sect. 2, Vol. I, Chapter 7, pp. 107–135. Williams & Wilkins, Baltimore, Maryland.

HARRIS, J. W. (1965). "The Red Cell." Harvard Univ. Press, Cambridge, Massachusetts.

HARVEY, W. (1958). "Exercitatio anatomica de motu cordis et sanguinis in animalilius" (English translation by C. D. Leake, 4th Ed. Thomas, Springfield, Illinois).

HAYNES, R. H. (1960). Physical basis of the dependence of blood viscosity on tube radius. *Amer. J. Physiol.* **198**, 1193–1200.

HAYNES, R. H. (1962). The viscosity of erythrocyte suspensions. *Biophys. J.* **2**, 95–103.

HOLMAN, E. (1954). On circumscribed dilation of an artery immediately distal to a partially occluding band: Poststenotic dilation. *Surgery* **36**, 3–24.

HOOKE, ROBERT (1664). "Micrographia" (reprinted by Dover, New York, 1960).

HYMAN, C. (1944). Filtration across the vascular wall as a function of several physical factors. *Amer. J. Physiol.* **142**, 671–685.

HYMAN, C., RAPOPORT, S. I., SAUL, A. M., and MORTON, M. E. (1952). Independence of capillary filtration and tissue clearance. *Amer. J. Physiol.* **168**, 674–679.

IBERALL, A. S. (1964). Study of the general dynamics of the physicalchemical systems in mammals. *NASA Contract. Rep. NASA CR*-**129**.

INTAGLIETTA, M., and ZWEIFACH, B. W. (1966). Indirect method of measurement of pressure in blood capillaries. *Circ. Res.* **19**, 199–205.

JACOBS, R. B. (1953). On the propagation of a disturbance through a viscous liquid flowing in a distensible tube of appreciable mass. *Bull. Math. Biophys.* **15**, 395–409.

JONES, E., CHANG, I. D., and ANLIKER, M. (1968). Effects of viscosity and external constraints on wave transmission in blood vessels. Biomech. Lab. Rep. No. Sudaar 344. Stanford University.

JOUKOWSKY, N. W. (1900). Ueber den hydraulischen Stoss in Wasserleitungsröhen. *Mem. Acad. Imp. Sci. St. Petersburg* [8] **9**, No. 5.

KAUDERDER, H. (1958). "Nichtlineare Mechanik." Springer, Berlin.

KENNER, T. (1971). Flow and pressure in the arteries. *In* "Biomechanics, Its Foundations and Objectives" (Y. C. Fung *et al.*, eds.). Prentice-Hall, Englewood Cliffs, New Jersey.

KING, A. L. (1947a). On a generalization of the Poiseuille law. *Amer. J. Phys.* **15**, 240–242.

KING, A. L. (1947b). Waves in elastic tubes: Velocity of the pulse waves in large arteries. *J. Appl. Phys.* **18**, 595–600.

KLIP, W. (1962). "Velocity and Damping of the Pulse Wave." Nijhoff, The Hague.

KLIP, W. (1967). Formulas for phase velocity and damping of longitudinal waves in thick-walled viscoelastic tubes. *J. Appl. Phys.* **38**, 3745–3755.

KLIP, W., and KLIP, D. A. B. (1964). Phase velocity and damping of torsional waves in thin-walled tubes of infinite length. *In* "Pulsatile Blood Flow" (E. O. Attinger, ed.), Chapter 10, pp. 323–330. McGraw-Hill, New York.

KORTEWEG, D. J. (1878). Ueber die Fortpflanzungesgeschwindigkeit des Schalles in elastischen Rohren. *Ann. Phys. Chem.* [3] **5**, 525–542.

KROVETZ, L. J. (1965). The effect of vessel branching on haemodynamic stability. *Phys. Med. & Biol.* **10**, 417–427.

KUCHAR, N. R., and OSTRACH, S. (1966). Flows in the entrance regions of circular elastic tubes. *In* "Biomedical Fluid Mechanics Symposium," pp. 45–69. Am. Soc. Mech. Eng., New York.

LAMB, H. (1897–1898). On the velocity of sound in a tube, as affected by the elasticity of the walls. *Phil. Soc. Mem. & Proc. Manchester, Lit. A* **42**, 1–16.

LAMBERT, J. W. (1956). Fluid flow in a nonrigid tube. Doctoral Dissertation, Ser. No. 19, 418. Univ. Microfilms, Inc., Ann Arbor, Michigan.

LAMBERT, J. W. (1958). On the nonlinearities of fluid flow in nonrigid tubes. *J. Franklin Inst.* **266**, 83–102.

LAMBOSSY, P. (1950). Aperçu historique et critique sur le problème de la propagation des ondes dans un liquide compressible enfermé dans un tube élastique. *Helv. Physiol. Pharmacol. Acta* **8**, 209–227.

LAMBOSSY, P. (1951). Aperçu historique et critique sur le problème de la propagation des ondes dans un liquide compressible enfermé dans un tube élastique. *Helv. Physiol. Pharmacol. Acta* **9**, 145–161.

LANDIS, E. M. (1927). The relation between capillary pressure and the rate at which fluid passes through the walls of single capillaries. *Amer. J. Physiol.* **82**, 217–238.

LANDOWNE, M. (1958). Characteristics of impact and pulse wave propagation in brachial and radial arteries. *J. Appl. Physiol.* **12**, 91–97.

LEE, J. S. (1966). The pressure-flow relationship in long-wave propagation in large arteries. *In* "Biomechanics" (Y. C. Fung, ed.), pp. 96–120. Am. Soc. Mech. Eng., New York.

LEE, J. S. (1969). Slow viscous flow in a lung alveoli model. *J. Biomech.* **2**, 187–198.

LEE, J. S., and FUNG, Y. C. (1968). Experiments on blood flow in lung alveoli models. Pap. No. 68-WA/BHF-2, Ann. Meeting. Am. Soc. Mech. Eng., New York.

LEE, J. S., and FUNG, Y. C. (1969a). Stokes' flow around a circular cylindrical post confined between two parallel plates. *J. Fluid Mech.* **37**, Part 4, 657–670.

LEE, J. S., and FUNG, Y. C. (1969b). Modeling experiments of a single red blood cell moving in a capillary blood vessel. *Microvascular Res.* **1**, 221–243.

LEE, J. S., and FUNG, Y. C. (1970). Flow in locally constricted tubes at low Reynolds number. *J. Appl. Mech.* **37**, 9–16.

LEE, J. S., FRASHER, W. G., and FUNG, Y. C. (1967). Two-dimensional finite-deformation on experiments on dog's arteries and veins. Tech. Rep. No. AFOSR 67-1980. University of California, San Diego, California.

LEE, J. S., FRASHER, W. G., and FUNG, Y. C. (1968). Comparison of elasticity of an artery *in vivo* and *in excision*. *J. Appl. Physiol.* **25**, 779–801.

LEW, H. S. (1969). Formulation of statistical equation of motion of blood. *Biophys. J.* **9**, 235–245.

LEW, H. S., AND FUNG, Y. C. (1969a). Flow in an occluded circular cylindrical tube with permeable wall. *Z. Angew. Math. Physik* **20** (Fasc. 5), 750–766.

LEW, H. S., AND FUNG, Y. C. (1969b). The motion of the plasma between the red blood cells in the bolus flow. *J. Biorheol.* **6**, 109–119.

LEW, H. S., AND FUNG, Y. C. (1969c). On the low Reynolds number entry flow into a circular cylindrical tube. *J. Biomechan.* **2**, 105–119.

LEW, H. S., AND FUNG, Y. C. (1970a). Entry flow into blood vessels at arbitrary Reynolds number. *J. Biomechan.* **3**, 23–38.

LEW, H. S., AND FUNG, Y. C. (1970b). Plug effect of erythrocytes in capillary blood vessels. *Biophys. J.* **10**, 80–99.

LI, C. S. (1971). Peristalsis in cylindrical tubes. *J. Biomech.* **3**, 513–523.

LIGHTHILL, M. J. (1968). Pressure-forcing of tightly fitting pellets along fluid-filled elastic tubes. *J. Fluid Mech.* **35**, Part 1, 113–143.

LING, S. C., ATABEK, H. B., FRY, D. L., PATEL, D. J., and JANICKI, J. S. (1968). Application of heated-film velocity and shear probes to hemodynamic studies. *Circ. Res.* **23**, 789–802.

MCCUTCHEON, E. P., and RUSHMER, R. F. (1967). Korotkoff sounds: An experimental critique. *Circ. Res.* **20**, 149–161.

MCDONALD, D. A. (1960). "Blood Flow in Arteries." Williams & Wilkins, Baltimore, Maryland.

McDonald, D. A. (1965). The flow properties of blood as a factor in the stability of pulsatile flow. *Proc. Int. Congr. Rheol., 4th,* 1963 Part 4, pp. 205–213.

McDonald, D. A. (1968). Regional pulse wave-velocity in the arterial tree (dog). *J. Appl. Physiol.* **24,** 73–78.

McDonald, D. A., and Gessner, U. (1966). Wave attenuation in viscoelastic arteries. *Proc. Int. Conf. Hemorheol. 1st,* pp. 113–125. Pergamon, Oxford.

McDonald, D. A., and Taylor, M. G. (1959). The hydrodynamics of the arterial circulation. *Progr. Biophys. Biophys. Chem.* **9,** 107–173.

Macfarlane, R. G., and Robb-Smith, A. H. T., eds. (1961). "Functions of the Blood." Academic Press, New York.

Malpighi, Marcello (1661). "DePulmonibus" [translated in *Proc. Roy. Soc. Med.* **23,** Part 1, 1–14 (1930)].

Martin, J. D. (1966). An extension to the theory of wave reflections. *In* "Biomedical Fluid Mechanics Symposium," pp. 70–77. Am. Soc. Mech. Eng., New York.

Mauro, A. (1957). Nature of solvent transfer in osmosis. *Science* **126,** 252–253.

Maxwell, J. A., and Anliker, M. (1968). The dissipation and dispersion of small waves in arteries and veins with viscoelastic wall properties. *Biophys. J.* **8,** 920–950.

Merrill, E. W., and Wells, R. E. (1961). Flow properties of biological fluids. *Appl. Mech. Rev.* **14,** 663–673.

Merrill, E. W., Gilliland, E. R., Cokelet, G., Shin, H., Britten, A., and Wells, R. E. (1963a). Rheology of blood and flow in the microcirculation. *J. Appl. Physiol.* **18,** 255–260.

Merrill, E. W., Cokelet, G. C., Britten, A., and Wells, R. E. (1963b). Non-Newtonian rheology of human blood-effect of fibrinogen deduced by "subtraction." *Circ. Res.* **8,** 48–55.

Merrill, E. W., Benis, A. M., Gilliland, E. R., Sherwood, T. K., and Salzman, E. W. (1965). Pressure-flow relations of human blood in hollow fibers at low flow rates. *J. Appl. Physiol.* **20,** 954–967.

Meschia, G., and Setnikar, I. (1958). Experimental study of osmosis through a collodion membrane. *J. Gen. Physiol.* **42,** 429–444.

Miekisz, S. (1961). Non-linear theory of viscous flow in elastic tubes. *Phys. Med. Biol.* **6,** 103–111.

Moens, A. I. (1878). "Die Pulskwive." Brill, Leiden, Netherlands.

Mollo-Christensen, E. (1966). Wave propagation, dispersion and energy transport in arterial blood flow. *Appl. Mech., Proc. Int. Congr., 11th,* 1964 pp. 1138–1144.

Morgan, G. W., and Ferrante, W. R. (1955). Wave propagation in elastic tubes filled with streaming liquid. *J. Acoust. Soc. Amer.* **27,** 715–725.

Morgan, G. W., and Kiely, J. P. (1954). Wave propagation in a viscous liquid contained in a flexible tube. *J. Acoust. Soc. Amer.* **26,** 323–328.

Morkin, E., Collins, J. A., Goldmann, H. S., and Fishman, A. P. (1965). Pattern of blood flow in the pulmonary veins of the dog. *J. Appl. Physiol.* **20,** 1118–1128.

Müller, V. A. (1935). *In* "Handbuch der biologische Arbeitsmethoden" (E. Ahderhalden, ed.), 5th ed., Part 8, pp. 1–237. Urban & Schwarzenberg, Berlin.

Müller, V. A. (1951). Ueber die Abhängigkeit der Fortpflanzungsgeschwindigkeit und der Dämpfung der Druckwellen in dehnbaren Röhren von deren Wellenlänge. *Helv. Physiol. Pharmacol., Acta* **9,** 162–176.

Müller, V. A. (1959). Die Mehrschichtige Rohrwand als Modell für die Aorta. *Helv. Physiol. Pharmacol. Acta* **17,** 131–145.

Nicoll, P. A. (1954). The anatomy and behavior of the vascular systems in nereis virens and nereis limbata. *Biol. Bull.* **106,** 69–82.

NICOLL, P. A., and WEBB, R. L. (1946). Blood circulation in the subcutaneous tissue of the living bat's wing. *Ann. N. Y. Acad. Sci.* **46**, 697–711.

NOORDERGRAAF, A., JAGER, G. N., and WESTERHOFF, N., eds. (1963). "Circulatory Analog Computers." North-Holland Publ., Amsterdam.

NOORDERGRAAF, A., VERDOUW, P. D., VAN BRUMMELEN, A. G. W., and WIEGEL, F. W. (1964). Analog of the arterial bed. *In* "Pulsatile Blood Flow" (E. O. Attinger, ed.), Chapter 22, pp. 373–387. McGraw-Hill, New York.

OKA, S., AND MURATA, T. (1970). A theoretical study of the flow of blood in a capillary with permeable wall. *Jap. J. Appl. Phys.* **9**, 345–352.

PAPPENHEIMER, J. R., and SOTO-RIVERA, A. (1948). Effective osmotic pressure of the plasma proteins and other quantities associated with the capillary circulation in the hindlimbs of cats and dogs. *Amer. J. Physiol.* **152**, 471–491.

PATEL, D. J., and FRY, D. L. (1966). Longitudinal tethering of arteries in dogs. *Circ. Res.* **19**, 1011–1021.

PATEL, D. J., DEFREITAS, F. M., GREENFIELD, J. C., JR., and FRY, D. L. (1963). Relationship of radius to pressure along the aorta in living dogs. *J. Appl. Physiol.* **18**, 1111.

PATEL, D. J., GREENFIELD, J. C., JR., and FRY, D. L. (1964). *In vivo* pressure-length-radius relationship of certain blood vessels in man and dog. *In* "Pulsatile Blood Flow" (E. O. Attinger, ed.), Chapter 17, pp. 293–305. McGraw-Hill, New York.

PATEL, D. J., JANICKI, J. S., and CAREW, T. E. (1969). Static anisotropic elastic properties of the aorta in living dogs. *Circ. Res.* **25**, 765–780.

PETERSON, L. H. (1962). Properties and behavior of the living vascular wall. *Physiol. Rev.* **42**, Suppl. 5, Part 2, 309–327.

PETERSON, L. H., JENSEN, R. E., and PARNELL, J. (1960). Mechanical properties of arteries *in vivo. Circ. Res.* **8**, 622–639.

POISEUILLE, J. L. M. (1828). "Recherche sur la force du coeur aortique." Didot, Paris.

POISEUILLE, J. L. M. (1840). Recherches expérimentals sur le mouvement des liquides dans les tubes de très petits diamètres. *C. R. Acad. Sci.* **11**, 961–967 and 1041–1048.

PONDER, E. (1948). "Hemolysis and Related Phenomena." Grune & Stratton, New York.

PROTHERO, J., and BURTON, A. C. (1961). The physics of blood flow in capillaries. 1. The nature of the motion. *Biophys. J.* **1**, 565–579.

PROTHERO, J., and BURTON, A. C. (1962a). 2. The capillary resistance to flow. *Biophys. J.* **2**, 199–212.

PROTHERO, J., and BURTON, A. C. (1962b). 3. The pressure required to deform erythrocytes in acid-citrate-dextrose. *Biophys. J.* **2**, 213–222.

RESAL, H. (1876). Sur les petits mouvements d'un fluid incompressible dans un tuyau élastique. *C. R. Acad. Sci.* **82**, 698–699.

ROACH, M. R. (1963). An experimental study of the production and time course of poststenotic dilation in the femoral and carotid arteries of adult dogs. *Circ. Res.* **13**, 537–551.

RODBARD, S. (1959). Physical forces and the vascular lining. *Ann. Intern. Med.* **50**, 1339–1351.

RUBINOW, S. I., and KELLER, J. B. (1968). Hydrodynamic aspects of the circulatory system. *In* "Hemorheology (A. L. Copeley, ed.), pp. 149–156. Pergamon Press, Oxford.

RUCH, T. C., and PATTON, H. D., eds. (1965). "Physiology and Biophysics," 19th ed. Saunders, Philadelphia, Pennsylvania.

RUDINGER, G. (1966). Review of current mathematical methods for the analysis of blood flow. *In* "Biomedical Fluid Mechanics Symposium," pp. 1–33. Am. Soc. Mech. Eng., New York.

RUDINGER, G. (1970). Shock waves in mathematical models of the aorta. *J. Appl. Mech.* **37**, 34–37.

SARPKAYA, T. (1966). Experimental determination of the critical Reynolds number for pulsating Poiseuille flow. *Trans. ASME* Pap. No. 66-FE-5.

SCHARFSTEIN, H., GUTSTEIN, W. H., and LEWIS, L. (1963). Changes of boundary layer flow in model systems: Implications for initiation of endothelial injury. *Circ. Res.* **13**, 580–584.

SCHMIDT-NIELSEN, K., and TAYLOR, C. R. (1968). Red blood cells: Why or why not? *Science* **162**, 274–275.

SCHOLANDER, P. F., HARGENS, A. R., AND MILLER, S. L. (1968). Negative pressure in the interstitial fluid of animals. *Science* **161**, 321–328.

SCHÖNENBERGER, F., and MÜLLER, A. (1960). Ueber die Vaskularisation der Rinderaortenwand. *Helv. Physiol. Pharmacol. Acta* **18**, 136–150.

SEGRÉ, G., and SILBERBERG, A. (1962). Behaviour of macroscopic rigid spheres in Poiseuille flow. Part 1. Determination of local concentration by statistical analysis of particle passages through crossed light beams. Part 2. Experimental results and interpretation. *J. Fluid Mech.* **14**, 115–135 and 136–157.

SESHADRI, V., and SUTERA, S. P. (1968). Concentration changes of suspensions of rigid spheres flowing through tubes. *J. Colloid. Interface Sci.* **27**, 101–110.

SESHADRI, V., AND SUTERA, S. P. (1970). Apparent viscosity of coarse, concentrated suspensions in tube flow. *Trans. Soc. Rheology* **14**, 351–373.

SHAPIRO, A. H., JAFFRIN, M. Y., and WEINBERG, S. L. (1969). Peristaltic pumping with long wavelengths at low Reynolds number. *J. Fluid Mech.* **37**, 799–825.

SKALAK, R. (1966). Wave propagation in blood flow. *In* "Biomechanics" (Y. C. Fung, ed.), pp. 20–40. Am. Soc. Mech. Eng., New York.

SKALAK, R. (1971). Mechanics of the microcirculation. *In* "Biomechanics: Its Foundations and Objectives" (Y. C. Fung, *et al.*, eds.). Prentice-Hall, Englewood Cliffs, New Jersey.

SKALAK, R., and STATIS, T. (1966). A porous tapered elastic tube model of a vascular bed. *In* "Biomechanics" (Y. C. Fung, ed.), pp. 68–81. Am. Soc. Mech. Eng., New York.

SKALAK, R., WIENER, F., MORKIN, E., and FISHMAN, A. P. (1966). The energy distribution in the pulmonary circulation, Part I. Theory. Part II. Experiments. *Phys. Med. & Biol.* **11**, 287–294 and 437–449.

SKALAK, R., BRANEMARK, P.-I., and EKHOLM, R. (1970). Erythrocyte adherence and diapedesis. *Angiology* **21**, 224–239.

SOBIN, S. S., TREMER, H. M., and FUNG, Y. C. (1970). Morphometric basis of the sheet-flow concept of the pulmonary alveolar microcirculation in the cat. *Circ. Res.* **26**, 397–414.

STARLING, E. H. (1896). On the absorption of fluid from the connective tissue spaces. *J. Physiol. (London)* **19**, 312–326.

STEHBENS, W. E. (1959). Turbulence of blood flow. *Quart. J. Exp. Physiol.* **44**, 110–117.

STREETER, V. L., KEITZER, W. F., and BOHR, D. F. (1963). Pulsatile pressure and flow through distensible vessels. *Circ. Res.* **13**, 3–20.

STREETER, V. L., KEITZER, W. F., and BOHR, D. F. (1964). Energy dissipation in pulsatile flow through distensible tapered tube. *In* "Pulsatile Blood Flow" (E. O. Attinger, ed.), Chapter 8, pp. 149–177. McGraw-Hill, New York.

SUTERA, S. P., AND HOCHMUTH, R. M. (1968). Large scale modeling of blood flow in the capillaries. *Biorheology* **5**, 45–73.

SUTERA, S. P., SESHADRI, V., CROCE, P. A., AND HOCHMUTH, R. M. (1970). Capillary blood flow, II. Deformable model cells in tube flow. *Microvascular Res.* **2**, 420–433.

SVANES, K., and ZWEIFACH, B. W. (1968). Variations in small blood vessel hematocrits produced in hypothermic rats by micro-occlusion. *Microvascular Res.* **1**, 210–220.

TAYLOR, M. G. (1957). An approach to an analysis of the arterial pulse wave. I. Oscillations in an alternating line. II. Fluid oscillations in an elastic pipe. *Phys. Med. & Biol.* **1**, 258–269 and 321–329.

TAYLOR, M. G. (1959). An experimental determination of the propagation of fluid oscillations in a tube with a visco-elastic wall. *Phys. Med. & Biol.* **4**, 63–82.

TAYLOR, M. G. (1965). Wave-travel in a non-uniform transmission line in relation to pulses in arteries. *Phys. Med. & Biol.* **10**, 539–550.

TAYLOR, M. G. (1966a). Use of random excitation and spectral analysis in the study of frequency-dependent parameters of the cardiovascular system. *Circ. Res.* **18**, 585–595.

TAYLOR, M. G. (1966b). Input impedance of an assembly of randomly branching elastic tubes. *Biophys. J.* **6**, 29–51.

TAYLOR, M. G. (1968). The influence of the viscous properties of blood and the arterial wall upon the input impedance of the arterial system. *In* "Hemorheology" (A. L. Copley, ed.), pp. 143–148. Pergamon Press, Oxford.

TEITEL, P. (1965). Disk-sphere transformation and plasticity alteration of red blood cells. *Nature (London)* **204**, 409–410.

TEXON, M. (1957). A hemodynamic concept of artherosclerosis with particular reference to coronary occlusion. *Arch. Intern. Med.* **99**, 418.

TEXON, M. (1967). Mechanical factors involved in atherosclerosis. *In* "Atherosclerotic Vascular Disease" (A. N. Brest and J. H. Moyer, eds.), Art. No. 3, Meredith Publ., New York.

UCHIDA, S. (1956). The pulsating viscous flow superposed on the steady laminar motion of incompressible fluid in a circular pipe. *Z. Angew. Math. Phys.* **7**, 403–421.

VAN CITTERS, R. L. (1960). Longitudinal waves in the wall of fluid-filled elastic tubes. *Circ. Res.* **8**, 1145–1148.

VAN DER POL, B., and VAN DER MARK, J. (1929). The heartbeat considered as a relaxation oscillation, and an electrical model of the heart. *Arch. Nelr. Physiol.* **14**, 418–443. [Note: The 1st paper on relaxation oscillation by Van der Pol was *Phil. Mag.* **51**, 978 (1926), in which heart beat was considered.]

WANG, H., and SKALAK, R. (1968). Viscous flow in a cylindrical tube containing a line of spherical particles. Tech. Rep. ONR Proj. NR 062-393. Columbia University, New York.

WEBER, E. H. (1850). Ueber die Anwendung der Wellenlehre auf die Lehre vom Kreislaufe des Blutes und insbesondere auf die Pulslehre. *Ber. Verh. Kgl. Saechs. Ges. Wiss., Math.-Phys. Kl.*

WEBER, E. H., and WEBER, W. (1825). "Wellenlehre auf Experimente gegründet; oder, Über die wellentropfbarer Flüssigkeiten mit Anwendung auf die Schall- und Lichtwellen." Fleischer, Leipzig.

WEBER, W. (1866). Theorie der durch Wasser order andere incompressible Flussigkeiten in elasischen Röhren fortgepflanztan Wellen. *Ber. Verh. Kgl. Saechs. Ges. Wiss., Math.-Phys. Kl.* **18**, 353–357.

WEISS, G. H. (1964). On the theory of blood flow in tapered arteries. *Biorheology* **2**, 153–158.

WETTERER, E., AND KENNER, T. (1968). Grundlagen der Dynamik des Arterienpulses. Springer, Berlin.

WHITMORE, R. L. (1963). Hemorheology and hemodynamics. *Biorheology* **1**, 201–220.

WIEDERHIELM, C. A. (1965). Distensibility characteristics of small blood vessels. *Fed. Proc., Fed. Amer. Soc. Exp. Biol.* **24**, 1075–1084.

WIEDERHIELM, C. A., WOODBURY, J. W., KIRK, S., and RUSHMER, R. F. (1964). Pulsatile pressures in the microcirculation of frog's mesentery. *Amer. J. Physiol.* **207**, 173–176.

WIENER, F. (1964). Mechanics of the pulmonary circulation. Doctoral Thesis, Columbia University, New York.

WIENER, F., MORKIN, E., SKALAK, R., and FISHMAN, A. P. (1966). Wave propagation in the pulmonary circulation. *Circ. Res.* **19**, 834–850.

WITZIG, K. (1914). Ueber erzwungene Wellenbewegungen zäher, inkompressibler Flüssigkeiten in elastischen Röhren. Inaugural Dissertation, Universitat Bern, K. J. Wyss, Bern.

WOMERSLEY, J. R. (1955a). Method for the calculation of velocity. Rate of flow and viscous drag in arteries when the pressure gradient is known. *J. Physiol. (London)* **127**, 553–563.

WOMERSLEY, J. R. (1955b). Oscillatory motion of a viscous liquid in a thinwalled elastic tube. I. The linear approximation for long waves. *Phil. Mag.* [7] **46**, 199–221.

WOMERSLEY, J. R. (1957). An elastic tube theory of pulse transmission and oscillatory flow in mammalian arteries. WADC Rep. TR 56-614. Wright Air Development Center, Dayton, Ohio.

WOMERSLEY, J. R. (1958). Oscillatory flow in arteries. I. The constrained elastic tube as a model of arterial flow and pulse transmission. II. The reflection of the pulse wave at junctions and rigid inserts in the arterial system. *Phys. Med. & Biol.* **2**, 178–187 and 313–323.

WYLIE, E. B. (1966). Flow through tapered tubes with nonlinear wall properties. *In* "Biomechanics" (Y. C. Fung, ed.), pp. 82–95. Am. Soc. Mech. Eng., New York.

YELLIN, E. L. (1966a). Hydraulic noise in submerged and bounded liquid jets. *In* "Biomedical Fluid Mechanics Symposium," pp. 209–221. Am. Soc. Mech. Eng., New York.

YELLIN, E. L. (1966b). Laminar-turbulent transition process in pulsatile flow. *Circ. Res.* **19**, 791–804.

YIN, F., and FUNG, Y. C. (1969). Peristaltic transport in a circular cylindrical tube. *J. Appl. Mech.* **36**, 579–587.

YIN, F., AND FUNG, Y. C. (1971). Comparison of theory and experiment in peristaltic transport. *J. Fluid Mech.* **47**, 93–112.

YOUNG, T. (1808). Hydraulic investigations, subservient to an intended Croonian lecture on the motion of the blood. *Phil. Trans. Roy. Soc. London* **98**, 164–186.

YOUNG, T. (1809). On the functions of the heart and arteries. *Phil. Trans. Roy. Soc. London* **99**, 1–31.

ZIEN, T. F., and OSTRACH, S. (1971). A long wave approximation to peristaltic motion. *J. Biomech.* **3**, 63–70.

ZWEIFACH, B. W. (1961). "Functional Behavior of the Microcirculation." Thomas, Springfield, Illinois.

ZWEIFACH, B. W., and INTAGLIETTA, M. (1968). Mechanics of fluid movement across single capillaries in the rabbit. *Microvascular Res.* **1**, 83–101.

Two-Dimensional Shock Structure in Transonic and Hypersonic Flow

MARTIN SICHEL

Department of Aerospace Engineering, The University of Michigan, Ann Arbor, Michigan

I. Introduction

In many flows viscous stresses are important only within very thin regions such as boundary layers and shock waves. Hence, in order to determine heat transfer and skin friction at the surface of bodies immersed in the fluid and to establish criteria for flow separation the details of the flow within the boundary layer must be established. The jump in pressure, density, velocity, and temperature across shock layers are usually the only quantities needed to establish the influence of such layers upon the rest of the flow field, and the one-dimensional inviscid conservation equations are sufficient to determine these jumps, provided gradients of the flow variables upstream and downstream of the shock are sufficiently small (Hayes, 1958). Generally then, unlike the boundary layer case, there is no need to consider the detailed structure of the shock if only the jump or Rankine–Hugoniot conditions are of interest.

There are, however, a number of situations where treatment of shock waves as locally one-dimensional Rankine–Hugoniot discontinuities fails. When a weak shock is adjacent to a curved wall, application of the R–H (Rankine–Hugoniot) conditions leads to a discontinuity in streamline curvature where the shock touches the wall (Emmons, 1946). Use of the R–H condition at the triple point in the analysis of the Mach reflection of weak oblique shocks leads to results which disagree drastically with experiment (Sternberg, 1959). The transonic flow near the throat of a C–D (converging-diverging) nozzle as the flow changes from purely subsonic to subsonic-supersonic flow cannot be described by a theory based only on inviscid flow with imbedded R–H shock waves (Emmons, 1946; Guderley, 1962). The pressure rise across weak shock waves terminating pockets of supersonic flow, such as arise in the transonic flow past wings and bodies, is observed to be less than the Rankine–Hugoniot value (Sinnott, 1960; Holder, 1964).

All the above described problems arise in transonic flows; however, the use of the one-dimensional R–H shock theory is also inconsistent in certain hypersonic flow problems. In the analysis of hypersonic blunt body flow at low Reynolds numbers consistent results can be obtained only if the influence of curvature upon the shock pressure rise is taken into account (Sedov *et al.*, 1953; Probstein and Kemp, 1960). Further, measurements of the flow near the leading edge of a flat plate in hypersonic flow (McCroskey *et al.*, 1967) provide a clear indication that R–H conditions no longer hold when the shock Reynolds number is low.

In the cases described above the assumption that the shock is locally one-dimensional breaks down. The flow within the shock is then two-dimensional, and it is necessary to compute the two-dimensional structure of such shock waves in order to establish the value of the shock pressure rise. Since the flow within the shock will, in these cases, depend on the upstream and downstream values of the flow variables and their gradients, the analysis of such problems is far more difficult than when it is possible to use the one-dimensional R–H conditions. The two-dimensional shock waves described above are generally very thin when compared to the characteristic length of the problem under consideration; nevertheless, the flows within these shocks can have a large influence on the overall flow field through their effect on the change in pressure, velocity, density, and temperature across the shock.

Flows in which two-dimensional shock structure is important form the subject of this article. The objective is to indicate where such problems arise and to show what work has been done in the theory and, in a few cases, in experimental measurements of such two-dimensional shocks. To start, then, a number of problems in which the use of the one-dimensional R–H conditions fail are described in detail in Section II. Two-dimensional shock structures frequently arise in transonic flow and can be described by a viscous-transonic equation. This equation is derived and its mathematical properties and similarity rules are discussed in Section III. Application of this equation to external and internal flows is considered in Sections IV and V.

Two-dimensional shock layers in supersonic and hypersonic flow are discussed in Section VI. Finally current problems in the study of two-dimensional shock layers are discussed in Section VII.

II. Flows with Two-Dimensional Shock Waves

Most compressible flow problems can be analyzed successfully by division into regions of inviscid flow separated by shock discontinuities across which the Rankine–Hugoniot conditions hold. The need to consider two-dimensional shock layers is signalled by the failure of such analyses to provide results which agree with observation or are physically meaningful. It is this failure which has also provided the impetus for the development of theories to deal with two-dimensional shock layers. To show the role of two-dimensional shock layers in compressible flows several important problems in which the inviscid R–H shock type analysis breaks down are therefore discussed in detail below.

A. Nozzle Flow

With decreasing back pressure the flow through a converging-diverging nozzle changes from a symmetrical one with subsonic velocities upstream and downstream of the throat to a flow in which the velocity accelerates to supersonic values downstream of the throat. The transition between these two classes of flow has formed the subject of many investigations. Many important features of such transitional flows are adequately explained by a one-dimensional or hydraulic theory with normal or oblique shocks in the supersonic portion of the nozzle located to satisfy the downstream boundary condition on pressure; however, to resolve the details of the flow near the nozzle throat, which is intimately related to the wall curvature, solutions of the two-dimensional or axisymmetric gasdynamic equations must be investigated.

In the asymmetrical case, when the flow accelerates from subsonic to supersonic velocities, a series solution of the exact potential equation of the form

$$\phi = a_1 x + a_2 x^2 + a_3 x^3 + c_3 x y^2 + a_4 x^4 + c_4 x^2 y^2 + \cdots \tag{2.1}$$

was found by Meyer in 1908. The calculations are straightforward and the solution adequately describes the flow near the nozzle throat. Extended versions of this series have been used for accurate computation of the flow through nozzles of arbitrary shape (Hall and Sutton, 1964), and such accelerating flows are often called "Meyer" flows.

Breakdown of the electrolytic tank method of obtaining compressible flow solutions when the maximum speed reaches the velocity of sound led Taylor (1930) to study series solutions of the form (2.1) when the flow is symmetrical with respect to the nozzle throat. Such symmetrical flows are sometimes called "Taylor" flow. Carrying terms up to fifth order and assuming symmetry with respect to both the nozzle axis and the nozzle throat the series (2.1) becomes

$$\phi = a_1 x + c_3 x y^2 + a_3 x^3 + a_5 x^5 + c_5 x^3 y^2 + c_6 x y^4. \tag{2.2}$$

Taylor's calculations showed the development of pockets of supersonic flow near the nozzle wall; however, above a certain value of the speed a_1 at the center of the nozzle, real solution for the coefficients of (2.2) no longer exist. For a ratio $h/R = 1/4$, where h is the half height of the nozzle and R the wall radius of curvature there are no solutions for $a_1 > 0.855a$, where a is the speed of sound.

Görtler (1939) showed that the series employed by Taylor tends to diverge as the velocity near the throat approaches the sonic value. Görtler extended Taylor's solution by carrying the series to higher order and relaxing the requirement of symmetry with respect to the nozzle throat; however, an analytical description of the complete transition from the Taylor to the Meyer flow was not found to be possible. Görtler (1939), using the inviscid equations, proved the theorem that if at some point $(x_0, 0)$ on the axis of a nozzle, $u = a$, and $\partial u/\partial x = 0$, and if the solution is analytic in some neighborhood K of this point, then for all y in K

1. $u(x_0, y) \equiv a$,
2. $(\partial u/\partial y)(x_0, y) \equiv 0$,
3. $v(x_0, y) \equiv 0$.

Since such a maximum at the sonic point must always occur during the transition from Taylor to Meyer flow, Görtler concluded that an analytical description of the Taylor–Meyer transition based on the inviscid equations is impossible for nozzles with finite wall curvature at the throat; rather, the transition must be accompanied by the formation of viscous shock waves.

Emmons (1946) used the method of relaxation to obtain numerical solutions of the inviscid gas dynamic equations for the flow near a nozzle throat. Emmons (1946) postulated that the transition from the symmetrical Taylor to the asymmetrical Meyer type of flow starts with the formation of shocks within the supersonic pockets which form near the nozzle wall, a postulate borne out by the calculations.

For the nozzle considered by Emmons (1946), the flow was similar to that in a venturi with supersonic pockets near the wall below a peak centerline Mach number, M, of 0.812. Above this Mach number, shock waves terminating the supersonic pockets had to be introduced in order to permit the elimination of residuals in the relaxation calculations, and for sufficiently large M the shock waves within the two supersonic pockets joined at the nozzle center. Even then there are difficulties for the shock wave when it first appears is already of finite length and strength rather than developing gradually. A second difficulty is that a sharp expansion must be introduced immediately behind the shock in order to avoid a discontinuity in streamline curvature where the shock touches the wall. This problem is discussed in greater detail in Section II, B below.

Instead of considering approximate solutions of the full equations of compressible inviscid flow as above, another approach is to study exact solutions of the approximate inviscid small disturbance transonic equation (Guderley, 1962), which can be written as

$$U_{YY} - (U^2)_{XX} = 0, \tag{2.3}$$

with

$$\epsilon U = (\bar{u}/\bar{a}^*) - 1,$$

$$X = (1/2)(\gamma + 1)\bar{x}/\bar{L}, \qquad Y = (\gamma + 1)/2^{3/2}\epsilon^{1/2}\bar{y}/\bar{L}.$$

Here $\bar{a}^*$ is the critical speed of sound, $\bar{L}$ is a characteristic length, and ϵ is a small parameter characterizing the deviation of $\bar{u}/\bar{a}^*$ from unity. Barred quantities are dimensional. With the transformation

$$U = Z(S) + 2\sigma^2 Y^2, \qquad S = X + \sigma Y^2, \tag{2.4}$$

where σ is an arbitrary constant, Tomotika and Tamada (1950) found that (2.3) collapses to the ordinary dlfferential equation

$$ZZ'' + (Z' - 2\sigma)(Z' + \sigma) = 0, \tag{2.5}$$

with the implicit solution

$$(Z - 2\sigma S)^2(Z + \sigma S) = 2\alpha^3/\sigma^3. \tag{2.6}$$

The flow described by (2.4) is symmetrical with respect to the X axis, and so any two steamlines equal distances above and below the X axis can be considered as the wall of a nozzle.

The nature of the solution (2.6) is determined by the value of the arbitrary constant α. With $\alpha = 0$ two solutions are $Z = -\sigma S$ and $Z = 2\sigma S$, with the latter being identical to the first few terms of the Meyer (1908) solution. Here, as in (2.4), σ is an arbitrary parameter which establishes the velocity gradient on the nozzle axis for these two special solutions while $Z \equiv U(X, 0)$ is the velocity distribution on the axis with $Z > 0$ and $Z < 0$ corresponding to supersonic and subsonic flow, respectively. The subsonic branch of the solutions corresponding to $\alpha < 0$ represent symmetrical Taylor flow with a different solution curve for each α. These solutions are shown in Fig. 1.

As $\alpha \rightarrow 0$ the maximum velocity on the axis approaches the sonic value and the Taylor solution curves approach 1-P-4. In the limit the velocity gradient becomes discontinuous at the sonic point. The Tomotika–Tamada (1950) solution does not permit a continuous transition from the limiting solution 1-P-4 to the Meyer solution 1-P-2, as is evident in the phase plane (Fig. 2), where the sonic line, $Z = 0$, is a barrier such that subsonic solutions can never become supersonic and vise versa except for the two special solutions $Z = -\sigma S$ and

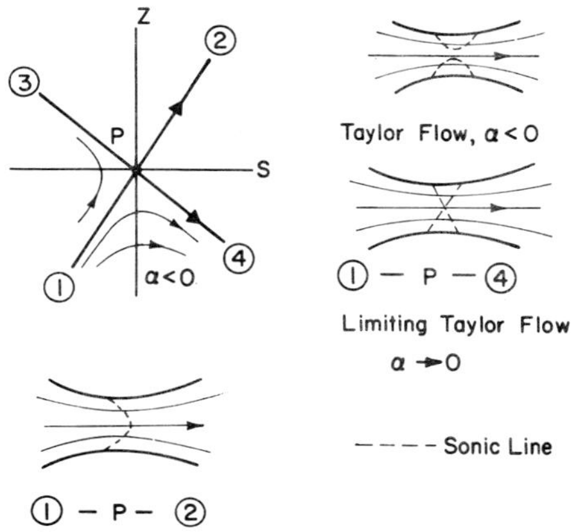

FIG. 1. The Tomotika–Tamada (1950) solution for transonic nozzle flow.

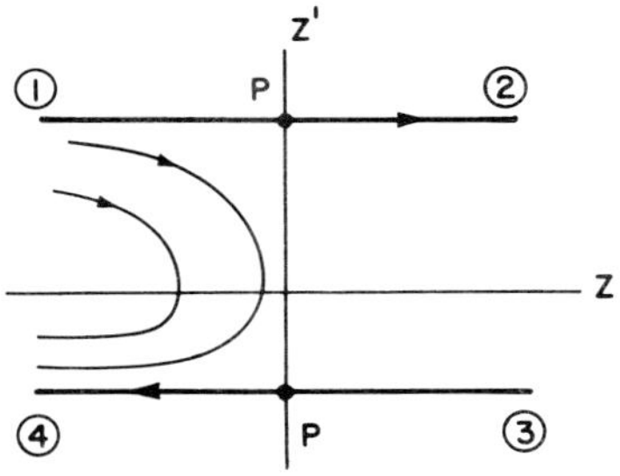

FIG. 2. The Tomotika–Tamada solution in the phase plane.

$Z = 2\sigma S$. The sonic point is a singularity for from (2.5) it is clear that only the two special solutions with $Z' = -\sigma$, 2σ can pass through the sonic point with finite Z'' or curvature.

A large family of exact solutions of the transonic small disturbance equation for two-dimensional nozzle flow has also been developed starting from the hodograph equations (Fal'kovitch and Chernov, 1966; Germain, 1964, for example); however, the planar self-similar solution of Ryzhov (1963) is of greatest interest since he considered both the transition from Taylor to Meyer flow and the formation of shock waves. Ryzhov used the criterion that shock waves must appear wherever limit lines along which the acceleration is infinite occur in the inviscid solution, and nozzle flows with shock waves are investigated. A weakness of this analysis is the use of the inviscid equations in the neighborhood of those flows where limit lines appear.

In concluding the discussion of the transition from Taylor to Meyer flow it appears appropriate to quote from the footnote on page 68 of Guderley's book (1962) on transonic flow:

> If the nozzle is symmetrical with respect to the throat, then as long as the flow is subsonic along the entire length, there exists a symmetry also in the flow field. This is certainly no longer true when the nozzle is acting as a DeLaval nozzle. One would expect that a study of the transition from one behavior to the other would provide an insight into the phenomenona of transonic flow. A direct analysis of nozzles has, however, *not in fact led* to this hoped-for result.

B. Shock Wave Adjacent to a Curved Wall

As indicated in Section II, A above, Emmons (1946) found it necessary to introduce shock waves in his numerical study of transonic flow through nozzles. However, in Emmons' solution the Mach number rises discontinuously downstream of the shock wave; the Mach numbers at all relaxation net points at the wall downstream of the shock fell closely on a smooth curve which when extrapolated to the shock gave this discontinuous behavior. In later calculations for transonic flow past a symmetrical airfoil Emmons (1948) also found it necessary to introduce shock waves to compute flows above a certain free stream Mach number. In these calculations there is also a sharp increase in Mach number downstream of the shock, but because of the finer grid spacing used in the airfoil calculations the rise is no longer discontinuous. It is interesting to note that such shocks followed by a rapid Mach number rise or pressure drop were also observed to occur outside the boundary layer by Ackeret, Feldmann, and Rott (1946).

This rapid pressure drop observed both in the course of numerical computations and experimentally is related to a singularity which arises at the point where a shock touches a curved surface, as shown in Fig. 3. If the flow follows the wall, the streamlines at the wall will have the curvature $1/R_1$ of the wall. To maintain this curvature there must be a pressure gradient

$$(\partial p/\partial r)_1 = (\rho_1 q_1^2)/R_1 \tag{2.7}$$

immediately upstream of the shock. If the flow is to remain attached to the wall. the shock must be normal at the point where it touches the surface. If the wall curvature is continuous then immediately downstream of the shock

$$(\partial p/\partial r)_2 = (\rho_2 q_2^2)/R_1 , \tag{2.8}$$

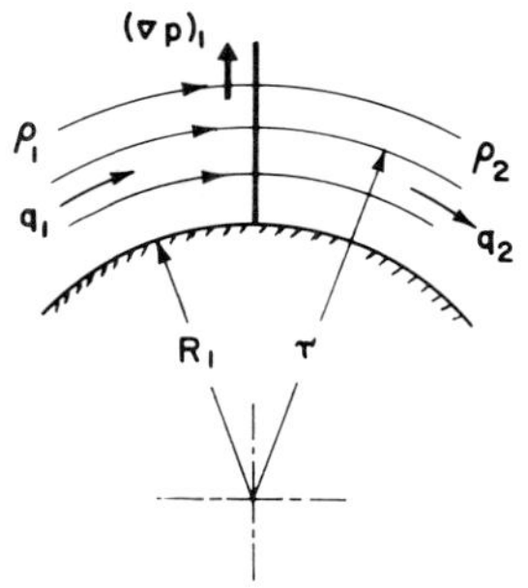

FIG. 3. Normal shock adjacent to a curved surface.

but it turns out that $(\partial p/\partial r)_2$ is determined by $(\partial p/\partial r)_1$ and the Rankine–Hugoniot conditions and does not in general satisfy (2.8). As shown by Emmons (1946), Lin and Rubinov (1948), and Tsien (1947) the streamline curvature is discontinuous across the shock with

$$\frac{R_2}{R_1} = \frac{M_1^2}{1 + \frac{1}{2}(\gamma - 1)M_1^2}\left(M_1^2 - \frac{\gamma + 3}{2}\right). \tag{2.9}$$

With $\gamma = 1.4$, $R_2 = R_1$ only for $M_1 = 1.66$ and Emmons indicates that for $M_1 = 1.075$ the pressure rise across the shock is just equal to the drop across the rapid expansion downstream of the shock.

The strange behavior at the foot of the shock wave apparently represents the adjustment of the flow to this sudden change in curvature across the shock. Lin and Rubinov (1948) suggest that the above difficulty might be resolved if the shock is permitted to have an infinite curvature where it touches the wall for then (2.9) is replaced by an inequality. Gadd (1960), Zierep (1958), and Oswatitsch and Zierep (1960) introduced a singularity at the foot of the shock to represent the sudden adjustment of the flow to the discontinuity in streamline curvature. In these theories, however, infinite velocity and pressure gradients occur immediately downstream of the point where the shock touches the wall.

In the theories described above, it is taken for granted that the Rankine–Hugoniot conditions must be satisfied across the shock. However, it is unlikely that infinite shock curvature and waves followed by regions with high gradients are consistent with the existence of locally one-dimensional shocks across which the R–H conditions hold. Emmons (1946) states:

> It appears that not only is the perfect fluid without shocks an insufficient mathematical theory to cover the practical phenomena arising in connection with the flow of compressible fluids but that even the extended theory which includes compression shock

discontinuities, as in this report, is not adequate to describe the facts properly. It is probable that a sufficiently general mathematical theory (and presumably a fluid with friction would provide such) could give smooth transitions from one type of solution to another. There is, however, no reason to assume that such solutions would be either unique or stable.

C. The Shock–Boundary Layer Interaction

The problem of the shock adjacent to a curved wall is unrealistic since a bondary or shear layer region will always lie between the shock and the solid surface. The flow in the region where the shock approaches the wall will be dominated by the interaction between the free stream shock and the boundary layer. Particularly in the case of the transonic interaction problem the use of R–H shocks outside the boundary layer does not in general yield results in agreement with experiment. It is generally observed that the pressure rise across a weak shock wave adjacent to a curved wall is less than the Rankine–Hugoniot value (Holder, 1964; Sinnot, 1960), and this effect appears to be related both to the singular behavior near the foot of the shock discussed in Section II, B above and to the flow within the boundary layer (Lighthill, 1950).

Perhaps the simplest transonic shock–boundary layer interaction problem showing the need to consider two-dimensional shock layers arises when a weak shock wave moves past a surface, for example the wall of a shock tube. The boundary layer induced on the wall by this shock results in shock attenuation and curvature. The absence of a boundary layer upstream of the shock distinguishes this shock–boundary interaction problem from the classical one in which the shock interacts with an already established boundary layer and makes this problem more tractable to analysis.

Hartunian (1961) considered the influence of the boundary layer on the shape of shock waves of arbitrary strength but because of the singularity at the leading edge of the boundary layer his solution is not valid in a small region near the foot of the shock. If the shock is very weak, and hence relatively thick, the flow can be divided into an outer region dominated by the shock layer and an inner shear layer governed by the diffusion equation (Sichel, 1962)

$$\partial u/\partial x = du_e/dx + \partial^2 u/\partial y^2, \tag{2.10}$$

as shown in Fig. 4a.

The shear layer flow is thus driven by the external velocity $u_e(x)$ and it is reasonable to assume that an oblique shock stands outside the shear layer. Then $u_e(x)$ will be determined by Taylor's weak shock solution

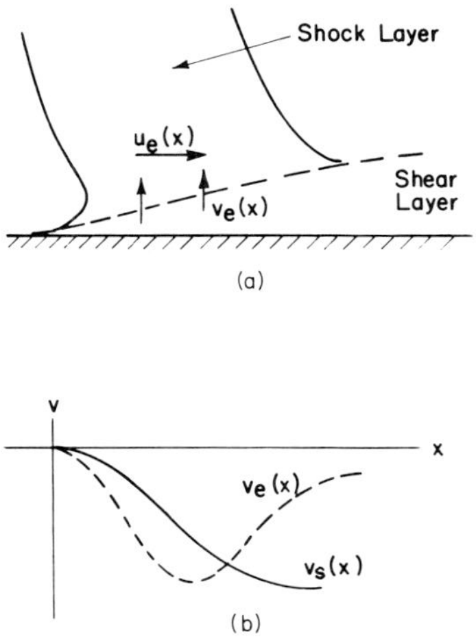

FIG. 4. Weak shock–boundary layer interaction in a shock tube.

(1910), and of course there is no longer any leading edge singularity. The y velocity $v_e(x)$ at the outer edge of the shear layer, which is now a function of $u_e(x)$, should match $v_s(x)$, the y velocity within the oblique shock. From Fig. 4b, which shows $v_s(x)$ and $v_e(x)$ when the maximum values of these velocities are equal, it is evident that, in fact, such matching does not occur. It is thus concluded that there must be a region immediately outside the shear layer where the shock structure is two-dimensional. An approximate solution for the flow in the interaction region, based on matching the maximum values of $v_e(x)$ and $v_s(x)$ (Fig. 4b) and upon an inviscid solution for the flow downstream of the shock and outside the boundary layer, yields the result that the pressure downstream of the shock actually overshoots the Rankine–Hugoniot value. This failure to satisfy the R–H condition is a characteristic feature of problems involving the interaction of a transonic shock wave with a boundary layer.

In the classical interaction problem an established boundary layer already exists at the point where the shock touches the wall. In transonic flow the pressure rise measured at the wall is invariably less than the R–H value and at present can only be predicted by empirical relations (Sinnott, 1960). At the wall the pressure is observed to rise gradually to its final value; however, shocklike compressions followed by equally sharp expansions are observed in the free stream (Ackeret *et al.*, 1946) when the boundary layer is turbulent. This behavior is like that computed by Emmons (1946) and plays a key role in establishing the shock pressure rise at the wall (Sinnott, 1960). Clearly the shock structure immediately outside the boundary layer is no longer one-dimensional and a rational theory for determining the shock pressure rise, which must also show how the pressure variations across the shock are smeared out by the boundary layer, thus requires a consideration of two-dimensional shock layers. When the boundary layer is laminar the free stream flow usually consists of one or more lambda shocks, and pressure contours plotted by Ackeret *et al.* (1946) also make it unlikely that the shock layers making up the lambda shock will be one-dimensional.

D. The Mach Reflection of Weak Shock Waves

A singularity closely related to that when a shock stands adjacent to a curved wall, but considerably more complex, occurs at the triple shock intersection which arises in the Mach reflection of shock waves. At the triple point the condition that the pressure and direction behind the Mach stem and behind the incident and reflected shocks must be equal, imposed in the von Neumann solution, is usually imposed on the flow, and all three shocks are treated as R–H discontinuities. Sternberg (1959) pointed out that for weak incident shocks theories based on the above treatment fail to agree with experiment. By using the electric tank analogy for compressible flow Sternberg was able to show that this discrepancy was real and not due to the inability to observe the details of the flow in the triple point region.

In those cases in which the von Neumann theory fails to agree with experiment the shock waves are observed to be curved near the triple point. In the hodograph plane the von Neumann solutions coincide with the intersection of the shock polars for the Mach stem and the reflected shock, and from this it can be shown (Guderley, 1947) that in the case of a weak incident shock, the streamlines on the two sides of the triple point will either converge toward one another or diverge if the shock curvature remains finite. When the steamlines converge this contradiction can be resolved only if the shock curvature becomes infinite at the triple point. This singular behavior is remarkably similar to that of a shock adjacent to a curved wall discussed above. In some cases the shock polars fail to intersect. Guderley (1947) has introduced a solution which involves a Prandtl–Meyer expansion in the neighborhood of the triple point, but shock curvatures then also become excessive near the triple point.

Sternberg (1959) properly points out that the shock structure will no longer be one-dimensional when the radius of curvature of the shock becomes of the same order as the shock thickness. Near the triple point

Sternberg (1959) indicates that "there must be a finite shock zone which provides a continuous transition inside the shocks between R–H waves in the upper and lower domains" and this situation is illustrated in Fig. 5. Sternberg coined the name "non-Rankine–Hugoniot shock wave" for this region of adjustment.

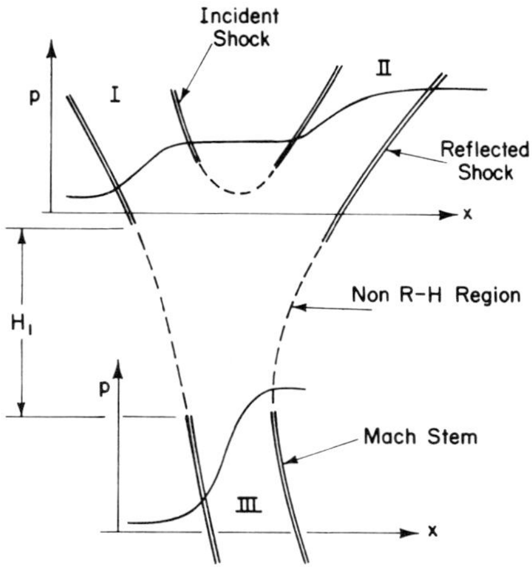

FIG. 5. Triple point region in a Mach reflection.

A control volume analysis of the non-R–H region shows that the height H_1 is greater than the shock thickness by almost an order of magnitude so that this region can support an appreciable pressure change. The introduction of the non-R–H zone frees the flow at the downstream edge from confinement to the ordinary shock polar. Although the height H_1 of the non-R–H region is much smaller than the characteristic dimension of the Mach reflection region, Sternberg (1959) shows that the insertion of this region causes an appreciable change in the overall flow field because of the change in the triple point boundary conditions.

An interesting question raised by Sternberg (1959) is whether, in the limit as $\mu \rightarrow 0$, the triple point solution returns to the solution based on R–H shock waves. As already mentioned in the Introduction, the flow in the non-R–H shock layer is two-dimensional and depends on the downstream flow, and a boundary condition in addition to that needed for R–H shocks is introduced. This additional boundary condition not only complicates the analysis but also suggests that the solution with R–H shocks is not the proper $\mu \rightarrow 0$ limit since then one boundary condition would be lost.

Although Sternberg (1959) did not find the detailed structure of the flow in the non-R–H shock layer his analysis clearly indicates that use of the R–H shock polar cannot be assumed valid everywhere and that even though the extent of two-dimensional shock layers may be minute relative to the characteristic dimension of the problem at hand, these layers may have an appreciable influence on the overall flow. An analogous situation arises with respect to boundary layers where the processes within a very thin region through their effect on separation can have an important influence on the overall flow.

E. Two-Dimensional Shocks in Hypersonic Flow

So far the discussion of shock waves with a two-dimensional internal structure has been restricted to problems in transonic flow. There are, however, several fundamental problems in hypersonic flow where consideration of non-R–H shock structure becomes crucial. Sedov *et al.* (1953) were among the first to recognize that in the supersonic flow past blunt bodies at low Reynolds numbers, the R–H conditions may not be satisfied across the detached shock upstream of the body. From the one-dimensional Navier–Stokes equations it can readily be shown (Hayes, 1958) that R–H conditions will be satisfied across shock waves provided all gradients vanish upstream and downstream of the wave, or at least are negligible compared to gradients of fluid parameters inside the wave. Now, however, Sedov *et al.* (1953) observed that in the supersonic flow of a gas past bodies of a small size the bow shock will be highly curved so that gradients of velocity density, pressure, etc., downstream of the shock may be quite large, large enough so that the R–H conditions no longer remain valid.

A very detailed investigation of hypersonic blunt body flow at low Reynolds numbers has been made by Probstein and Kemp (1960). At high Reynolds numbers $\mathbf{R}_\infty(=\rho_\infty U R_b/\mu_\infty)$ there is a region of inviscid flow, the shock layer, downstream of the thin R–H bow shock and there is a thin boundary layer at the body surface as shown Fig. 6. As $\mathbf{R}_\infty$ decreases the boundary layer thickens until at $\mathbf{R}_\infty \approx 100$ it occupies the entire region between the shock and the body. This flow is sometimes called the viscous layer regime. At still lower Reynolds numbers the shock thickness increases and the shock structure begins to merge with the viscous layer. Once the viscous region extends to the shock, i.e., $\mathbf{R}_\infty \lesssim 100$, the viscous shear and heat conduction behind the shock are sufficiently high that it no longer is possible to treat the shock as an R–H discontinuity (Probstein and Kemp, 1960). As in the case of the

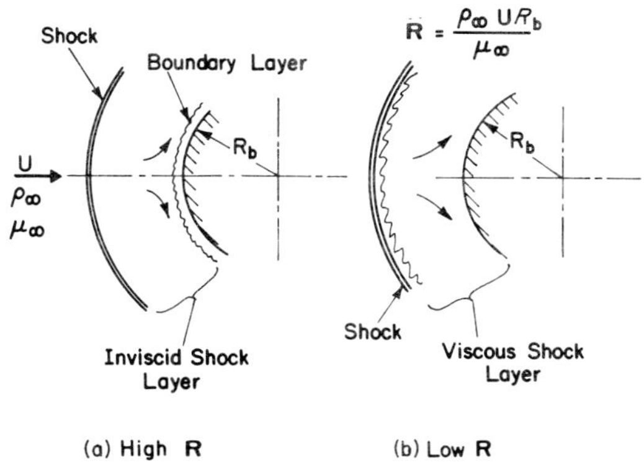

FIG. 6. Blunt body flow at (a) high and (b) low **R**.

Mach reflection discussed in Section II, D above, the conditions across the shock no longer depend solely upon the flow upstream but, rather, are coupled to the flow downstream of the shock wave. The merging of the shock layer with the shock structure and the departure from the R–H conditions is verified by the measurements of Ahouse and Harbour (1969), although merging seems to start for $\mathbf{R}_\infty < 400$ rather than the value of 100 indicated by the theory.

In the high velocity flow past a sharp edged semi-infinite flat plate the presence of the boundary layer at the surface of the plate induces a shock wave which in turn has a large influence upon the pressure and heat transfer distribution at the plate surface (Hayes and Probstein, 1959). The distance, x, from the leading edge of the flat plate is the appropriate characteristic length characterizing the flow at any position along the plate and so $\mathbf{R}_x$, the Reynolds number based on x, ranges from zero at the leading edge to very large values downstream. Far downstream the flow still behaves like the classical boundary layer on a flat plate, but closer to the leading edge the boundary layer induced shock becomes stronger and interacts first weakly, then strongly with the boundary layer. These regimes of flow have been termed the strong and weak interaction regimes (Hayes and Probstein 1959). Still closer to the leading edge the boundary layer and shock structure merge (Pan and Probstein, 1966) so that, as in the blunt body problem, the Rankine–Hugoniot conditions across the shock must be modified. The different regimes of flow are shown in Fig. 7.

This behavior, suggested by the analysis of Pan and Probstein (1966), was verified by the careful experimental measurements of McCroskey *et al.* (1966) and by Vidal and Bartz (1969). But while the analysis of Pan

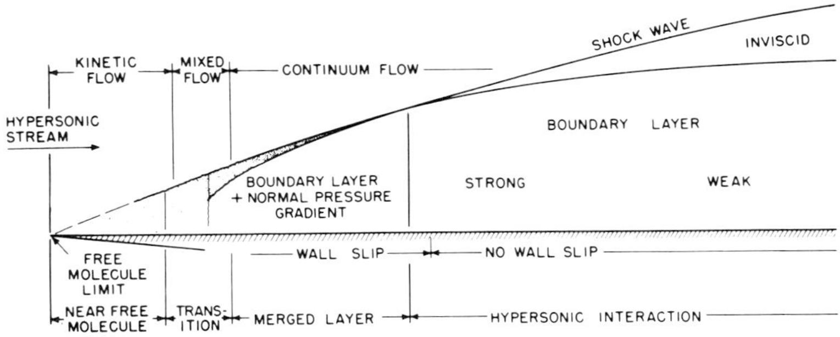

FIG. 7. Hypersonic rarefied flow past a semi-infinite sharp leading edge flat plate (Reprinted from Pan and Probstein, 1966 by permission of Cornell University Press).

and Probstein (1966) resulted in only about a 5% change from the R–H conditions across the shock, the measurements (McCroskey *et al.*, 1966) indicated changes of the order of 50%. In fact one of the main conclusions of McCroskey *et al.* (1966) was that the shock structure plays the dominant role in establishing the surface pressure distribution near the leasing edge. The analysis of the non-Rankine–Hugoniot shock layer also plays a dominant role in an improved analysis of leading-edge flow by Shorenstein and Probstein (1968). The occurrence of two-dimensional shock structure is not restricted to semi-infinite flat plates but is also observed to occur in the hypersonic flow past sharp-nosed slender bodies in general (Vas *et al.*, 1969).

III. The Viscous-Transonic Equation

A. DERIVATION

As mentioned in Section II above the need to consider two-dimensional shock wave structure arises in certain transonic flows and in hypersonic flow at low Reynolds numbers. In the transonic case it is possible to formulate such two-dimensional shock structure problems in terms of a viscous transonic equation, which is essentially the inviscid transonic small disturbance equation modified by the inclusion of a viscous term. This approach to the problem represents an extension to two dimensions of Taylor's treatment (1910) of the structure of weak normal shock waves. The derivation and properties of the viscous-transonic (V–T) equation are considered below.

A simple qualitative argument showing that viscous terms may be important in certain transonic flows has been given by Szaniawski (1962). In transonic small disturbance flow it can be shown (Guderley, 1962) that the perturbation potential ϕ' satisfies the approximate equation

$$-(\gamma + 1)\phi_x' \phi_{xx}' + \phi_{yy}' = \Psi, \tag{3.1}$$

where Ψ represents the dissipative terms. Now $-(\gamma + 1)\phi_x'\phi_{xx}' \approx (1 - M^2)\phi_{xx}'$, and when the magnitudes of $(1 - M^2)\phi_{xx}'$, ϕ_{yy}' and Ψ in Eq (3.1) are compared it is usually found that Ψ is negligible. However, as $M \to 1$, $(1 - M^2)\phi_{xx}'$ vanishes whereas the magnitude of Ψ may not necessarily change; consequently there may be domains $\mathscr{D}$ where $(1 - M^2\phi_{xx}' \sim O(\Psi)$ so that the dissipative terms Ψ cannot be neglected. The interior of a shock wave with either a one- or two-dimensional structure is, of course, an important example of such a domain. The thickness of weak normal shocks is of the order $\bar{\mu}''/\bar{\rho}^*\bar{a}^*(M_1 - 1)$ and so can be quite large as the upstream Mach number $M_1 \to 1.0$. Hence, the region in which convective and dissipative effects are comparable may be extensive in transonic flow. Even when $\mathscr{D}$ is very small compared to the region under consideration, the effect of the flow within $\mathscr{D}$ upon pressure temperature and velocity changes across the region $\mathscr{D}$ may, as already indicated, exert a large influence on the overall flow.

A viscous-transonic (V–T) equation was first derived by Liepmann, Ashkenas, and Cole (1948), and was later rederived independently by Ryzhov and Shefter (1964), Sichel (1962, 1963) and Szaniawski (1962, 1963). In the derivation presented here steady flow of a perfect gas with constant specific heats, viscosities, and thermal conductivity will be considered. Variation of the thermodynamic properties is taken into account by Sichel (1963), Szaniawski (1962), and Ryzhov and Shefter (1964). However, since only small perturbations from uniform flow are considered, there is no material change in the results. The derivations cited above consider perturbations with respect to a uniform sonic flow, but a more general equation is obtained if perturbations are taken with respect to a uniform free stream flow in the x or axial direction, with velocity $\bar{U}$ near the sonic value but not necessarily equal to it (Sichel, 1970). In what follows barred quantities are dimensional.

From the method of characteristics (Guderley, 1962; Sichel, 1963) for inviscid flow or from the transonic Hugoniot conditions across shocks it follows that perturbations in the dimensionless quantities $\bar{p}/\bar{\rho}_\infty \bar{U}^2$, $\bar{\rho}/\bar{\rho}_\infty$, $\bar{T}/\bar{T}_\infty$, and $\bar{a}/\bar{a}_\infty$ will be of the same order as the deviation of the dimensionless x or axial velocity component $\bar{u}/\bar{U}$ from the undisturbed value of unity. Thus if $[(\bar{u}/\bar{U}) - 1] \sim O(\epsilon)$ where $\epsilon \ll 1$, the flow parameters can be expanded as

$$\begin{aligned} u &= (\bar{u}/\bar{U}) = 1 + \epsilon u^{(1)} + \cdots, \\ \rho &= (\bar{\rho}/\bar{\rho}_\infty) = 1 + \epsilon \rho^{(1)} + \cdots, \\ p &= (\bar{p}/\bar{\rho}_\infty U^2) = (1/\gamma M_\infty{}^2) + \epsilon p^{(1)} + \cdots, \\ T &= \bar{T}/\bar{T}_\infty = 1 + \epsilon T^{(1)} + \cdots, \\ \bar{a} &= \bar{a}/\bar{a}_\infty = 1 + \epsilon a^{(1)} + \cdots. \end{aligned} \tag{3.2}$$

The transverse velocity is expanded as

$$v = \bar{v}/\bar{U}_\infty = \lambda \epsilon v^{(1)} + \cdots. \tag{3.3}$$

The factor $\lambda \ll 1$, and if the disturbance is caused by a slender body it will be shown below that $\lambda \sim O(\epsilon^{1/2})$. In a one-dimensional flow, e.g., a normal shock, $\lambda = 0$. From the method of characteristics or the oblique shock relations (Sichel, 1963; Guderley, 1962) it can be shown that if $\bar{L}$ is the characteristic length of the problem under consideration then the appropriate stretchings of the $\bar{x}$ and $\bar{y}$ coordinates are

$$x = \bar{x}/\bar{L}, \qquad y = \bar{y}\lambda/\bar{L}. \tag{3.4}$$

The stretch factor λ, thus takes into account the difference in the characteristic x and y dimensions which arises in transonic flow. The expansion coefficients $u^{(1)}$, $v^{(1)}$, $\rho^{(1)}$, etc., are presumed to be $O(1)$.

In deriving the inviscid transonic small disturbance equations (Guderley, 1962) the expansions (3.2), (3.3), and the stretched coordinates (3.4) are substituted in the equations for inviscid flow. In the present case (3.2), (3.3), and (3.4) are directly substituted into Navier–Stokes equations:

$$\frac{\partial(\bar{\rho}\bar{u})}{\partial\bar{x}} + \frac{\partial(\bar{\rho}\bar{v})}{\partial\bar{y}} + \frac{k\,\bar{\rho}\bar{v}}{\bar{y}} = 0, \tag{3.5a}$$

$$\begin{aligned} \bar{\rho}\bar{u}\frac{\partial\bar{u}}{\partial\bar{x}} + \bar{\rho}\bar{v}\frac{\partial\bar{u}}{\partial\bar{y}} = &-\frac{\partial\bar{p}}{\partial\bar{x}} + 2\frac{\partial}{\partial\bar{x}}\left(\bar{\mu}\frac{\partial\bar{u}}{\partial\bar{x}}\right) + \frac{\partial}{\partial\bar{y}}\left[\bar{\mu}\left(\frac{\partial\bar{u}}{\partial\bar{y}} + \frac{\partial\bar{v}}{\partial\bar{x}}\right)\right] \\ &+ \frac{\partial}{\partial\bar{x}}\left[(\bar{\mu}'' - 2\bar{\mu})\left(\frac{\partial\bar{u}}{\partial\bar{x}} + \frac{\partial\bar{v}}{\partial\bar{y}}\right)\right], \end{aligned} \tag{3.5b}$$

$$\bar{\rho}\bar{u}\frac{\partial \bar{v}}{\partial \bar{x}} + \bar{\rho}\bar{v}\frac{\partial \bar{v}}{\partial \bar{y}} = -\frac{\partial \bar{p}}{\partial \bar{y}} + 2\frac{\partial}{\partial \bar{x}}\left[\bar{\mu}\left(\frac{\partial \bar{v}}{\partial \bar{x}} + \frac{\partial \bar{u}}{\partial \bar{y}}\right) + 2\frac{\partial}{\partial \bar{y}}\left(\bar{\mu}\frac{\partial \bar{v}}{\partial \bar{y}}\right)\right]$$

$$+ \frac{\partial}{\partial \bar{y}}\left[(\bar{\mu}'' - 2\bar{\mu})\left(\frac{\partial \bar{v}}{\partial \bar{y}} + \frac{\partial \bar{u}}{\partial \bar{x}}\right)\right], \tag{3.5c}$$

$$\bar{\rho}\bar{C}_{\mathrm{p}}\left(\bar{u}\frac{\partial \bar{T}}{\partial \bar{x}} + \bar{v}\frac{\partial \bar{T}}{\partial \bar{y}}\right) - \left(\bar{u}\frac{\partial \bar{p}}{\partial \bar{x}} + \bar{v}\frac{\partial \bar{p}}{\partial \bar{y}}\right) = \frac{\partial}{\partial \bar{x}}\left(\bar{k}\frac{\partial \bar{T}}{\partial \bar{x}}\right) + \frac{\partial}{\partial \bar{y}}\left(\bar{k}\frac{\partial \bar{T}}{\partial \bar{y}}\right) + \Phi, \tag{3.5d}$$

$$\bar{p} = \bar{\rho}R\bar{T}. \tag{3.5e}$$

The integer $k = 1$ for axisymmetric flow with $\bar{x}$ as the axis of symmetry and $\bar{y}$ as the radius; $k = 0$ for two-dimensional plane flow. In (3.5b,c) $\bar{\mu}'' = \frac{4}{3}\bar{\mu} + \bar{\mu}'$ is what Hayes (1958) calls the longitudinal viscosity, and $\bar{\mu}'$ is the bulk viscosity, while Φ in the energy equation (3.5d) is the viscous dissipation.

Introducing (3.2), (3.3), and (3.4) into the Navier–Stokes equations (3.5) and equating coefficients of the lowest power in ϵ yields the following set of partial differential equations for the first order expansion coefficients:

$$\rho_x^{(1)} + u_x^{(1)} = 0, \quad \text{(continuity)} \tag{3.6a}$$

$$u_x^{(1)} + p_x^{(1)} = 0, \quad (x \text{ or axial momentum}) \tag{3.6b}$$

$$v_x^{(1)} + p_y^{(1)} = 0, \quad (y \text{ or transverse momentum}) \tag{3.6c}$$

$$T_x^{(1)} - (\gamma - 1)M_\infty{}^2 p_x^{(1)} = 0, \quad \text{(energy)} \tag{3.6d}$$

$$\gamma M_\infty{}^2 p_x^{(1)} = \rho_x^{(1)} + T_x^{(1)}. \quad \text{(state)} \tag{3.6e}$$

Assuming uniform flow upstream, we can integrate (3.6a), (3.6b), and (3.6d) to yield

$$\begin{aligned} \rho^{(1)} + u^{(1)} &= 0, \\ u^{(1)} + p^{(1)} &= 0, \\ T^{(1)} - (\gamma - 1)M_\infty{}^2 p^{(1)} &= 0, \end{aligned} \tag{3.7}$$

which are identical to the relations between $\rho^{(1)}$, $u^{(1)}$, $p^{(1)}$, and $T^{(1)}$ in a leftward propagating acoustic wave. Using (3.7) the first order y momentum equation (3.6c) reduces to the irrotationality condition

$$u_y^{(1)} - v_x^{(1)} = 0. \tag{3.8}$$

Since entropy changes are of a higher order in tansonic flows and stagnation enthalpy is assumed to be constant, irrotationality of the first order flow could also have been derived from the Crocco relation.

Equations (3.6) are five equations for $\rho^{(1)}$, $u^{(1)}$, $v^{(1)}$, $T^{(1)}$, and $p^{(1)}$; however, with $(M_\infty{}^2 - 1) \ll 1$ it is readily shown that the energy and state equations (3.6d) and (3.6e) are identical so that the set (3.6) is redundant. The equations (3.6) are therefore insufficient for obtaining solutions for the first order V–T flow. The second order equations, obtained by equating coefficients of the next higher power in ϵ, are

$$\rho_x^{(2)} + u_x^{(2)} + (\rho^{(1)}u^{(1)})_x + \frac{\lambda^2}{\epsilon}\left(v_y^{(1)} + \frac{k}{y}v^{(1)}\right) = 0, \quad \text{(continuity)} \tag{3.9a}$$

$$u_x^{(2)} + p_x^{(2)} = \frac{\bar{\mu}''}{\epsilon\rho_\infty \bar{U}\bar{L}} u_{xx}^{(1)}, \quad (x \text{ momentum}) \tag{3.9b}$$

$$T_x^{(2)} - (\gamma - 1)M_\infty{}^2 p_x^{(2)} = -(\gamma - 1)M_\infty{}^2 u^{(1)}u_x^{(1)} + (\bar{\mu}''/\bar{\rho}_\infty \bar{U}\bar{L}Pr''\epsilon)T_{xx}^{(1)}, \quad \text{(energy)} \tag{3.9c}$$

$$T_x^{(2)} - \gamma M_\infty{}^2 p_x^{(2)} + \rho_x^{(2)} = -\gamma M_\infty{}^2 u^{(1)}u_x^{(1)} + (1 - M_\infty{}^2)u_x^{(1)} + (u^{(1)} + T^{(1)})u_x^{(1)}. \quad \text{(state)} \tag{3.9d}$$

In deriving the second order equations (3.9) it was assumed that $\lambda \sim O(\epsilon^{1/2})$ or smaller. If λ were larger than this, say $O(\epsilon^{1/4})$, then the expansions (3.2) would be inconsistent. The transverse shear terms in the x-momentum equation are of $O(\lambda^2\, \epsilon\bar{\mu}/\bar{p}_\infty \bar{U}\bar{L})$ while the compressive viscous terms are of $O(\epsilon\bar{\mu}''/\bar{\rho}_\infty \bar{U}\bar{L})$. In transonic flow $\bar{U} \simeq \bar{a}_\infty$, so that the inverse Reynolds numbers $\bar{\mu}/\bar{\rho}_\infty \bar{U}\bar{L}$, and $\bar{\mu}''/\bar{\rho}_\infty \bar{U}\bar{L}$ are of $O(\bar{L}_c/\bar{L})$ where $\bar{L}_c = \bar{\mu}/\bar{\rho}_\infty \bar{a}_\infty$ is a viscous length of the order of a mean free path. Unless $\bar{L}_c/\bar{L} \ll 1$, the continuum theory being used here will be inapplicable. Consequently the viscous terms are of a higher order and it is assumed here that $\bar{L}_c/\bar{L} \sim O(\epsilon)$ or less. Also, $\lambda \ll 1$ reflects the slow y variation so characteristic of transonic flow with the consequence that the transverse shear terms are of a higher order than the compressive viscous terms.

The second order equations (3.9) are redundant in $\rho^{(2)}$, $u^{(2)}$, $T^{(2)}$, and $p^{(2)}$, and upon eliminating these second order expansion coefficients the following equation relating $u^{(1)}$ and $v^{(1)}$ is obtained (Sichel, 1963, 1970; Ryzhov, 1965)

$$\frac{\bar{\mu}''}{\bar{\rho}_\infty \bar{U}\bar{L}\epsilon}\left(1 + \frac{\gamma - 1}{Pr''}\right) u_{xx}^{(1)} + \frac{(1 - M_\infty{}^2)}{\epsilon} u_x^{(1)} - (\gamma + 1)\, M_\infty{}^2 u^{(1)}u_x^{(1)} + \frac{\lambda^2}{\epsilon}\left(v_y^{(1)} + \frac{k}{y}v^{(1)}\right) = 0. \tag{3.10}$$

Together with the irrotationality condition (3.8) this equation is sometimes called the V-T equation. Pr'' is a Prandtl number based on the longitudinal viscosity $\bar{\mu}''$. Since the first order flow is irrotational, introduction of the perturbation potential Φ such that $u^{(1)} = \Phi_x$, $v^{(1)} = \Phi_y$ reduces Eq. (3.8) and (3.10) to the V-T potential equation

$$\frac{\bar{\mu}''}{\bar{\rho}_\infty \bar{U} \bar{L} \epsilon}\left(1 + \frac{\gamma - 1}{Pr''}\right) \Phi_{xxx} + \frac{(1 - M_\infty{}^2)}{\epsilon} \Phi_{xx}$$
$$- (\gamma + 1) M_\infty{}^2 \Phi_x \Phi_{xx} + \frac{\lambda^2}{\epsilon}\left(\Phi_{yy} + \frac{k}{y} \Phi_y\right) = 0. \tag{3.11}$$

From the above method of derivation, in which the first order terms satisfy the acoustic relations, the V-T flow appears as an acoustic wave whose structure is determined by a balance between higher order viscous and convective effects.

B. Boundary Conditions and Similarity

Given (3.11) there remains the problem of relating λ and ϵ to the boundary conditions. First, two-dimensional flow past a body aligned with the x axis and with the surface coordinate

$$\bar{y} = \bar{y}_{\mathrm{w}}(\bar{x}) = \bar{\tau} f(\bar{x}/\bar{L}) \tag{3.12}$$

will be considered. In (3.12) $\bar{\tau}$ is the maximum thickness, and the body shape function $f(\bar{x}/\bar{L}) \sim O(1)$. It can be shown that the mean surface approximation is applicable to the V-T equation and that, as in the inviscid case, specification of the transverse velocity should be sufficient. The boundary condition at the body thus will be

$$\lambda \epsilon v^{(1)}(x, 0) = \lambda \epsilon \Phi_y(x, 0) = (d\bar{y}_{\mathrm{w}}/d\bar{x}) = \tau (df/dx), \tag{3.13}$$

where $\tau = \bar{\tau}/\bar{L}$ is a thickness ratio. Also, $u^{(1)}$ and $v^{(1)}$ must remain finite as $y \to \infty$. The pressure coefficient, to the first order, is given by

$$C_{\mathrm{p}} = (\bar{p} - \bar{p}_\infty)/\tfrac{1}{2}\bar{\rho}_\infty \bar{U}^2 = -2\epsilon \Phi_x . \tag{3.14}$$

Since both $v^{(1)}$ and $f'(x)$ should be $O(1)$ it appears reasonable, following Ferrari and Tricomi's discussion (1968) of inviscid transonic similarity, to let

$$\lambda \epsilon = \tau. \tag{3.15}$$

Using (3.15) the V-T equation (3.11) can be written in the form

$$\frac{\bar{\mu}''\epsilon^2}{\bar{\rho}_\infty \bar{U}\bar{L}\tau^2}\left(1 + \frac{\gamma - 1}{Pr''}\right)\Phi_{xxx} + \frac{(1 - M_\infty{}^2)\epsilon^2}{\tau^2}\Phi_{xx} - \frac{(\gamma + 1) M_\infty{}^2\epsilon^3}{\tau^2}\Phi_x\Phi_{xx} + \Phi_{yy} = 0. \tag{3.16}$$

In V-T flow the four terms in (3.16) should be of the same order and Φ_{xxx}, Φ_{xx}, $\Phi_{yy} \sim O(1)$, if the expansions (3.2) and (3.3) and the stretched coordinates (3.4) are appropriate. Thus the coefficients in (3.16) should also be $O(1)$ and to simplify (3.16) it becomes convenient, as in the inviscid case, to choose ϵ so that

$$\frac{(\gamma + 1) M_\infty{}^2\epsilon^3}{\tau^2} = 1 \qquad \text{or} \qquad \epsilon = \frac{\tau^{2/3}}{(\gamma + 1)^{1/3} M_\infty^{2/3}}. \tag{3.17}$$

The requirement that the nonlinear term in (3.16) be of $O(1)$ thus leads to the relation (3.17) between ϵ and τ, which is identical to that obtained in inviscid transonic flow, and from (3.15) it then follows that $\lambda \sim O(\epsilon^{1/2})$. The coefficient of Φ_{xx} in (3.16) then becomes

$$\frac{(1 - M_\infty{}^2)\epsilon^2}{\tau^2} = \frac{(1 - M_\infty{}^2)}{\tau^{2/3} M_\infty^{4/3}(\gamma + 1)^{2/3}} = \chi_\infty, \tag{3.18}$$

where χ_∞ is now the usual inviscid transonic similarity parameter (Ferrari and Tricomi, 1968). The coefficient of the viscous term in (3.16) becomes

$$\frac{\bar{\mu}''\epsilon^2}{\bar{\rho}_\infty \bar{U}\bar{L}\tau^2}\left(1 + \frac{\gamma - 1}{Pr''}\right) = \frac{\bar{\mu}''}{\bar{\rho}_\infty \bar{U}\bar{L}\tau^{2/3} M_\infty^{4/3}(\gamma + 1)^{2/3}}\left(1 + \frac{\gamma - 1}{Pr''}\right) = \chi_{\mathrm{V}}, \tag{3.19}$$

where χ_{V} is now a viscous transonic similarity parameter. The V-T equation can thus be written in the form

$$\chi_{\mathrm{V}}\Phi_{xxx} + \chi_\infty\Phi_{xx} - \Phi_x\Phi_{xx} + \Phi_{yy} = 0, \tag{3.20}$$

with the boundary conditions

$$\Phi_y(x, 0) = f'(x) \qquad \Phi_y, \Phi_x \text{ bounded as } |y|, |x| \to \infty, \tag{3.21}$$

and the pressure coefficient is

$$C_{\mathrm{p}} = \frac{2^{2/3}}{(\gamma + 1)^{1/3} M_\infty^{2/3}}\Phi_x(x, y; \chi_\infty, \chi_{\mathrm{V}}). \tag{3.22}$$

Equations (3.20), (3.21), and (3.22) provide the basis for viscous-transonic similarity.

Denoting $\bar{U}\bar{L}\bar{\rho}_\infty/\bar{\mu}''$, the Reynolds number based on longitudinal viscosity, by $\mathbf{R}$, the newly defined V-T similarity parameter χ_V can also be expressed in the form

$$\chi_\text{V} = \frac{1 + (\gamma - 1)/Pr''}{\mathbf{R}\tau^{2/3}M_\infty^{4/3}(\gamma + 1)^{2/3}} = \frac{1 + (\gamma - 1)/Pr''}{\epsilon\mathbf{R}(\gamma + 1)}. \tag{3.23}$$

The quantity

$$\frac{\bar{\mu}''}{\bar{\rho}_\infty}\left(1 + \frac{\gamma - 1}{Pr''}\right) = \frac{\bar{\mu}''}{\bar{\rho}_\infty} + \frac{(\gamma - 1)\bar{\kappa}}{\bar{\rho}_\infty \bar{C}_\text{p}} = \delta, \tag{3.24}$$

where δ is what Lighthill (1956) calls the "Diffusivity of sound," and is the combination of transport properties governing the attenuation of sound waves. Thus it is also possible to write

$$\chi_\text{V} = \delta/\bar{U}\bar{L}\tau^{2/3}M_\infty^{4/3}(\gamma + 1)^{2/3} = \bar{\eta}/\bar{L}, \tag{3.25}$$

where $\bar{\eta}$ is a length of the same order of magnitude as the thickness of a weak shock wave with upstream Mach number M_1 such that $(M_1 - 1) \sim O(\epsilon)$. Finally, in terms of the viscous length $\bar{L}_\text{c}$ defined above

$$\chi_\text{V} = (L_\text{c}/L)\left(1 + \frac{\gamma - 1}{Pr''}\right)\Big/\tau^{2/3}M_\infty^{7/3}(\gamma + 1)^{2/3}. \tag{3.26}$$

With $\chi_\text{V} \ll 1$, (3.20) reduces to the inviscid transonic small disturbance equation except, possibly, for very thin regions where gradients are large so that $\chi_\text{V}\Phi_{xxx} \sim O(1)$. The singular perturbation nature of the V-T equation (3.20) may account for some of the difficulties of inviscid transonic theory. A key question is, clearly, whether or not solutions of the V-T equation approach the inviscid solution as $\chi_\text{V} \to 0$. If perturbations are with respect to the sonic velocity or if the undisturbed velocity is sonic, $\chi_\infty = 0$ and the second term of (3.20) drops out. It is readily shown that at the sonic point

$$u^{(1)} = u_*^{(1)} = \chi_\infty. \tag{3.27}$$

Equation (3.22) implies that the pressure coefficient C_p varies as $\tau^{2/3}$ for fixed χ_∞ and χ_V at least to the first order in $(1 - M_\infty{}^2)$ or ϵ. For a fixed τ it is clear that whereas C_p depends only on M_∞ in the inviscid case, C_p will now also depend upon $\mathbf{R}$. In actual airfoil flows the presence of the boundary layer also introduces a $\mathbf{R}$ dependence. When both a boundary layer and a non-R–H shock are present the nature of the

R effect can therefore be expected to be very complex. The influence of the individual physical variables $\bar{L}$, M_∞, $\bar{\tau}$, and $\bar{L}_c$ upon χ_∞ and χ_V has been discussed by Sichel (1970).

The condition $\lambda^2 \sim O(\epsilon)$ imposed by the boundary condition (3.15) and (3.17) is not the only possibility (Ryzhov, 1965). If $\lambda^2 \ll \epsilon$ while $\epsilon \sim O(\mathbf{R}^{-1})$, then taking $M_\infty = 1$ the V-T equation becomes

$$\chi_V \Phi_{xxx} - \Phi_x \Phi_{xx} = 0. \tag{3.28}$$

This equation describes the structure of weak normal shock waves. Since no boundaries are involved here ϵ cannot be related to any thickness ratios; rather, an appropriate choice is $\epsilon = (\bar{u}_1/\bar{a}^*) - 1$, where $\bar{u}_1$ is the velocity upstream of the shock and $\bar{a}^*$ is the critical speed of sound.

Another possibility (Ryzhov, 1965) is $\lambda^2 \sim O(\mathbf{R}^{-1})$, $\epsilon \ll \mathbf{R}^{-1}$. Then assuming a sonic free stream and letting $\lambda^2 = \mathbf{R}^{-1}[1 + (\gamma - 1)/Pr'']$, (3.11) becomes

$$\Phi_{xxx} + \Phi_{yy} + (k/y)\Phi_y = 0. \tag{3.29}$$

Equation (3.29) is linear and Ryzhov shows that (3.29) is sufficient to describe viscous transonic flow at large distances from axisymmetric bodies. Actually with $\lambda^2 \sim O(\mathbf{R}^{-1})$ the expansion (3.2) and (3.3) and the stretching (3.4) must be replaced by

$$\begin{aligned} u &= \bar{u}/a^* = 1 + \epsilon u^{1)} + \epsilon\,\delta u^{(2)} + \cdots, \\ \rho &= \bar{\rho}/\rho^* = 1 + \epsilon\rho^{(1)} + \epsilon\,\delta\rho^{(2)}, \\ v &= \bar{v}/a^* = \epsilon\lambda v^{(1)} + \epsilon\lambda\,\delta v^{(2)} + \cdots, \\ x &= \bar{x}/L, \qquad y = \lambda y/L, \end{aligned} \tag{3.30}$$

and with $\delta \sim O(\epsilon^{1/2})$, (3.30) is different from the usual expansion and stretching used in tansonic flow.

C. The Variation of Entropy

Using the relation

$$T\,dS = C_p\,dT - (dp/\rho), \tag{3.31}$$

the energy equation (3.5d) can also be written as an equation for the entropy change. Keeping only the largest terms the dimensionless entropy $S = \bar{S}/\bar{C}_p$, with $\lambda^2 = \epsilon$, satisfies

$$Pr'' S_x = \epsilon^2(\epsilon\mathbf{R})^{-1} T^{(1)}_{xx}. \tag{3.32}$$

In those regions where $\epsilon \sim O(\mathbf{R}^{-1})$, that is, where $\chi_V \sim O(1)$, (3.32) implies that

$$S = \epsilon^2 S^{(2)} + \epsilon^3 S^{(3)} + \cdots \tag{3.33}$$

is the appropriate expansion for S. Equation (3.32) is an entropy transport equation which yields the result $S_1^{(2)} = S_2^{(2)}$ when integrated across a weak shock, where subscripts 1 and 2 refer to upstream and downstream conditions. From the equation for $S^{(3)}$

$$S^{(3)}{}_x = \frac{1}{Pr''}(\epsilon\mathbf{R})^{-1}(T^{(2)}{}_{xx} + T^{(1)}{}_{yy}) + (\gamma - 1)(\epsilon\mathbf{R})^{-1}\left(1 + \frac{\gamma - 1}{Pr''}\right)(u^{(1)}{}_x)^2, \tag{3.34}$$

it follows that entropy production is of the third order, since $(u_x^{(1)})^2$ is always positive. These results regarding entropy are in agreement with weak shock theory, as indeed they should be.

D. Properties of the V-T Equation

As indicated above when $\lambda^2 \ll \epsilon$ but $(\epsilon\mathbf{R}) \sim O(1)$ the V-T equation reduces to (3.28). One integration of (3.28) yields

$$\chi_V u^{(1)}{}_x - \tfrac{1}{2}(u^{(1)})^2 = C. \tag{3.35}$$

In the case of a weak normal shock, choosing $\epsilon = (u_1/a^*) - 1$ means that $u^{(1)}(-\infty) = +1$, so that the constant of integration $C = -1/2$. Equation (3.35), a Riccati equation, has the solution

$$u^{(1)}(x) = \Phi_x(x) = -\tanh(x/2\chi_V), \tag{3.36a}$$

which is Taylor's solution (1910) for the structure of a weak shock. In addition (3.35) has the solutions

$$u^{(1)} = +1, \tag{3.36b}$$

$$u^{(1)} = -1, \tag{3.36c}$$

$$u^{(1)} = -\coth(x/2\chi_V). \tag{3.36d}$$

The solutions (3.36b) and (3.36c) are trivial in the present case while (3.36d) diverges at $x = 0$. As will be seen below, integration of the V-T equation often leads to a Riccati equation with several solutions, some of which may be divergent. The structure of a transonic oblique shock at an angle $\alpha\epsilon^{1/2}$ with the y axis can be shown to be a similarity solution of the V-T equation (3.20) with $\chi_\infty = 0$, with the velocity components $u^{(1)}$ and $v^{(1)}$ given by (Sichel, 1963)

$$\begin{aligned} u^{(1)}(x, y) &= \frac{\alpha^2}{\gamma+1} - \left(1 - \frac{\alpha^2}{\gamma+1}\right) \tanh\left(1 - \frac{\alpha^2}{\gamma+1}\right) \frac{(x+\alpha y)}{2\chi_V}, \\ v^{(1)}(x, y) &= -\alpha\left(1 - \frac{\alpha^2}{\gamma+1}\right)\left[1 + \tanh\left(1 - \frac{\alpha^2}{\gamma+1}\right) \frac{(x+\alpha y)}{2\chi_V}\right]. \end{aligned} \tag{3.37}$$

For both supersonic ($\Phi_x > \chi_\infty$) and subsonic ($\Phi_x < \chi_\infty$) flow the V-T potential equation (3.20) is parabolic with the threefold characteristic $y = \text{const}$. This property is in distinct contrast to the inviscid transonic equation

$$(\chi_\infty - \Phi_x)\,\Phi_{xx} + \Phi_{yy} = 0, \tag{3.38}$$

which changes from elliptic to hyperbolic as the velocity passes from sub- to supersonic values. The V-T equation is, thus, in some ways simpler than the inviscid equation.

A limited uniqueness theorem for the two-dimensional case gives some indication of properly posed boundary conditions. Given a rectangular domain $\mathscr{R}$ ($x_1 < x < x_2$, $y_1 < y < y_2$), it can be shown (Sichel, 1963) that if Φ_x is specified on $x = x_1$ and Φ is specified on the entire boundary of $\mathscr{R}$, then the solution of Eq. (3.20) is unique provided $\Phi_x < 0$ in $\mathscr{R}$. Thus only one condition may be specified on the boundaries parallel to the undisturbed flow just as in the inviscid case. Since the mean surface approximation remains valid in a V-T flow (Sichel, 1962) a tentative conclusion, which has already been used in Section III, B above, is that slender-body boundary conditions will be identical to those used in inviscid flow. The V-T equation cannot satisfy the no slip condition at the wall, and the complete viscous problem still requires the inclusion of a boundary layer. Even though the V-T equation is parabolic it is interesting to note that Φ must be specified over a closed boundary as for second order elliptic equations. The high order of the V-T equation is responsible for this result.

E. Higher Order Equations

Higher order equations for $u^{(2)}$, $v^{(2)}$, etc., which are linear and non-homogeneous have been formulated by Sichel (1961), Szaniawski (1963), and Ryzhov (1965). To second order the flow remains irrotational so that

$$u^{(2)}{}_y - v^{(2)}{}_x = 0. \tag{3.39}$$

Obtaining meaningful solutions of the higher order equations is a formidable task; however, in the one-dimensional case it has been possible to solve both the second and third order equations for corrections to the Taylor (1910) weak shock solution (Sichel, 1960; Szaniawski, 1966). Comparison with several exact solutions for the shock structure indicates that the higher order terms do improve the accuracy of the Taylor solution. While this result does not constitute a convergence proof for the expansions (3.2) and (3.3), which may also be asymptotic, it does provide some confidence that the expansion scheme is a reasonable one.

IV. The Nozzle Problem

A. Nozzle Solutions

As indicated in Section II above, the inviscid theory fails to explain the transition from the Taylor to the Meyer type of flow in converging diverging nozzles. It would appear to be more consistent to begin the investigation of such transitional flows with the V-T equation and then to study the behavior of the viscous solutions as $\mathbf{R} \to \infty$. This problem has been considered by Szaniawski (1964a,b), and in great detail by Kopystyński and Szaniawski (1965).

Kopystyński and Szaniawski used a double expansion in x and y such that

$$\begin{aligned} \bar{u}/a^* &= 1 + \epsilon u^{(1)} + \epsilon^2 u^{(2)} + \cdots \\ &= 1 + \epsilon \left[U_{10}(x) + \frac{1}{2!} U_{12}(x) y^2 + \cdots \right] \\ &\quad + \epsilon^2 \left[U_{20}(x) + \frac{1}{2!} U_{22}(x) y^2 + \cdots \right], \qquad (4.1) \\ \bar{v}/a^* &= \epsilon^{3/2} \left[V_{11}(x) y + V_{13}(x) \frac{y^3}{3!} + \cdots \right] \\ &\quad + \epsilon^{5/2} \left[V_{21}(x) y + V_{23}(x) \frac{y^3}{3!} + \cdots \right]. \end{aligned}$$

If the nozzle half height $\bar{l}$ is used as the characteristic length $\bar{L}$ of the problem,

$$\bar{y}/\bar{L} = \bar{y}/\bar{l} = \tilde{y} \sim O(1) \qquad (4.2)$$

and $y = \epsilon^{1/2}\tilde{y}$. Then the leading terms of the expansion (4.1) become

$$\bar{u}/a^* = 1 + \epsilon U_{10}(x) + \cdots, \qquad \bar{v}/a^* = \epsilon^2 V_{11}(x)\, \tilde{y}. \qquad (4.3)$$

Substituting the values of $u^{(1)}$ and $v^{(1)}$ from (4.3) into the V-T equation (3.10) with $M_\infty = 1$ then yields the equation

$$\chi_V U'' - UU' + V' = 0, \tag{4.4}$$

where the subscripts of U_{10} and V_{11} have been suppressed and χ_v corresponds to the second form given in (3.23). If the relative magnitudes of x and y are retained the stretching (4.2) implies that $x = \bar{x}/\bar{l} \sim O(\epsilon^{1/2})$. Kopystyński and Szaniawski considered flow through a nozzle throat with contour

$$\pm \bar{y}_W/\bar{l} = 1 + \tfrac{1}{2}\epsilon^2 f^2(x) \tag{4.5}$$

to be consistent with the expansion (4.1). $f(x)$ is an increasing function of x with $f(0) = 0$, $f'(0) \neq 0$. Using (4.5) to evaluate $V(x)$, (4.4) finally becomes

$$\chi_V U'' = UU' - ff'. \tag{4.6}$$

When $\chi_v = 0$, (4.6) has the two solutions

$$U = \pm f(x) \tag{4.7}$$

and the problem is to determine under what conditions solutions of (4.6) approach the inviscid solution (4.7) as $\chi_v \rightarrow 0$. Applying the theory of Vasileva (1963a,b) Kopystyński and Szaniawski showed that, under appropriate conditions, the viscous solutions will converge to (4.7) everywhere except in the neighborhood of the sonic point $U = 0$.

The stretched variables

$$x = \sqrt{2/f'(0)}\chi_V^{1/2}\xi$$

and

$$U = \sqrt{2\chi_V f'(0)}\psi(\xi) + R(\chi_V, \xi) \tag{4.8}$$

$$\lim_{\chi_V \rightarrow 0} R(\chi_V, \xi) = 0$$

are introduced to examine the structure of the flow near the sonic point in greater detail. Then expanding $f(x)$ in a Taylor series,

$$f = \sqrt{2f'(0)}\,\chi_V^{1/2}\xi + O(\chi_V^{3/2}\xi^2),$$

(4.6) after one integration reduces to the following Riccati equation for ψ:

$$(d\psi/d\xi) - \psi^2 = -\xi^2 + A, \tag{4.9}$$

where A is an arbitrary constant of integration. As already mentioned In Section III V-T flows are closely related to properties of various forms of the Riccati equation. The transformation $\psi = 2\xi - (v'/v)$ changes (4.9) to the second order linear equation

$$v'' - 2\xi v' + (A - 1)v = 0, \tag{4.10}$$

with solutions expressible in terms of confluent hypergeometric functions so that

$$\psi(\xi, A, \psi_0) = -\xi + \frac{(1 + A)\,\xi F(\tfrac{1}{4}(1 - A), \tfrac{3}{2}, \xi^2) + \psi_0 F(-\tfrac{1}{4}(1 + A), \tfrac{1}{2}, \xi^2)}{F(\tfrac{1}{4}(1 - A), \tfrac{1}{2}, \xi^2) - \psi_0 \xi F(\tfrac{1}{4}(3 - A), \tfrac{3}{2}, \xi^2)}, \tag{4.11}$$

where ψ_0 is the value of ψ at $\xi = 0$.

The solution (4.11) possesses a number of interesting properties. ψ diverges whenever the denominator of (4.11) vanishes and this condition defines curves $\psi_0 = \Psi(A)$ which are the boundary between continuous and divergent solutions in the ψ–A plane as shown in Fig. 8.

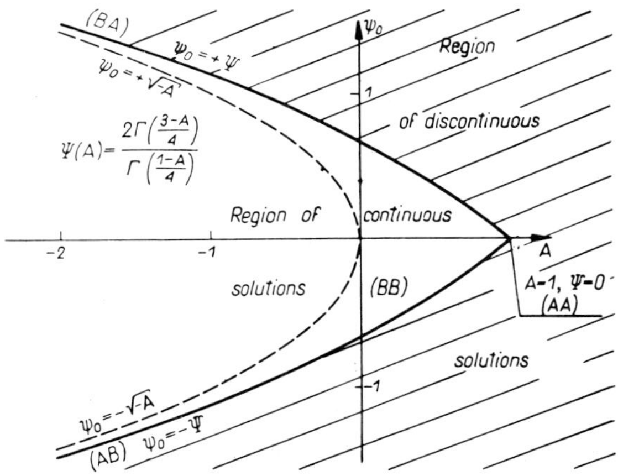

FIG. 8. Domains of continuous and discontinuous nozzle solutions in the ψ_0–A plane (from Kopystyński and Szaniawski, 1965).

Divergent solutions are eliminated from consideration on physical grounds. Solutions $\psi = \psi(\xi, A, \psi_0)$ with $-\Psi(A) < \psi_0 < \Psi(A)$, $A < 1$, have the asymptotic behavior $\lim_{\xi \to \pm\infty} \psi = -\xi$ but contain an interesting dissipative layer near the sonic point as shown in Fig. 9, which shows solutions for $A = -1$. The precise significance of this "layer" is not entirely clear. The flow either decelerates rapidly and then re-accelerates

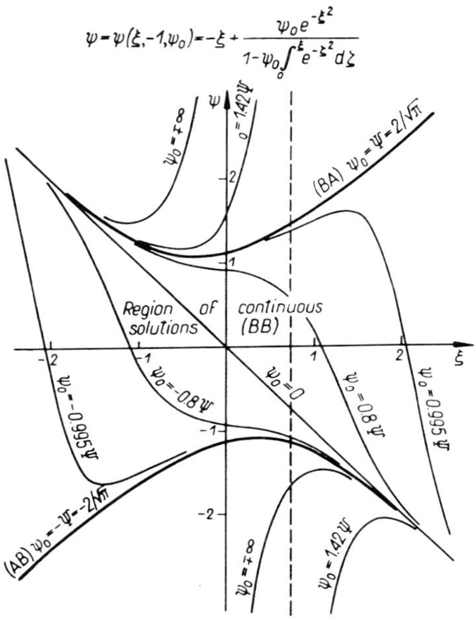

FIG. 9. Nozzle solutions for $A = -1$ (from Kopystyński and Szaniawski, 1965).

or accelerates gradually and then rapidly decelerates through a shocklike compression. Rapid shock compressions followed by expansions have, as discussed in Section II above, been observed when shocks are adjacent to an airfoil surface (Ackeret *et al.*, 1946), but the boundary conditions are then different from the nozzle problem. Still, the flow through an individual streamtube may be quite similar to the flow through a nozzle.

Choosing $\psi_0 = -\Psi(A)$ yields solutions with asymptotes $\lim_{\xi\to-\infty} \psi = +\xi$ and $\lim_{\xi\to+\infty} \psi = -\xi$. These solutions, shown for different values of A in Fig. 10, provide a reasonable picture of the transition from Taylor to Meyer flow and indicate the initial stages of shock formation downstream of the nozzle throat. In the limiting case $A = 1$, the only converging solution is $\psi = \xi$. The mirror image of this solution is obtained when $\psi_0 = +\Psi(A)$. The problem of matching the viscous solutions above to the inviscid solutions upstream and downstream of the throat is also discussed.

The nozzle solutions of Kopystyński and Szaniawski (1965) have several limitations. The solutions in Fig. 10 are nonunique for all of them approach $\psi = \pm\xi$ as $|\xi| \to \infty$. This is a disturbing result, since the solutions should, in some way, be sensitive to the downstream

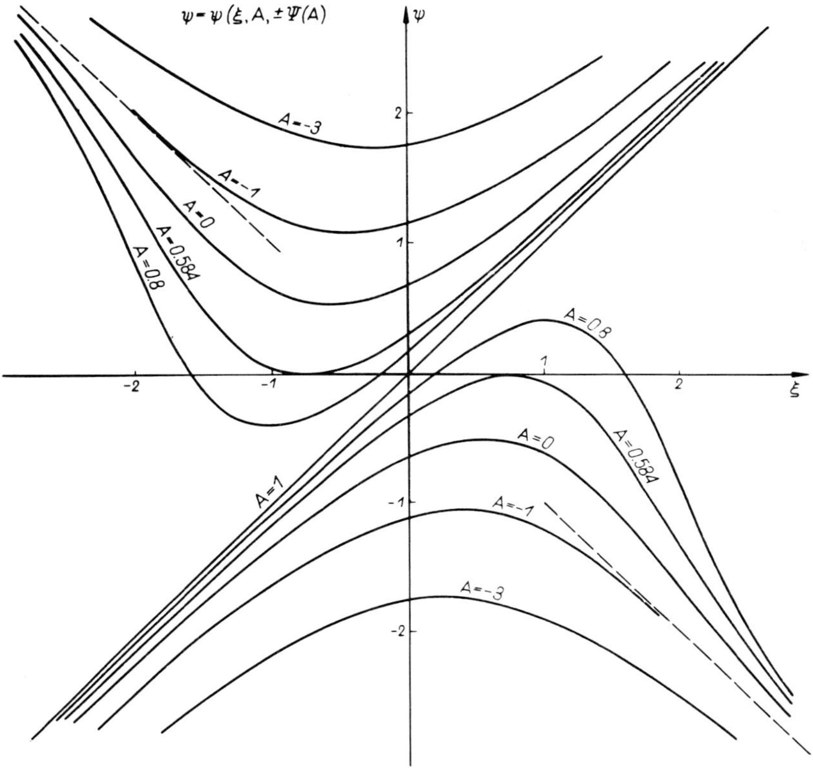

FIG. 10. Nozzle solutions with $\psi_0 = -\Psi(A)$ (from Kopystyński and Szaniawski, 1965).

boundary conditions. Because of the expansion scheme used the velocity or U is a function of x only, that is, the flow is quasi one-dimensional and isotachs, or lines of constant velocity, will all be normal to the x axis.

A different approach was used by Sichel (1966), Sichel and Yin (1967), and Ryzhov (1968), who found that inviscid nozzle-type similarity solutions, such as those found by Tomotika and Tamada (1950) and Tomotika and Hasimoto (1950), can also be developed in the viscous case. Differentiating the V-T equation (3.11) with respect to x yields the following equation for $u^{(1)}$

$$\chi_{\mathrm{V}} u^{(1)}{}_{xxx} - (\tfrac{1}{2} u^{(1)^2})_{xx} + u^{(1)}{}_{\tilde{y}\tilde{y}} + (k/\tilde{y})\, u^{(1)}{}_{\tilde{y}} = 0, \tag{4.12}$$

with $\lambda = \epsilon^{1/2}$, $M_\infty = 1$, and $\tilde{y} = (\gamma + 1)^{1/2} y$. Substitution of the transformation

$$u^{(1)} = 2Z(S) + 4\sigma^2 \tilde{y}^2, \qquad S = x + \sigma \tilde{y}^2 \tag{4.13}$$

reduces (4.12) to the ordinary differential equation

$$\chi_V Z''' - 2ZZ'' - 2(Z' - \omega_1\sigma)(Z' - \omega_2\sigma) = 0, \tag{4.14}$$

with $\omega_1, \omega_2 = [(k+1)/2]\{1 \pm [1 + 8(k+1)]^{1/2}\}$, where $k = 0, 1$ for two-dimensional and axisymmetric flows, respectively. From (4.13) it is evident that the flow is even with respect to the x axis and so can be considered a nozzle flow by choosing one of the stream tubes as the nozzle wall. The function $2Z$ is also $u^{(1)}(x, 0)$, the velocity on the nozzle axis, and the present solution takes the variation of $u^{(1)}$ with both x and y into account. It should be remarked that to the present order of approximation the fluid velocity is given by

$$(\bar{q}/a^*) = q = (u^2 + v^2)^{1/2} = [(1 + \epsilon u^{(1)})^2 + \epsilon^3 v^{(1)^2}]^{1/2} = \epsilon u^{(1)} + O(\epsilon^2).$$

The arbitrary parameter σ in (4.13) can be related to the nozzle geometry since the nozzle wall radius of curvature is inversely proportional to σ^2 (Sichel and Yin, 1967). Ryzhov (1968) used a similarity transformation more general than (4.13) in that azimuthal variation of the velocity u is considered, and he also examined the case of nozzle flow with a swirl or azimuthal velocity component.

Equation (4.14) is identical to that obtained by Tomotika and Hasimoto (1950) except for the viscous term $\chi_V Z'''$. The special solutions $Z = \omega_1\sigma S$, and $Z = \omega_2\sigma S$ satisfy both (4.14) and the inviscid equation, and the nozzle velocity distribution corresponding to $Z = \omega_1\sigma S$ is identical to that obtained from the first three terms of Meyer's original series solution (1908). With $\chi_V \ll 1$, (4.14) presents a singular perturbation problem. Equation (4.14) may be integrated once to yield, upon letting $Z = -\zeta$,

$$\chi_V \zeta'' + 2\zeta\zeta' + 2\sigma(\omega_1 + \omega_2)\,\zeta = -2\omega_1\omega_2\sigma^2 S + C_1, \tag{4.15}$$

an equation which is closely related to one considered by Cole (1968), and is of the same form as the nozzle equation used by Ryzhov (1968). Taking

$$Z = g(\tilde{S}) + \chi_V g_1(\tilde{S}) + \cdots, \tag{4.16}$$

with

$$\tilde{S} = (S - S_0)/\chi_V,$$

and keeping only the largest terms, substitution in (4.14) leads to the inner equation

$$g''' - 2(gg')' = 0, \tag{4.17}$$

with the Taylor weak shock structure (1910) as a solution. The constant

S_0 merely identifies the location of the shock layer. Equation (4.14) can thus describe a nozzle flow with an imbedded shock layer. Higher order terms of (4.16) need to be evaluated to establish the influence of two-dimensionality on the shock structure. The transformation

$$Z = \chi_V^{1/2} f(\hat{S}) + \chi_V f_1(\hat{S}) + \cdots, \qquad \hat{S} = (S - S_0)/\chi_V^{1/2}, \tag{4.18}$$

which Cole (1968) calls a "distinguished limit" leads to the inner equation

$$f''' - 2ff'' - 2(f' - \omega_1\sigma)(f' - \omega_2\sigma) = 0. \tag{4.19}$$

In (4.18) the requirement that $Z \sim O(1)$ is relaxed. Equation (4.19) was investigated in detail by Sichel (1966), Sichel and Yin (1967), and Ryzhov (1968). Rather than using (4.18), these authors arrived at this equation by assigning a value of unity to χ_V in (4.14), a procedure equivalent to choosing the weak-shock thickness η as the chracteristic length of the problem. From the stretching (3.4) it then follows that the nozzle half height $\bar{l} \sim O(\bar{\eta}/\epsilon^{1/2})$, which is an extremely small length except for very small ϵ, or for large $\bar{\eta}$ such as might occur at low densities. The end result is thus the same as in the distinguished limit described above. With $\chi_V = 1$ the behavior of the solutions of (4.14) in the (Z'', Z', Z) phase space can be established by studying the two-dimensional trajectories obtained when Z is held constant. These trajectories are dominated by singularities which occur where the inviscid solutions $Z = \omega_1\sigma S$ and $Z = \omega_2\sigma S$ pierce the $Z =$ const. planes. In these planes the point $Z' = \omega_1\sigma$, $Z'' = 0$, corresponding to the Meyer solution (1908), will be a saddle point for all Z while the point $Z' = \omega_2\sigma$, $Z'' = 0$ will be an unstable node, an unstable spiral point, a stable spiral point, or a stable node, respectively, for Z corresponding to the ranges

$$Z > [2\sigma(\omega_1 - \omega_2)]^{1/2}, \qquad [2\sigma(\omega_1 - \omega_2)]^{1/2} > Z > 0,$$
$$0 > Z > -[2\sigma(\omega_1 - \omega_2)]^{1/2}, \qquad -[2\sigma(\omega_1 - \omega_2)]^{1/2} > Z. \tag{4.20}$$

Thus any solution starting near the inviscid accelerating solution, $Z = \omega_1\sigma S$, will diverge from it for all Z. Some of these solutions pass through a maximum and then asymptotically approach the inviscid decelerating solution $Z = \omega_2\sigma S$.

Numerical solutions of (4.14) with $\chi_V = 1$, representing stages in the transition from the Taylor to the Meyer type of flow and similar to those found by Kopystyński and Szaniawski (1965), are shown in Figs. 11a–c for $\sigma = 0.1$, 0.5, and 1.0. As the maximum of Z increases beyond the

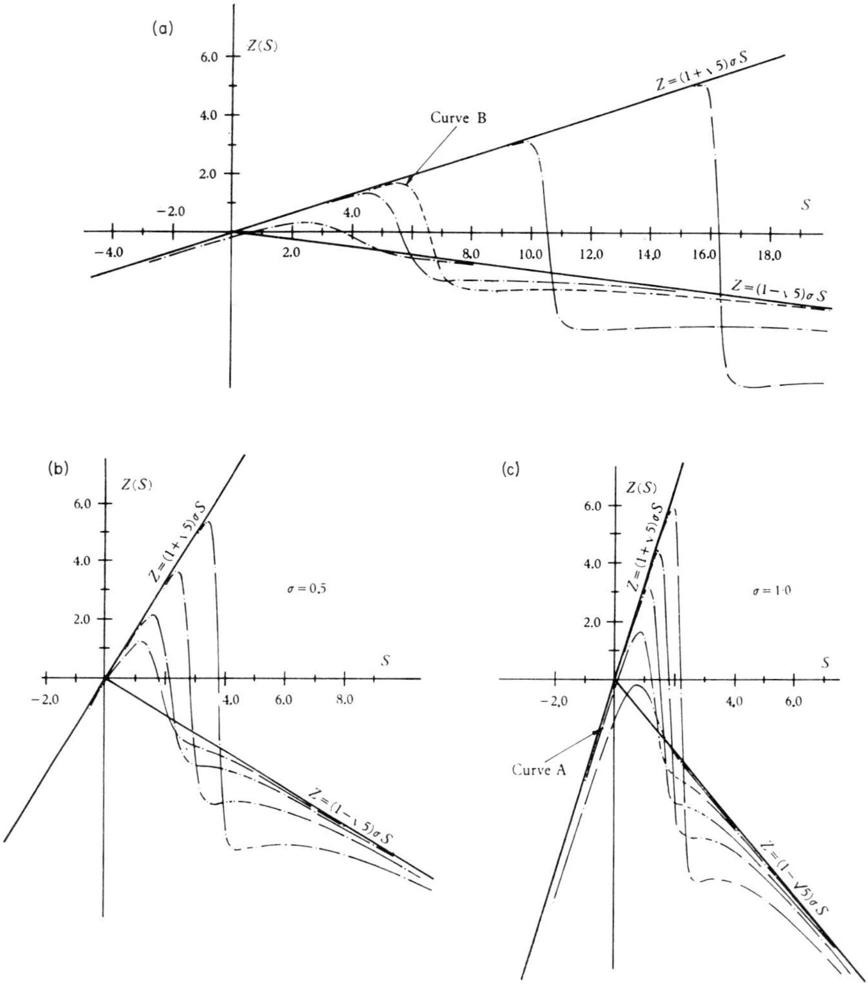

FIG. 11. Axisymmetric V-T nozzle solutions for different values of the parameter σ: (a) $\sigma = 0.1$; (b) $\sigma = 0.5$; (c) $\sigma = 1.0$ (from Sichel and Yin, 1967, with permission of Cambridge University Press).

sonic value $Z = 0$ the deceleration to subsonic velocity becomes steeper, resembling the transition through a shock wave. Although (4.14) is essentially an "inner" equation it is remarkable that the solution approaches the inviscid solution for accelerating and decelerating nozzle flow as $S \to \pm\infty$. A detailed discussion of the asymptotic behavior of the solution as $S \to \pm\infty$ has been given by Sichel and Yin (1967) and by Ryzhov (1968).

The similarity solution, described above, is indirect in that the shape of the nozzle wall cannot be specified in advance. Rather, it is necessary to accept one of the streamlines produced by the similarity solution as the nozzle contour. For a given σ each of the solutions for Z produces a different nozzle contour, and as the centerline velocity maximum increases undulations occur in the nozzle wall streamlines. Typical nozzle flow fields showing streamlines and constant velocity lines are shown in Fig. 12.

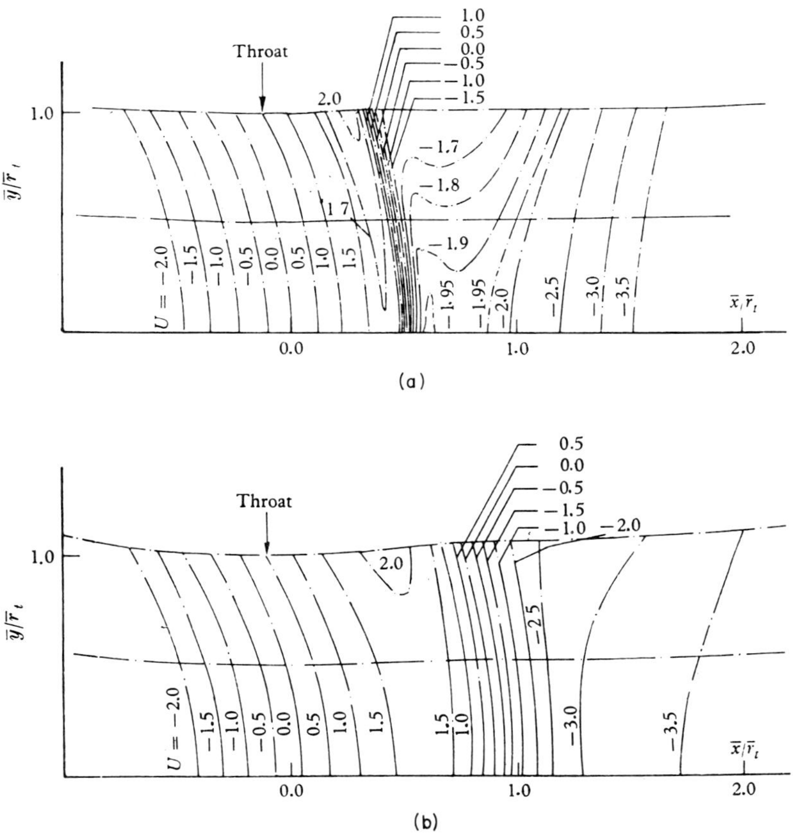

FIG. 12. Streamlines and isotachs for axisymmetric V-T nozzle flow (from Sichel and Yin, 1967, with permission of Cambridge University Press).

B. Source and Source-Vortex Flows

Exact solutions of the equations of two-dimensional inviscid compressible flow for source and source-vortex or spiral flow, which are closely related to the nozzle solutions, contain limiting circles at or near the sonic velocity where the acceleration becomes infinite (Taylor, 1930; von Mises, 1958; Oswatitsch, 1956). Solutions do not exist inside these limiting circles, and the inviscid theory is clearly inapplicable when the velocity gradients become very large. Viscous source flow using the full Navier–Stokes equations was therefore investigated by Wu (1955), Sakurai (1958), and Levey (1954, 1959). Wu and Sakurai were able to find closed form source solutions valid in the region of transonic flow, and it has been possible to show that these solutions are also a similarity solution of the V-T equation (Sichel and Yin, 1969a). With the transformation

$$u^{(1)} = 2f(S), \qquad S = x + \lambda\tilde{y}, \tag{4.21}$$

which was first introduced by Tomotika and Tamada (1950), the two-dimensional V-T equation in the form of (4.12) becomes

$$\chi_{\mathrm{V}} f''' - (ff')' + \lambda^2 f'' = 0. \tag{4.22}$$

The variable S is equal to the dimensionless radial distance of any point from the sonic circle, i.e.,

$$S = (\bar{r} - \bar{r}^*)/\bar{L}. \tag{4.23}$$

The arbitrary parameter λ determines the circulation and is zero for source flow, and x and $\tilde{y} = (\gamma + 1)^{1/2} y$ can be shown to be equivalent to the potential ϕ and the stream function ψ, respectively (Sichel and Yin, 1969a).

Choosing $\chi_{\mathrm{V}} = 1$ is, as in the nozzle problem, equivalent to using the shock thickness $\bar{\eta}$ as a characteristic lenth $\bar{L}$ for the region $\bar{r} - \bar{r}^*$. Equation (4.22) can be integrated twice and, with $\chi_{\mathrm{V}} = 1$, yields

$$f' - f^2 + \lambda^2 f = -C_1 S + C_2\,. \tag{4.24}$$

The integration constant C_1 can be shown to be

$$C_1 = 4\bar{\eta}/\bar{r}^*\epsilon^2, \tag{4.25}$$

and so is inversely proportional to the sonic radius which in turn depends upon the source strength. By considering the Navier-Stokes equations in cylindrical coordinates (Sichel and Yin, 1969a) it can be shown that the V-T equation is valid only as long as $\bar{r}^* \sim O(\bar{\eta}/\epsilon^2)$ or $C_1 \sim O(1)$. With $\bar{r}^* \ll \bar{\eta}/\epsilon^2$ the balance between convection and dissipation is no longer the mechanism determining the flow while with $\bar{r}^* \gg \bar{\eta}/\epsilon^2$, $C_1 \ll 1$ and the solution of (4.24) reduces to the structure of a weak normal or oblique shock. The integration constant C_2 in (4.24) determines the origin of the coordinate S and has been chosen zero for convenience.

The condition $\bar{r}^* \sim O(\bar{\eta}/\epsilon^2)$ implies that $\epsilon \sim O(\mathbf{R}^{-1/3})$, with $\mathbf{R} = \bar{\rho}^*\bar{a}^*\bar{r}^*/\bar{\mu}^{*\prime\prime}$, and that $\bar{\eta}/\bar{r}^* \sim O(\mathbf{R}^{-2/3})$. Thus the expansion and stretching used here can also be written in the form

$$\begin{aligned} \bar{u}/\bar{a}^* &= 1 + \mathbf{R}^{-1/3}u^{(1)} + \cdots, \\ \bar{r}/\bar{r}^* &= 1 + \mathbf{R}^{-2/3}S, \end{aligned} \tag{4.26}$$

and (4.26) is identical to the expansion and stretching used by Wu (1955) and Sakurai (1958).

Without the viscous term f', the solution of (4.24) becomes

$$S + (\lambda^4/4C_1) = (1/C_1)[f - (\lambda^2/2)]^2, \tag{4.27}$$

with a limit circle at $S = -(\lambda^4/4C_1)$ where $df/ds \to \infty$. In the transonic regime (4.27) is identical to Taylor's solution (1930) for compressible source-vortex flow and has an accelerating or supersonic branch and a decelerating or subsonic branch. The inviscid solution provides no mechanism for transition from one branch to the other.

The solution of the viscous equation (4.24) can be found in closed form in terms of Airy functions by first transforming (4.24) to a second order linear equation, and is

$$\begin{aligned} \hat{f} &= -C_1^{1/3}\frac{KA_i'(C_1^{1/3}\xi) + B_i'(C_1^{1/3}\xi)}{KA_i(C_1^{1/3}\xi) + B_i(C_1^{1/3}\xi)}, \\ \hat{f} &= f - \tfrac{1}{2}\lambda^2, \qquad \xi = S + (\lambda^4/4C_1). \end{aligned} \tag{4.28}$$

A_i and B_i are Airy functions of the first and second kind while the prime denotes differentiation. The solution (4.28) is universal in that it is independent of the parameter λ and is shown in Fig. 13 for various values of the constant of integration K. The inviscid solution in the form $\hat{f} = \pm C_1{}^{1/2}\xi^{1/2}$ is also displayed. For large values of K the solution f first approaches the supersonic branch of the inviscid solution, then after passing through a shocklike compression f approaches the subsonic branch of the inviscid solution, and it can be shown from the property of Airy functions that as $\xi \to \infty$, $\hat{f} \to -C_1^{1/2}\xi^{1/2}$. It is clear that the inclusion of the viscous term in the equations for radial flow has

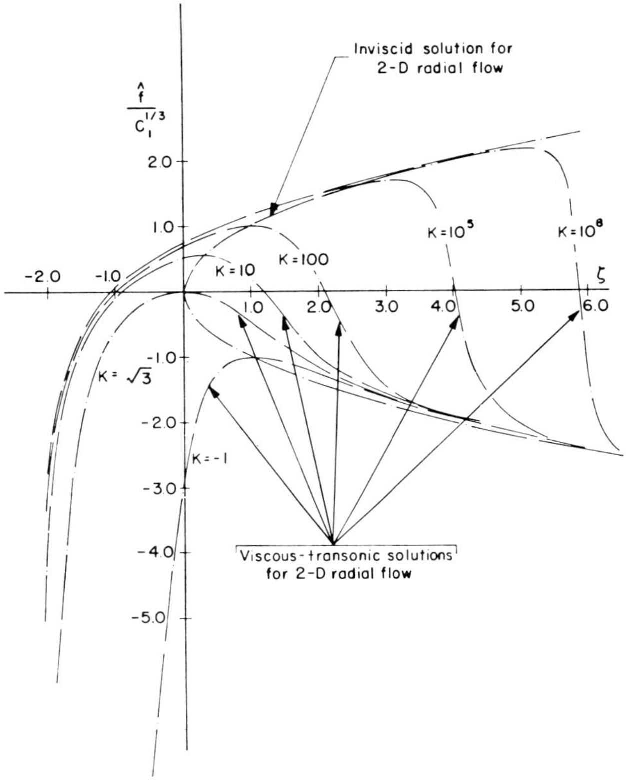

FIG. 13. Radial velocity for V-T spiral and radial flow (from Sichel and Yin, 1969a).

eliminated the singular behavior near the sonic radius. However the solution still diverges at some point inside the sonic circle where the denominator vanishes. With viscosity, a mechanism is now provided for transition from the supersonic to the subsonic branch of the inviscid solution. Perhaps the most striking result is that near the sonic radius the viscous solution differs drastically from the behavior of the inviscid solution.

Compressible radial flow in the presence of a gravitational field behaves like the flow in a converging–diverging nozzle in that the flow can accelerate from subsonic to supersonic values and vice versa, and such flows have been used as a model for the solar and stellar wind and for the accretion of interstellar gas by a star (Axford and Newman, 1967; Parker, 1965). As in the case of nozzle flows the inviscid theory fails to provide a smooth transition from flows subsonic throughout to accelerating or decelerating subsonic–supersonic flows. Consequently, Axford and Newman (1967) studied viscous transonic solutions for such radial flows with a gravitational field. In the case of spherically symmetric flow the appropirate expansion is such that $\epsilon \sim O(\mathbf{R}^{-1/2})$, which implies that $\bar{r}^* \sim O(\bar{\eta}/\epsilon)$. Upon introducing the stretched variable ξ, defined by $\bar{r} = \bar{r}^*(1 + \epsilon\xi)$, the conservation equations are finally reduced to the following Riccati equation for the radial velocity u:

$$\pm 2\left(\frac{4}{3} + \frac{\gamma - 1}{Pr}\right)\frac{du}{d\xi} = (\gamma + 1)\, u^2 + 4(\gamma - 1)\, u\xi - 2(3 - 2\gamma)\, \xi^2 + C. \quad (4.29)$$

The positive sign is for outflow, the negative for inflow, while C is a constant of integration. Equation (4.29) is very similar to (4.9), considered by Kopystyński and Szaniawski, and the solutions can be expressed in terms of parabolic cylinder functions. While some of the solutions of (4.29) diverge, there is also a family of solutions which provide a smooth shocklike transition from the accelerating to the decelerating branch of the inviscid solution just as in the nozzle problem.

V. External Flows

A. General Considerations

Determination of the jump conditions across a non-Rankine–Hugoniot shock wave terminating a region of supersonic flow is one of the key problems in the study of external V-T flow (Pearcey, 1964). While this important problem remains unsolved some progress in the study of V-T flow past bodies has been possible.

Some simple results can be obtained directly from the V-T equation (3.20). The oblique-shock similarity solution as presented in (3.37) depends on an obliquity parameter α, which has been taken as constant. In a curved shock, however, α will vary with y so that $\alpha = \alpha(y)$. Unless, however, $(d\alpha/dy) \sim O(\epsilon)$ or less, the solution (3.37) will fail to satisfy the V-T equation. Since $(d\alpha/dy)$ can be related to the shock radius of curvature, the above conditions imply that $\bar{\eta}/\bar{R}_s \sim O(\epsilon^2)$ or less, i.e., the ratio of shock thickness $\bar{\eta}$ to shock radius of curvature $\bar{R}_s$ must be of second order in ϵ for the conventional oblique-shock conditions to remain valid (Sichel, 1962).

Formal integration of the two dimensional V-T equation (3.20) with respect to x and with y held constant yields the result

$$\chi_V u^{(1)}{}_x + \chi_\infty u^{(1)} - \frac{1}{2} u^{(1)2} = -\int v^{(1)}{}_y \, dx + K, \tag{5.1}$$

where K is a constant of integration. For constant y the V-T equation thus behaves like a Riccati equation with the right-hand side dependent on the behavior of $v^{(1)}$. This result is significant since, as already shown in Section IV above, many solutions of the V-T equation reduce to a solution of a Riccati equation with different functions of the independent variable on the right-hand side. In the simplest case of the weak normal shock the right-hand side of (5.1) is constant. In the case of a uniform upstream flow, with $\epsilon = M_1{}^* - 1$, and χ_∞ chosen zero, (5.1) yields the expression

$$u^{(1)}(+\infty, y) = -\left[1 + 2\int_{-\infty}^{\infty} v^{(1)}{}_y \, dx\right]^{1/2} \tag{5.2}$$

for $u^{(1)}$ downstream of the shock, provided $u^{(1)}{}_x(+\infty, y) = 0$. In a normal shock $v^{(1)} = 0$ and then Eq. (5.2) yields the Rankine–Hugoniot result $u^{(1)}(+\infty, y) = -1$. In an oblique shock with the free stream in the $+x$ direction, $v_y^{(1)} < 0$ within the shock so that $u^{(1)}(+\infty, y) > -1$, as is actually the case. In the shock terminating a supersonic region it appears reasonable that $v^{(1)} > 0$ because of the wall boundary layer. Then if $v^{(1)}$ decays with increasing y, $v_y^{(1)} < 0$ and Eq. (5.2) predicts $u^{(1)}(\infty, y) > -1$, so that the shock pressure rise is less than the R–H value in agreement with the numerical results of Emmons (1946) and with experiment (Sinnott, 1960). In the shock tube $v_y^{(1)} > 0$ near the shock-induced boundary layer so that (5.2) implies $u^{(1)}(+\infty, y) < -1$, in agreement with the approximate calculations of Sichel (1962).

B. Influence of the Viscosity in the Far Field

A desirable objective of any preliminary study of the V-T equations would be to examine the effect of viscosity upon a known inviscid solution. In two-dimensional or axisymmetric flow past finite or semi-infinite bodies with sonic velocity at infinity, Frankl' (1947), Guderley and Yosihara (1951), and Fal'kovitch and Chernov (1966) have found that the asymptotic behavior of flow far from the body is described by self-similar solutions of the inviscid potential equation of the form

$$\Phi = \tilde{y}^{3n-2} F(\xi), \qquad \xi = x/\tilde{y}^n. \tag{5.3}$$

Investigation of the singularities of the ordinary differential equation for $F(\xi)$ shows that the solution with $n = 4/5$ represents the asymptotic flow past a finite two-dimensional profile, while $n = 4/7$ represents flow past finite bodies of revolution.

To examine the effect of viscosity upon these inviscid solutions Ryzhov and Shefter (1964) studied self-similar solutions of the V-T equation (3.20) of the same form as (5.3). For fixed ξ, and with $\chi_v = 1$, it then follows that with increasing y the terms of the V-T equation vary as

$$\phi_x \phi_{xx}, \qquad \phi_{\tilde{y}\tilde{y}} \sim O(\tilde{y}^{3n-4}), \qquad \phi_{xxx} \sim O(\tilde{y}^{-2}). \tag{5.4}$$

Thus when $n > 2/3$ the dissipative term ϕ_{xxx} decreases faster than the other terms of the V-T equation as $y \to \infty$, while for $n < 2/3$ the dissipative term will be comparable or greater than the other terms in the V-T equation with increasing $\tilde{y}$. Since $4/7 < 2/3 < 4/5$, Ryzhov and Shefter concluded that the two-dimensional inviscid solutions with $n = 4/5$ will not be affected by viscosity for large y but that viscosity may have important effects on the asymptotic solution for axysymmetric flow with $n = 4/7$.

A consideration of a Reynolds number defined by

$$\mathbf{R} = \bar{\rho}^* \bar{a}^* \bar{h} / \bar{\mu}^{*\prime\prime}, \tag{5.5}$$

with $\bar{h}$ denoting the horizontal distance between two parabolas $\xi = \text{const.}$, provides another explanation for these results. Since $u^{(1)} \sim O(\tilde{y}^{2n-2})$, $\bar{h} \sim O(\tilde{y}^n)$ it follows that $\mathbf{R} \sim O(\tilde{y}^{3n-2})$ as $\tilde{y} \to \infty$. Thus for $n < 2/3$, $\mathbf{R}$ decreases so that viscous effects will be important throughout the flow.

The freedom in the choice of n disappears in the V-T case, a similarity solution of the V-T equation (3.20) with $\chi_\infty = 0, \chi_v = 1$,being possible only for $n = 2/3$. Then (5.3) implies that

$$u^{(1)} = \tilde{y}^{-2/3} f(\xi), \qquad v^{(1)} = \tilde{y}^{-1} g(\xi), \tag{5.6}$$

and $f(\xi)$ satisfies the ordinary differential equation

$$f'' + [(4/9)\xi^2 - f] f' + (4/9)\, \xi f + (2/3)(k-1)(c - f\xi) = 0, \tag{5.7}$$

while

$$g = (2/3)(c - f\xi), \tag{5.8}$$

with c a constant of integration. Numerical solutions of (5.7) were compared to the inviscid similarity solution with $n = 2/3$. The $n = 2/3$

solutions correspond to the asymptotic behavior of sonic flow past the semi-infinite body

$$\tilde{y} = ((8/3)cx)^{1/2}. \tag{5.9}$$

The variation of f and g computed by Ryzhov and Shefter (1964) from (5.7) and (5.8) for $c = 1.2$, 0.0149 are compared to the corresponding inviscid solutions in Fig. 14a and b. For the smaller value of c, corresponding to the smaller of the two Reynolds numbers, the difference between the viscous and inviscid solutions is appreciable.

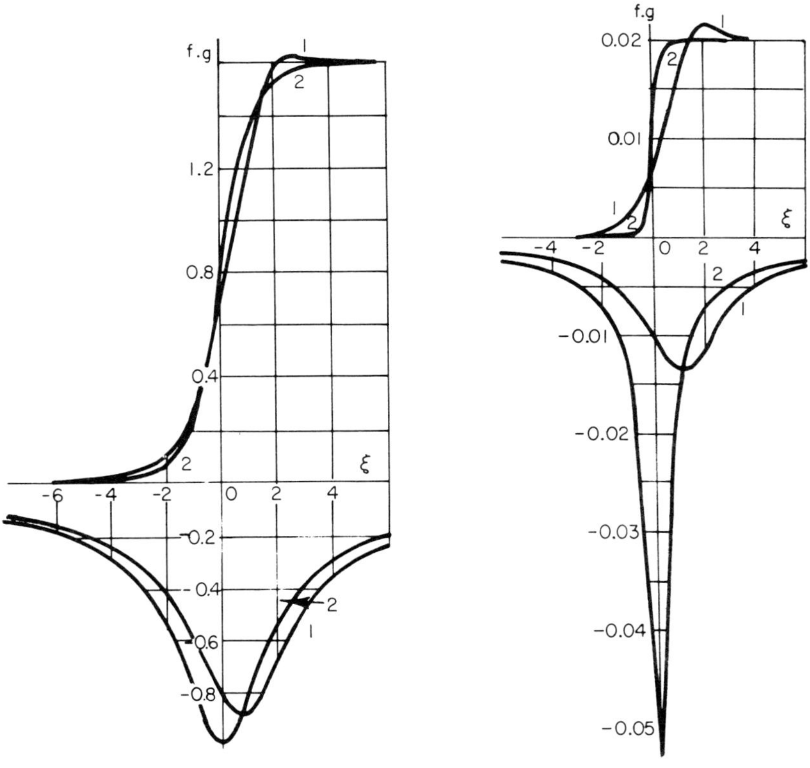

FIG. 14. Comparison of the viscous (curve 1) and inviscid (curve 2) solutions for the flow past a semi-infinite body: (a) $C = 1.2$; (b) $C = 0.0149$ (from Ryzhov and Shefter, 1964).

Another approach to the study of the asymptotic flow past axisymmetric bodies in a sonic stream (Ryzhov, 1965) is to assume that the perturbations die out sufficiently rapidly so that for large y the stretch factor λ and expansion parameter ϵ are given by

$$\lambda^2 \sim O(\mathbf{R}^{-1}), \qquad \epsilon \ll \mathbf{R}^{-1}, \tag{5.10}$$

where $\mathbf{R} = \bar{U}\bar{L}\bar{\rho}_\infty/\bar{\mu}''$. Then, as already discussed in Section III, the resultant equations will be linear. By carrying the expansion to second order, Diesperov and Ryzhov (1967) showed that subject to (5.10) the appropriate expansion scheme to use is that given in (3.30). The equations satisfied by $u^{(1)}$, $u^{(2)}$, $v^{(1)}$, and $v^{(2)}$ are then, with

$$\lambda^2 = \mathbf{R}^{-1}[1 + (\gamma - 1)/Pr''],$$

given by

$$u^{(1)}{}_y = v^{(1)}{}_x\,, \qquad u^{(1)}{}_{xx} + v^{(1)}{}_y + (1/y)\, v^{(1)} = 0, \tag{5.11}$$

and with the appropriate choice of δ (Diesperov and Ryzhov, 1967) the second order equations become

$$u^{(2)}{}_y = v^{(2)}{}_x\,,$$

$$u^{(2)}{}_{xx} + v^{(2)}{}_y + (1/y)\, v^{(2)} = u^{(1)}u^{(1)}{}_x + u^{(1)}{}_{xxx}\,. \tag{5.12}$$

The validity of the assumption (5.10) is a key question. The Reynolds number $\mathbf{R}$ based on $\bar{h}$ may be more appropriate to use in (5.10) than $\mathbf{R}$ based on $\bar{L}$. Then if $\mathbf{R}$ decreases with increasing y, (5.10) may well be a valid assumption.

Similarity solutions of both the first order and second order equations (5.11) and (5.12) representing source and doublet flows have been found by Ryzhov (1965) and by Diesperov and Ryzhov (1967). Introduction of

$$u^{(1)} = y^{-\tilde{n}} f_1(\xi), \qquad v^{(1)} = y^{-(3\tilde{n}+1)/3} g_1(\xi), \qquad \xi = xy^{-2/3}, \tag{5.13}$$

in Eq. (5.11) leads to the equations

$$f_1''' + \tfrac{4}{9}\xi^2 f_1'' + \tfrac{2}{3}(\tilde{n} + \tfrac{1}{3})\,\xi f_1' + \tilde{n}^2 f_1 = 0,$$

$$g_1 = [3/(3\tilde{n} - 2)](f_1'' + \tfrac{4}{3}\xi^2 f_1' + \tfrac{1}{3}\tilde{n}\xi f_1), \tag{5.14}$$

where $\tilde{n}$ plays the role of an eigenvalue. It should be noted that the similarity transformation (5.13) differs from (5.3) used by Guderley and Yosihara (1951) and Fal'kovitch and Chernov (1966). From (5.13) it follows that for fixed ξ

$$[u^{(1)}{}_{xx}\,, v^{(1)}{}_y\,, (1/y)\, v^{(1)}] \sim O[y^{-(\tilde{n}+4/3)}],$$

while

$$u^{(1)}u^{(1)}{}_x \sim O[y^{-(2\tilde{n}+2/3)}], \tag{5.15}$$

so that for $\tilde{n} > 2/3$ the nonlinear term vanishes faster than the other terms of the V-T equation consistent with (5.10). Comparing $u^{(1)}$ from (5.3) and (5.13) shows that the exponents n and $\tilde{n}$ are related by $\tilde{n} = -2(n-1)$ and are identical for $n = \tilde{n} = 2/3$.

The solution of (5.11) corresponding to source flow is that one for which $v^{(1)}(x, y)$ vanishes along the y axis everywhere except at the origin, and Ryzhov (1965) shows that this is the solution corresponding to $\tilde{n} = 4/3$, with $f_1(\xi)$ given by

$$\begin{aligned} f_1 &= c_1[F(2/3, 1/3; \eta) - (\Gamma^2(1/3)/2\Gamma^2(2/3))\, \eta^{2/3}F(4/3, 5/3; \eta)] \\ &= (\Gamma(1/3)/3\Gamma(2/3))\, c_1\Psi(2/3, 1/3, \eta), \\ \eta &= -(4/27)\xi^3. \end{aligned} \tag{5.16}$$

Γ is Euler's Gamma function, F is the confluent hypergeometric function, c_1 is proportional to the source strength, while Ψ is a function introduced by Tricomi *et al.* (1954). The function $g_1(\xi)$ is given by

$$\begin{aligned} g_1 &= 2(2)^{1/3}c_1[\eta^{1/3}F(5/3, 4/3; \eta) - (\Gamma^2(1/3)/6\Gamma^2(2/3))\, F(4/3, 2/3; \eta)] \\ &= -(2(2)^{1/3}\Gamma(1/3)/9\Gamma(2/3))\, c_1\Psi(4/3, 2/3; \eta). \end{aligned} \tag{5.17}$$

The source solution (or fundamental solution) of (5.11) can now be used to solve the boundary value problem

$$yv^{(1)} \to \begin{cases} 0 & \text{when } y \to 0, \quad x < 0, \\ R(x) & \text{when } y \to 0, \quad x > 0, \end{cases}$$

with the result that

$$\begin{aligned} u^{(1)} &= -\frac{1}{2^{2/3}3} D \int_0^\infty \Psi(2/3, 1/3; \eta_1) \frac{R(x_1)}{y^{4/3}}\, dx_1, \\ v^{(1)} &= 2^{2/3}D \int_0^\infty \Psi(4/3, 2/3; \eta_1) \frac{R(x_1)}{y^{5/3}}\, dx_1, \end{aligned} \tag{5.18}$$

with $\eta_1 = -\frac{4}{27}(x - x_1)^3/y^2$. D is a normalization constant given by

$$D = \left[\int_{-\infty}^{\infty} \eta_1^{-2/3}\Psi(4/3, 2/3; \eta_1)\, d\eta_1\right]^{-1}, \tag{5.19}$$

and the constant c_1 is now related to $R(x)$ by

$$c_1 = -\frac{9\Gamma(2/3)}{2^{2/3}\Gamma(1/3)} D \int_0^l R(x_1)\, dx_1 , \tag{5.20}$$

where it is assumed that $R(x) \neq 0$ in $0 \leqslant x \leqslant l$ but is zero everywhere else. The significance of the source solution (5.16) and (5.17) is now apparent for, provided the integral in (5.20) is finite, the flow due to the body described by $R(x)$ behaves like the source solution for large values of x_1 and y. For a closed body the integral in (5.20) vanishes so that $c_1 = 0$. Then the first contribution to the asymptotic flow comes from the doublet solution of (5.12). The doublet solution f_2 and g_2 corresponding to f_1 and g_1 can be shown to be (Ryzhov, 1969)

$$f_2 = df_1/d\xi, \qquad g_2 = dg_1/d\xi, \tag{5.21}$$

and in general the asymptotic behavior of $u^{(1)}$ and $v^{(1)}$ can be represented by the series

$$\begin{aligned} u^{(1)} &= \sum_{i=1}^{\infty} a_i f_i(\xi)\, y^{-2(i+1)/3}, \\ v^{(1)} &= \sum_{i=1}^{\infty} b_i g_i(\xi)\, y^{-(2i+3)/3}, \end{aligned} \tag{5.22}$$

where

$$f_n = d^{n-1}f_1/d\xi^{n-1}, \qquad g_n = d^{n-1}g_1/d\xi^{n-1}.$$

Equation (5.22) represents a summation of sources, doublets, quadrupoles, etc. The doublet solution is discussed in detail by Diesperov and Ryzhov (1967).

Comparison of the viscous source solution with the inviscid source solution of Guderley and Yosihara (1951) and Fal'kovitch and Chernov (1966) shows that

		Viscous	Inviscid
$\bar{h}$	$\sim$	$y^{2/3}$	$y^{4/7}$
$u^{(1)}$	$\sim$	$y^{-4/3}$	$y^{-6/7}$
$v^{(1)}$	$\sim$	$y^{-5/3}$	$y^{-9/7}$.

Viscosity thus smears out the asymptotic flow pattern over a larger horizontal distance and results in a more rapid attenuation of the disturbance with increasing y than in inviscid flow.

Ryzhov (1965) shows that the source solution (5.16) and (5.17) automatically takes account of the formation of a vortex trail behind the immersed body. In general, for flow past a semi-infinite body of revolution with contour

$$R(x) = \epsilon^{1/2}\bar{L}Ax^{\nu}, \qquad x \geqslant 0,$$

it can be shown that ν and n are related by

$$n = -(4/3)(\nu - 1). \tag{5.23}$$

With $\nu = 1/2$, $\tilde{n} = n = 2/3$ and then, according to Ryzhov and Shefter (1964) the nonlinear term $\phi_x\phi_{xx}$ and the term ϕ_{xxx} die out at the same rate with increasing y. In this special case Ryzhov (1965) suggests that the full V-T equation must be used. On the other hand for $1/2 < \nu \leqslant 1$, $n > 2/3$ and the viscous term decays faster than the others so that the inviscid equations can be used to compute the asymptotic behavior of the flow. Ryzhov (1965) suggests that the linearized equations are appropriate in the range $0 \leqslant \nu \leqslant (1/2)$.

The variation of the source functions f_1 and g_1 and of the doublet functions f_2 and g_2 with c_1 chosen to make $f_1 \sim 1/\xi^2 + \cdots$ with $\xi \to \infty$ is shown in Fig. 15a and b. The inviscid solutions for flow past finite bodies (Guderley and Yosihara, 1951; Fal'kovitch and Chernov, 1966) require the insertion of a shock discontinuity at soome appropriate

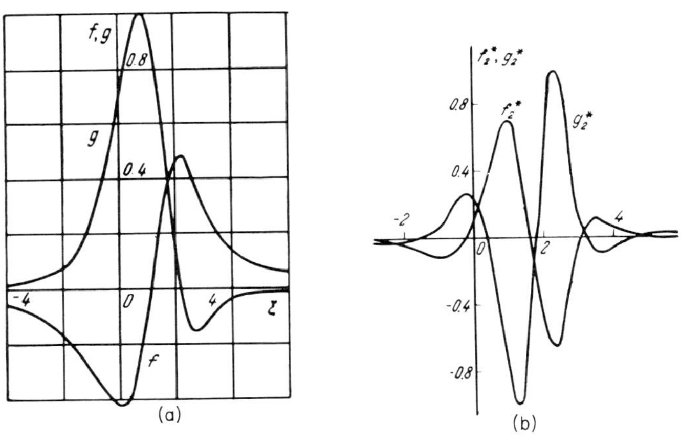

FIG. 15. (a) Variation of the source and (b) doublet function (from Ryzhov, 1965, and Diesperov and Ryzhov, 1967 with permission from *Prikl. Mat. Mekh.*, Pergamon Press).

$\xi = \xi_s$. The asymptotic behavior of a finite axisymmetric body for which $c_1 = 0$ is, according to Eq. (5.22), given by

$$u^{(1)} \sim y^{-2} f_2(\xi), \qquad v^{(1)} \sim y^{-7/3} g_2(\xi).$$

In Fig. 15b $f_2(\xi)$ first accelerates to supersonic velocity, decelerates rapidly to a subsonic value, and then re-accelerates. Thus the viscous dipole solution may well represent the behavior of a shock which has, at very large distances, decayed to some sort of viscous wave. Solutions with shock waves are discussed in Section V, C below. The source solution f_1 is dominated by a rapid expansion which has no resemblance to a shock wave.

The asymptotic behavior of viscous transonic flow has also been studied by Hubert (1968) using an expansion scheme originally applied by Euvrard (1967) to the inviscid problem. For two-dimensional flows Hubert (1968) uses the coordinate expansion

$$\Phi = x + y^{2/5} f^*(\zeta) + y^0 f^{**}(\zeta) + \cdots, \tag{5.24}$$

with

$$\zeta = x/(\gamma + 1)^{1/3} y^{4/5}.$$

Then $f^*(\zeta)$ is identical to the first order solution obtained by Euvrard (1967).

The mathematical behavior of the V-T potential equation is dominated by the terms ϕ_{xxx} and ϕ_{yy}. Therefore the two-dimensional linearized V-T equation

$$\phi_{xxx} + \phi_{yy} = 0, \tag{5.25}$$

as well as the axisymmetric equation (5.11) or (3.29), has been investigated, with the thought that the linearized solutions might provide the initial step in an iterative solution of the full nonlinear equation. Partial differential equations of the type (5.25) have been studied in general by Block (1912) and by Dezin (1958, 1959). The solution of the axisymmetric equation has already been discussed. Source-type solutions of (5.25) have been investigated in detail by Sichel (1961) while solutions of (5.25) on a finite interval using separation of variables are considered by Sichel, Yin, and David (1968). The source solution of (5.25) is given by

$$\begin{aligned} \Phi &= (y^{1/3}/\pi) f(\xi), \qquad \Phi_x = (y^{-1/3}/\pi) f'(\xi), \\ \Phi_y &= (y^{-2/3}/\pi)\, g(\xi), \qquad \xi = x y^{-2/3}, \end{aligned} \tag{5.26}$$

where f and g satisfy the differential equations

$$g'' + \tfrac{4}{9}\xi^2 g' + \tfrac{10}{9}\xi g = 0,$$
$$f'' + \tfrac{4}{9}\xi^2 f' - \tfrac{2}{9}\xi f = 0, \tag{5.27}$$

which are closely related to (5.14) obtained by Ryzhov. Using the asymptotic behavior of the source solution (5.26) it can be shown that the boundary value problem

$$\Phi_y(x, 0) = v^{(1)}(x, 0) = G(x) \tag{5.28}$$

has the following solution for $u^{(1)}(x, 0)$:

$$u^{(1)}(x, 0) = -\frac{1}{(\pi)^{1/2}} \int_x^\infty \frac{G(s)}{(s - x)^{1/2}}\, ds.$$

Thus $u^{(1)}(x, 0)$ depends only on the behavior of $v^{(1)}(x, 0) = G(x)$ downstream of the point x, and this strange "upstream wake" behavior as well as (5.25) are also encountered in some magnetohydrodynamic problems (Ludford, 1961). It is not clear whether such upstream influence will also arise in the nonlinear problem, although as indicated in Section II above in the case of two-dimensional shock structure the flow downstream of the shock will affect the upstream flow.

Using Fourier integrals Rae (1960) has carried out an extensive investigation of the linearized V-T equation for axisymmetric and two-dimensional flow and for subsonic, sonic, and supersonic free-stream flows. The linearized equations are derived by a straightforward linearization of the Navier–Stokes equation without any coordinate stretching and it is generally assumed that $\chi_V \ll 1$. In the subsonic case the solution consists of the inviscid linearized solution plus a viscous correction. In the sonic and supersonic cases the solutions are more complex but again show a more rapid decay than the corresponding inviscid solutions. For supersonic flow the viscous solutions show how the Mach waves at the leading and trailing edges of the body are smeared out and decay with increasing y. The sonic and supersonic solutions diverge in the limit $\chi_V \to 0$. Rae's linear theory makes no provision for the existence of shock waves. Axisymmetric V-T flows with $M_\infty > 1$ have been discussed by Ryzhov (1969) who showed that in characteristic coordinates the equations reduce to a form of Burger's equations (discussed by Lighthill, 1956). Viscous-transonic flow past lifting bodies, using the linearized equations, has been investigated by Ryzhov and Terent'ev (1967).

C. Solution for Flow with Shocks

As already mentioned above the Frankl-Guderley inviscid similarity solution (5.3) for a finite body in a sonic stream must have at least one discontinuity at some value $\xi = \xi_s$ corresponding to a compression shock (Guderley, 1962). A solution describing the structure of a weak shock wave in such a flow was developed by Szaniawski (1968), using the method of matched asymptotic expansions. A key question was whether or not viscosity influences the flow outside the shock structure in the limit as $\chi_V \to 0$ (or $\mathbf{R} \to \infty$). Unlike the linearized analyses described above the existence of a shock wave is assumed at the very outset.

The analysis starts with the V-T equation (3.10) with $M_\infty = 1.0$. Introduction of the new variables

$$
\begin{gathered}
u(x, \tilde{y}) = \tilde{u}(\xi, \eta), \qquad v(x, \tilde{y}) = (\gamma + 1)^{1/2}\tilde{v}(\xi, \eta), \\
\xi = x\tilde{y}^{-n}, \qquad \eta = \ln|\tilde{y}|, \qquad \text{and} \qquad \tilde{y} = (\gamma + 1)^{1/2}y
\end{gathered}
\tag{5.29}
$$

transforms the V-T equation to

$$
\begin{gathered}
\tilde{u}\tilde{u}_\xi + (n\xi\tilde{v}_\xi - \tilde{v}_\eta - k\tilde{v})\, e^{(n-1)\eta} - \chi_V\, e^{-n\eta}\tilde{u}_{\xi\xi} = 0, \\
\tilde{v}_\xi + (n\xi\tilde{u}_\xi - \tilde{u}_\eta)\, e^{(n-1)\eta} = 0.
\end{gathered}
\tag{5.30}
$$

Superscripts on $u^{(1)}$ and $v^{(1)}$ have been suppressed. The shock is presumed to lie on some surface $\xi = \xi_s$ and the flow is divided into two outer regions (—) and (+) upstream and downstream of ξ_s and an inner shock layer region in the neighborhood of ξ_s.

Substituting an outer expansion of the form

$$
\begin{aligned}
\tilde{u} &= \xi^2[f(\xi) + \chi_V\theta(\xi, \eta; \chi_V)]\, e^{2(n-1)\eta}, \\
\tilde{v} &= \xi^3[g(\xi) + \chi_V\omega(\xi, \eta; \chi_V)]\, e^{2(n-1)\eta},
\end{aligned}
\tag{5.31}
$$

in (5.30) leads to the following equations for $f(\xi)$ and $g(\xi)$:

$$
\begin{gathered}
\xi^3\{\xi(ff' + ng') + [2f^2 + (3 - k)g]\} = \chi_V(\xi^2 f)''\, e^{-(3n-2)\eta} \equiv A, \\
\xi(nf' + g') + (2f + 3g) = 0.
\end{gathered}
\tag{5.32}
$$

Introduction of an inner expansion

$$
\begin{aligned}
\tilde{u} &= \xi_s^2[F(\bar{\xi}) + \chi_V\Theta(\bar{\xi}, \eta; \chi_V)]\, e^{2(n-1)\eta}, \\
\tilde{v} &= \xi_s^3[G(\bar{\xi}) + \chi_V\Omega(\bar{\xi}, \eta; \chi_V)]\, e^{3(n-1)\eta}, \\
\bar{\xi} &= [(\xi - \xi_s)/\chi_V]\, e^{(3n-2)\eta},
\end{aligned}
\tag{5.33}
$$

in (5.30) leads to the following "inner" equations for $F(\xi)$ and $G(\xi)$:

$$\xi_s^{-2}F'' - (FF' + nG')$$
$$= (\chi_v/\xi_s)\{n\tilde{\xi}G'\, e^{-(3n-2)\eta} - [3(n-2)(\tilde{\xi}G)' + (k-1)G]\, e^{-(4n-3)\eta}\} = B,$$
$$G' + nF' = 2(\chi_v/\xi_s)(n-1)(\tilde{\xi}F)'\, e^{-(3n-2)\eta} = C. \tag{5.34}$$

In both (5.32) and (5.33) the higher order terms involving θ, ω, Θ, and Ω have been neglected. If now it is assumed that $f(\xi)$, $f'(\xi)$, $f''(\xi)$, $F(\xi)$, $F''(\xi)$, etc., remain $O(1)$ as $\chi_v \to 0$, then in this limit it becomes possible to neglect the terms A, B, and C on the right side of (5.32) and (5.33).

In this limit Eqs. (5.32) become identical to those which arise in the inviscid solution of Guderley and Yosihara (1951), Frankl' (1947), and Fal'kovitch and Chernov (1966), and the solution of (5.32) can be expressed in the parametric form (Szaniawski, 1967)

$$f = (1 - \tfrac{1}{3}k)\,\sigma(\sigma-2)/(\sigma-1)^2, \qquad g = -\tfrac{2}{3}(1 - \tfrac{1}{3}k)\,\sigma^2(\sigma-3)/(\sigma-1)^3,$$
$$\xi/\xi_s = [(\sigma-1)/(\sigma^{\pm}-1)](\sigma^{\pm}/\sigma)^{n/2}. \tag{5.35}$$

The values of the parameter σ upstream (σ_-) and downstream (σ_+) of the shock are given by

$$\sigma_{\mp} = [(2 - \tfrac{1}{2}n) \pm \sqrt{3}]/(1 - \tfrac{1}{2}n), \qquad 0 < \sigma < \sigma_-, \qquad \sigma_+ < \sigma < 0. \tag{5.36}$$

As discussed in Section V, B above $n = 4/5$ corresponds to plane and $n = 4/7$ to axisymmetric flow. The inner equations (5.34) with B and C neglected have the solution

$$F = n^2 - c_1 \tanh[c_1 \tfrac{1}{2}(\tilde{\xi} + c_0)\xi_s^2],$$
$$G = c_1 \tanh[c_1^{\frac{1}{2}}(\tilde{\xi} + c_0)\xi_s^2] + c_2\,, \tag{5.37}$$

and comparison with (3.37) shows that Eq. (5.37) describes the structure of an oblique shock wave. The constants c_0, c_1, and c_2 can be determined by matching with the outer or inviscid solution. After appropriate matching the viscous and inviscid solutions can be combined to form the composite solution

$$u = u_i + u_d\,, \qquad v = v_i + v_d\,, \tag{5.38}$$

where u_i and v_i are the inviscid and u_d and v_d the viscous contributions to the velocity. u_i and v_i are readily determined from (5.35) and are

$$u_i = (\xi_s y^{n-1})^2 \left[(1 - \tfrac{1}{3}k) \frac{\sigma_\mp}{(1 - \sigma_\mp)^2} \left(\frac{\sigma}{\sigma_\mp}\right)^{1-n} (\sigma - 2)\right],$$

$$v_i = (\gamma + 1)^{1/2}(\xi_s y^{n-1})^3 \left[\tfrac{2}{3}(1 - \tfrac{1}{3}k) \frac{\sigma_\mp{}^2}{(1 - \sigma_\mp)^3} \left(\frac{\sigma}{\sigma_\mp}\right)^{2-(3/2)n} (\sigma - 3)\right], \tag{5.39}$$

$$x = \tilde{y}^n \xi_s \frac{(\sigma - 1)}{\sigma_\mp - 1} \left(\frac{\sigma_\mp}{\sigma}\right)^{n/2},$$

and u_i and v_i are discontinuous at the shock wave $x_s = \xi_s y^n$. The velocities u_d and v_d from the inner solution are

$$u_d = (\xi_s y^{n-1})^2 E \left[\mp 1 - \tanh \frac{\tilde{y}^{3n-2}\xi_s{}^3}{\chi_v} \left(\frac{x}{x_s} - 1\right)\right],$$

$$v_d = (\gamma + 1)^{1/2}(\xi_s \tilde{y}^{n-1})^2\, nE \left[\pm 1 + \tanh \frac{\tilde{y}^{3n-2}\xi_s{}^3}{\chi_v} \left(\frac{x}{x_s} - 1\right)\right], \tag{5.40}$$

where

$$E = (1 - \tfrac{1}{3}k)(1 - \tfrac{1}{2}n)^2(3)^{1/2}/2.$$

The discontinuities in (5.39) and (5.40) cancel in (5.38) so that the composite solution is smooth. The composite solution (5.38) is shown in Fig. 16 in terms of the normalized velocities

$$U = u(\xi_s \tilde{y}^{n-1})^{-2}, \qquad V = v(\xi_s \tilde{y}^{n-1})^{-2}(\gamma + 1)^{-1/2},$$

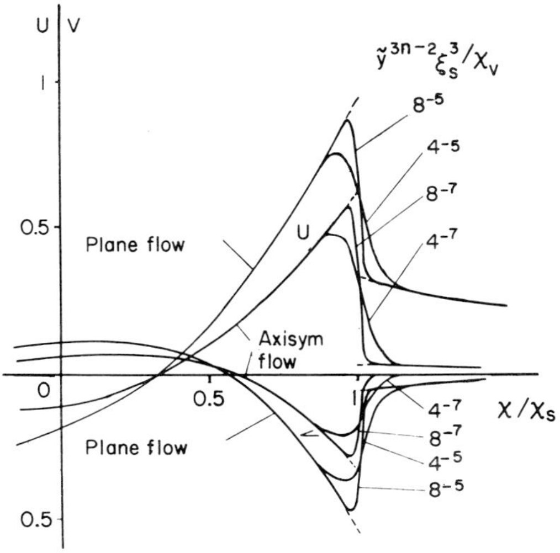

FIG. 16. Solutions for sonic flow with shock waves (from Szaniawski, 1968, with permission of *Acta Mechanica*, Springer-Verlag, New York).

for several values of $\tilde{y}^{3n-2}\xi_s{}^3\chi_V^{-1}$, for both plane flow ($n = 4/5$), and axisymmetric flow ($n = 4/7$).

Several features of the solution in Fig. 16 are worthy of note. The shock layer becomes thinner with decreasing χ_V (or increasing **R**). The pressure or velocity change across the shock is less than would occur if the shock were replaced by a Rankine–Hugoniot discontinuity connecting the upstream and donwstream branches of the inviscid solution at $x/x_s = 1$. Because of the factor $(1/x_s) \sim O(\tilde{y}^{-n})$ in (5.40), the shock thickness $\Delta \sim O(\tilde{y}^{2(1-n)}\chi_V^{-1})$, so that the shock becomes smeared out with increasing y. The thickness increases more rapidly in the axisymmetric case with $n = 4/7$. It is interesting to compare Fig. 16 with Fig. 15b which shows the doublet solutions of Diesperov and Ryzhov (1967). Here $f_2 \sim u$ and $g_2 \sim v$ and there is certainly some similarity between the two solutions, at least up to the rapid shocklike decrease in u or f_2. While the velocity is still supersonic downstream of the oblique shock in Szaniawski's solution (Fig. 16) the velocity drops to subsonic values in the solution of Diesperov and Ryzhov. However, the Diesperov–Ryzhov solution represents the flow at transverse distances y so large that the shock structure has become completely merged with the rest of the flow.

The entire analysis described above hinges on the initial assumption that A, B, and C in (5.32) and (5.34) can be neglected in the limit $\chi_V \to 0$, which is equivalent to assuming that viscosity has no effect outside the shock structure. Evaluation of A, B, and C using the solutions (5.35) and (5.37) shows that A, B, and C vanish as $\chi_V \to 0$, $|y| \to \infty$, only for $n > 3/4$, so that the above assumption holds only in the two-dimensional case. Thus in agreement with Ryzhov and Shefter (1964), Szaniawski finds that viscosity may also be important outside the shock layer in axisymmetric flow. However, these neglected terms only become important for very large values of $|x|$ and $|y|$, and by consideration of A, B, and C Szaniawski finds that the solutions (5.35) and (5.37) remain valid for

$$
\begin{aligned}
|x| &\ll \xi_s{}^7\chi_V^{-2} \simeq \xi_s{}^7\mathbf{R}^2, \\
|y| &\ll (\xi_s{}^3/\chi_V)^{7/2} \simeq (\xi_s{}^3\mathbf{R})^{7/2}.
\end{aligned}
\tag{5.41}
$$

The region defined by (5.41) may be quite large for $\mathbf{R} \gg 1$, but even then the viscous terms may have an important influence on the asymptotic behavior of the solution far from the body. Szaniawski's solution is of course an asymptotic one valid only at large distances from the body, but provides no information whatsoever about the details of the flow in the immediate neighborhood of the body surface. In the two-dimensional case the requirement that A, B, and C be negligible imposes the additional restriction for sufficiently small ξ_s and χ_V/ξ_s^3 that

$$|x| \gg (\chi_V/\xi_s^3)^2\xi_s, \qquad |y| \gg (\chi_V/\xi_s^3)^{5/2}.$$

The shock waves which arise in the asymptotic solution described above have, as is evident from (5.40), an essentially one-dimensional structure. It is the finite thickness of the shock layer which accounts for the difference between the actual shock pressure rise and that which would occur if the shock were replaced by a discontinuity. A solution for the two-dimensional structure of a weak curved shock wave was derived by Szaniawski (1969) by generalizing the oblique-shock solution (3.37). In terms of a potential $\tilde{\Phi}(x, \tilde{y})$ defined by

$$u^{(1)} = \tilde{\Phi}_x, \qquad v^{(1)} = (\gamma + 1)^{1/2}\tilde{\Phi}_{\tilde{y}},$$

and with $\lambda = \epsilon^{1/2}$, the oblique shock solution can be expressed in the form

$$\tilde{\Phi} = \theta(x, \tilde{y}) - 2\chi_V \ln \cosh[\omega(x, \tilde{y})/2\chi_V]. \tag{5.42}$$

For the oblique shock θ and ω are linear functions of x and $\tilde{y}$ of the form

$$\theta = \theta_0 + \theta_{10}x + \theta_{01}\tilde{y}, \qquad \omega(x, \tilde{y}) = \omega_0 + \omega_{10}x + \omega_{01}\tilde{y}, \tag{5.43}$$

with the relation $\theta_{10}\omega_{10}^2 = \omega_{01}^2$ imposed by the Rankine–Hugoniot conditions. Szaniawski's generalization consists of permitting θ and ω to be arbitrary functions of x and $\tilde{y}$. $\tilde{\Phi}$ in the V-T equation can then be replaced by θ and ω as dependent variables. The center of the shock is defined by $\omega(x, \tilde{y}) = 0$, or the equivalent relation $x = f(\tilde{y})$.

With uniform flow upstream θ and ω are related by $\omega = x - \theta$, and, in the limit $\chi_V \to 0$, θ is shown to satisfy

$$\theta_{xx} = -[\theta_{\tilde{y}\tilde{y}} + (k/\tilde{y})\theta_{\tilde{y}}] + (\theta_x^2)_x, \tag{5.44}$$

with the condition

$$\theta = x \quad \text{and} \quad \theta_x = \theta_{\tilde{y}}^2 + (2 - \theta_x)\theta_x^2 \quad \text{for} \quad x = f(\tilde{y}), \tag{5.45}$$

on the shock surface. Szaniawski considered a curved shock symmetrical with respect to the x axis as shown in Fig. 17, and described by the series

$$f(\tilde{y}) = \frac{\alpha_1}{2!}\tilde{y}^2 + \frac{\alpha_2}{4!}\tilde{y}^{+4} + \cdots. \tag{5.46}$$

The function $\theta(x, \tilde{y})$ is expanded in a double series in x and $\tilde{y}$ and the coefficients are evaluated by substitution in (5.44) and (5.45). With

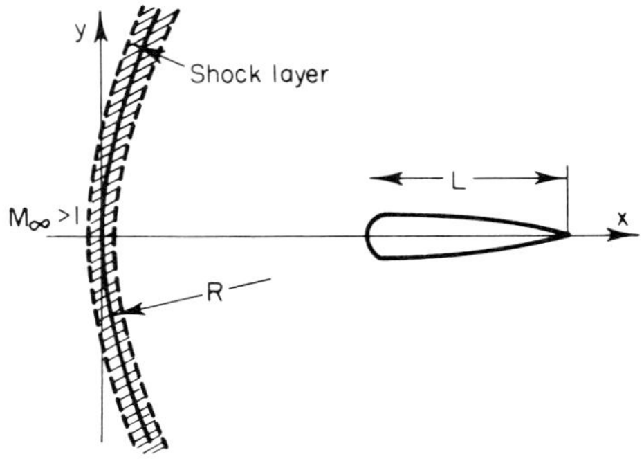

FIG. 17. Curved shock configuration (from Szaniawski, 1969).

$\alpha_2 = 0$, the final result for $u^{(1)}(x, 0)$, the velocity along the axis of symmetry is, in the two-dimensional case,

$$u^{(1)} = (-\alpha_1 x - \tfrac{1}{2}\alpha_1 x^2 - \cdots)\left[1 + \tanh\left(\frac{\alpha_1 x + \frac{1}{2}\alpha_1 x^2 + \cdots}{2\alpha_1 \chi_v}\right)\right] - \tanh\left(\frac{\alpha_1 x + \frac{1}{2}\alpha_1 x^2 + \cdots}{2\alpha_1 \chi_v}\right), \tag{5.47}$$

with a similar result for axisymmetric flow. Typical velocity profiles are shown in Fig. 18 for different values of $2\alpha_1\chi_v$.

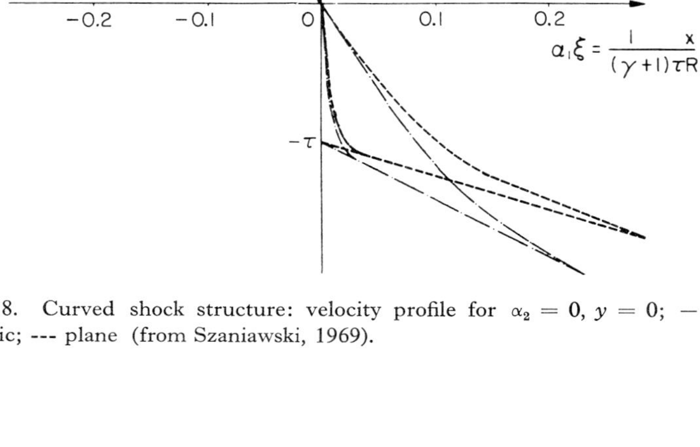

FIG. 18. Curved shock structure: velocity profile for $\alpha_2 = 0$, $y = 0$; — · — axisymmetric; --- plane (from Szaniawski, 1969).

The solution for $\theta(x, \tilde{y})$ must represent the first term of an asymptotic expansion in χ_v since θ is evaluated from the V-T equation in the limit $\chi_v \to 0$. $\theta(x, \tilde{y})$ is thus independent of χ_v and the influence of viscosity enters only through the factor $\alpha_1\chi_v$ in (5.47). From Eq. (5.46) for the shock shape it is readily shown that the parameter α_1 is related to the shock radius of curvature R_s through

$$\alpha_1 = \frac{1}{(\gamma+1)\epsilon}\frac{L}{\bar{R}_s}, \tag{5.48}$$

and since $\chi_v = \bar{\eta}/\bar{L}$ it follows that

$$\alpha_1\chi_v = \frac{1}{(\gamma+1)\epsilon}\frac{\bar{\eta}}{\bar{R}_s}. \tag{5.49}$$

When $\alpha_1\chi_v \ll 1$ the solution (5.47) reduces to a Taylor or one-dimensional shock, as described by (3.36a), followed by a decelerating flow, $u^{(1)} = (-1 - 2\alpha_1 x - \alpha_1 x^2 \cdots)$. For $\alpha_1\chi_v \sim O(1)$ the flow within the shock and the inviscid downstream flow merge together and the shock structure differs appreciably from the one-dimensional Taylor structure. It is interesting to note that the condition $\alpha_1\chi_v \ll 1$ for negligible curvature effect is, from (5.49), equivalent to the requirement $(\bar{\eta}/\bar{R}_s) \ll \epsilon$ consistent with the discussion of oblique shock structure in Section V, A above. Near the axis of symmetry of the curved shock considered here $v^{(1)} > 0$ which, as noted in Section V, A, implies that $u^{(1)}(+\infty, y) < -1$, a result borne out by the curves in Fig. 18.

Szaniawski's solution (1969) is indirect in that no attempt is made to satisfy the boundary conditions at the body which is responsible for the existence of the shock wave. The shock under consideration could represent the detached shock upstream of a blunt body at a free stream Mach number, M_∞, slightly greater than unity. As $(M_\infty - 1) \to 0$ the detachment distance as well as R_s will in this case increase. As $(M_\infty - 1) \to 0$ the shock structure will be influenced by curvature only if $\alpha_1\chi_v$ increases, which for fixed fluid properties is equivalent to $\lim_{\epsilon\to 0}(\epsilon^2 R_s) \to 0$. It is not clear whether this condition is satisfied but the above analysis is of great interest regardless since, as indicated in Section II above, there are many transonic problems in which the ratio $\bar{\eta}/\bar{R}_s$ may become of $O(\epsilon)$ or larger.

D. The Wavy-Wall Problem

An approximate solution of the V-T equation for the classical problem of two-dimensional flow past a wavy wall has been determined (Sichel,

1970; Sichel and Yin, 1969b) following the method applied by Hosokawa (1960a, b) to the inviscid problem. The appropriate boundary condition for flow past a wall with ordinate $\bar{y}_w = \bar{\tau} \sin 2\pi(\bar{x}/\bar{l})$ will be

$$v^{(1)}(x, 0) = \Phi_y(x, 0) = \omega \cos \omega x, \qquad \omega = 2\pi\bar{L}/\bar{l}, \tag{5.50}$$

where $\bar{\tau}$ is the wavy wall amplitude. The expansion parameter ϵ is chosen in accordance with the discussion in Section III, B above. The potential Φ is split into two parts $\Phi = \phi + g$ such that ϕ satisfies

$$\chi_V \phi_{xxx} + \chi_\infty \phi_{xx} - K\phi_x + \phi_{yy} = 0, \tag{5.51}$$

which is the V-T equation linearized by replacing ϕ_{xx} in the nonlinear term by a constant K. The potential ϕ is made to satisfy the wavy-wall boundary condition so that

$$\phi_y(x, 0) = \omega \cos \omega x = R(\omega e^{i\omega x}). \tag{5.52}$$

From the V-T equation (3.20) it follows that the nonlinear correction $g(x, y)$ then must satisfy

$$\begin{aligned} \chi_V g_{xxx} + \chi_\infty g_{xx} + (K - \phi_{xx})\,\phi_x - (g_x\phi_x)_x - g_x g_{xx} + g_{yy} &= 0, \\ g_y(x, 0) &= 0. \end{aligned} \tag{5.53}$$

A key assumption in the approximate analysis, which is discussed in detail by Hosokawa (1960b) and Sichel and Yin (1969b), is that $g_{yy}(x, 0) = 0$, so that (5.53) can be treated as an ordinary differential equation for g at the wall ($y = 0$), which is "driven" by the linear solution $\phi(x, y)$.

The solution of (5.51) and (5.52) is

$$\phi = R[-(\omega/m)^{1/2} \exp(-m^{1/2} y \cos \tfrac{1}{2}\beta) \exp i(\omega x - m^{1/2} y \sin \tfrac{1}{2}\beta - \tfrac{1}{2}\beta)], \tag{5.54}$$

with

$$m = [\chi_\infty^2 \omega^4 + (\omega^3 \chi_V + \omega K)^2]^{1/2},$$

$$\beta = \arctan(\omega^3 \chi_V + \omega K)/\chi_\infty \omega^2 \qquad 0 \leqslant \beta \leqslant \pi.$$

Following Hosokawa (1960b), the constant acceleration K is chosen as $\phi_{xx}(x, 0)$ at the accelerating sonic point x^* where $\phi_x(x^*, 0) = \chi_\infty$. The constants K and x^* are then determined by

$$\begin{aligned} \phi_x(x^*, 0; \chi_\infty, \chi_V) &= \chi_\infty, \\ \phi_{xx}(x^*, 0; \chi_\infty, \chi_V) &= K. \end{aligned} \tag{5.55}$$

In the limit, as $K \to 0$, $\chi_v \to 0$, the solution (5.54) reduces to the well-known linearized subsonic or supersonic wavy-wall solution, with exponential decay with increasing y in the subsonic case ($\chi_\infty > 0$), and with properties constant along characteristics in the supersonic case ($\chi_\infty < 0$). In the inviscid limit $\chi_v \to 0$, $K \neq 0$, the solution (5.54) decays exponentially with increasing y for either subsonic or supersonic flow, although the subsonic rate of decay is higher. In the general case the phase of the solution remains constant along the characteristic cuves $\omega x - m^{1/2} y \sin(\beta/2) = \text{const.}$ which now exist in both the subsonic and supersonic case. For sonic flow with $\chi_\infty = 0$, $\chi_v \to 0$ the linearized equation (5.51) reduces to the diffusion equation $-K\phi_x + \phi_{yy} = 0$ which has, for example, been used by Spreiter and Alksne (1958) to approximate the transonic equation when the free stream is sonic.

Combining (5.54) and (5.53) leads to the following ordinary differential equation for $\zeta(x) = g_x(x, 0) + \phi_x(x, 0) - \chi_\infty$, the deviation of $u^{(1)}(x, 0)$ from the critical value:

$$\zeta' - \frac{1}{2\chi_v}\zeta^2 = -\frac{\chi_\infty}{2\chi_v} + \frac{A_1}{\chi_v} + \frac{m^{1/2}}{\chi_v}\cos\left(\omega x + \frac{\beta}{2} - \frac{\pi}{2}\right), \tag{5.56}$$

where A_1 is a constant of integration. Without the harmonic term and with $A_1 = 0$ the Riccati equation (5.56) reduces to (3.35), the equation describing the structure of a weak normal shock. The transformation

$$\xi = \tfrac{1}{2}\omega x + \tfrac{1}{4}\beta + \tfrac{1}{4}\pi, \qquad \zeta = -\omega\chi_v(1/T)\, dT/d\xi \tag{5.57}$$

changes (5.56) to the Mathieu equation (McLachlan, 1947)

$$T'' + (a - 2q\cos 2\xi)T = 0, \tag{5.58}$$

with the parameters a and q related to the wavy-wall problem by

$$a = (2A_1 - \chi_\infty^2)/\omega^2\chi_v^2, \qquad q = m^{1/2}/\omega^2\chi_v^2. \tag{5.59}$$

It is readily shown that the appropriate solution of Eq. (5.58) is

$$T = A_2\, e^{\mu\xi} \sum_{r=-\infty}^{\infty} c_{2r}\, e^{2r\xi i} + A_3\, e^{-\mu\xi} \sum_{r=-\infty}^{\infty} c_{2r}\, e^{-2r\xi i}, \tag{5.60}$$

where A_2 and A_3 are constants of integration and $\mu(a, q)$ is the characteristic value of (5.58), which depends on the parameters a and q. The coefficients c_{2r} are determined from the recurrence relation

$$[a - (2r - i\mu)^2]\, c_{2r} - q(c_{2r+2} + c_{2r-2}) = 0. \tag{5.61}$$

The parameters χ_V and χ_∞ thus enter the solution in a rather complex way, for first the relation between m, χ_V, χ_∞, and β must be established by solving the transcendental equation (5.55). χ_V and χ_∞ then determine a and q which in turn establish the characteristic value μ and the coefficients c_{2r} in (5.60) through the recurrence relation (5.61).

From (5.57) it follows that ζ depends only on the ratio A_2/A_3 of the constants of integration. For a wavy wall in subsonic flow it is to be expected that $\lim_{\tau\to 0} \zeta = -\chi_\infty$, that is there is no disturbance as the wall amplitude approaches zero. This condition is satisfied only with $A_2/A_3 \to \infty$. Other possibilities are

$$A_2/A_3 = 1.0, \qquad \lim_{\tau\to 0} \zeta = -\chi_\infty \tanh[(\chi_\infty/\omega\chi_V)\xi];$$

$$A_2/A_3 = 0, \qquad \lim_{\tau\to 0} \zeta = \chi_\infty;$$

$$A_2/A_3 = -1.0, \qquad \lim_{\tau\to 0} \zeta = -\chi_\infty \coth[(\chi_\infty/\omega\chi_V)\xi].$$

But these were discarded on the grounds that they were not in accord with the problem under consideration. Once again some of the solutions of the Riccati equation (5.56) diverge. The appropriate wavy-wall solution for a subsonic free stream is thus

$$\zeta = -\omega\chi_V\mu + \frac{\omega\chi_V \sum_{r-1}^{\infty} 2r(\rho_{2r}/\rho_0)\sin(2r\xi + \phi_{2r})}{1 + \sum_{r=1}^{\infty} (\rho_{2r}/\rho_0)\cos(2r\xi + \phi_{2r})}, \tag{5.62}$$

where ρ_{2r}, ρ_0, ϕ_{2r} are related to c_{2r} by

$$c_{2r} = \rho_{2r}\, e^{i\phi_{2r}}, \qquad c_{-2r} = \rho_{2r}\, e^{-i\phi_{2r}}, \qquad c_0 = 2\rho_0\,.$$

The inviscid wavy-wall solutions of Hosokawa (1960b) are shown in Fig. 19a and the corresponding viscous solutions for $\chi_V = 0.61$ and $\chi_V = 0.33$ are shown in Figs. 19b and c. With $\tau = 0.01$, for example, these values of χ_V correspond to $\mathbf{R} = 28$ and 52 while the range $0 < \chi_\infty < 0.8$ corresponds to $1 > M_\infty > 0.970$. The constant A_1 has been chosen to make $\zeta(x^*) = 0$, consistent with the inviscid solutions of Hosokawa. It can be seen that in the viscous theory shock discontinuities terminating regions of supersonic flow are replaced by smooth compressions across which the Rankine–Hugoniot conditions are not necessarily satisfied. The transition to subsonic flow becomes steeper with decreasing χ_V or increasing $\mathbf{R}$ as is to be expected. With a sonic free stream $\chi_\infty = 0$, χ_V does not seem to affect the location of the

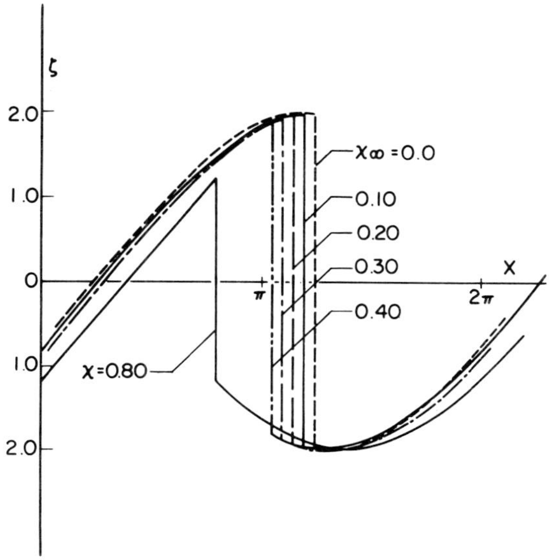

FIG. 19a. Inviscid wavy-wall solution of Hosokawa (from Sichel, 1970).

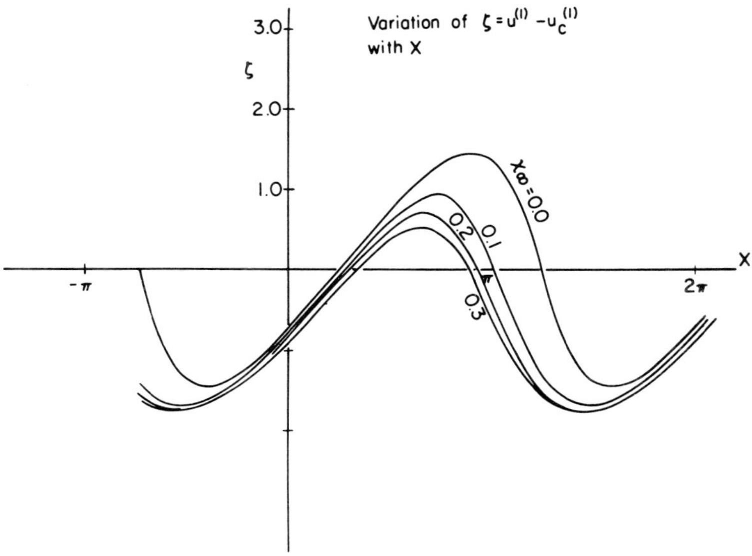

FIG. 19b. Viscous transonic wavy-wall solutions for $\chi_v = 0.61$, $\omega = 1.00$, $q = 3.0$ (from Sichel, 1970).

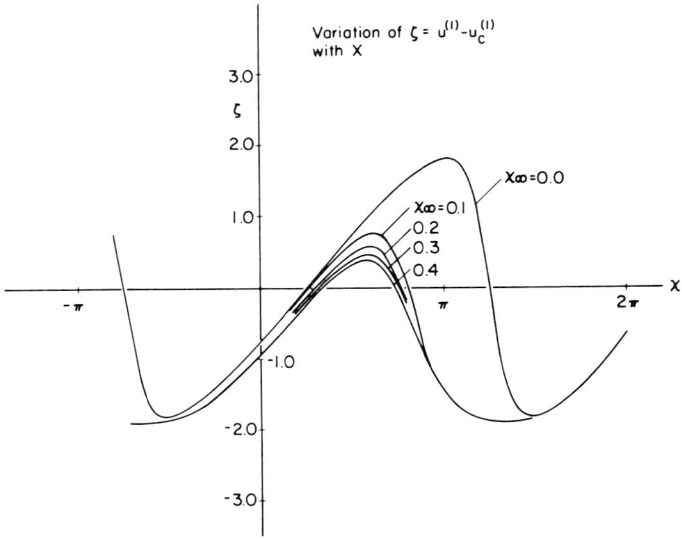

FIG. 19c. Viscous transonic wavy wall solutions for $\chi_v = 0.33$, $\omega = 1.00$, $q = 10.0$ (from Sichel, 1970).

compression wave as is quite apparent from Fig. 20, which compares sonic-inviscid and viscous solutions. In the limit $\chi_v \to 0$ the viscous solution, in fact, approaches the inviscid solution. With a subsonic free

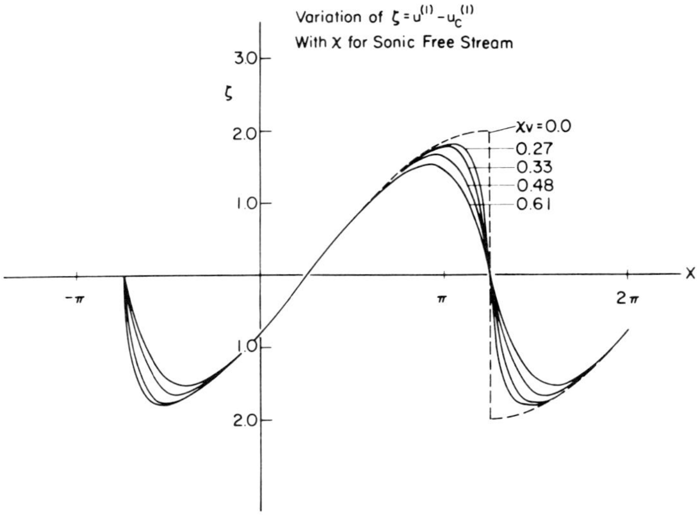

FIG. 20. Effect of **R** on wavy-wall flow with $M_\infty = 1.0$ (from Sichel, 1970).

stream, $\chi_\infty > 0$; on the other hand, viscosity shifts the compression upstream and shortens the supersonic region as compared to the inviscid solution, i.e., the viscous solution does not appear to approach the inviscid solution in the limit $\chi_v \to 0$.

The choice of A_1 to make $\zeta(x^*) = 0$ has an important influence on the behavior of the viscous solution and is a troublesome aspect of the wavy-wall analysis. The choice is based in part on the inference of Oswatitsch and Keune (1955) that the linearized solution should be valid near the accelerating sonic point where, by (5.55), the constant K coincides with the actual value of ϕ_{xx} from linearized solution. There is, however, no mathematically or physically rigorous justification for the choice $\zeta(x^*) = 0$.

VI. Curved Shock Waves in Hypersonic Flow

As already mentioned in Section II above it is essential to consider the deviation of the shock jump conditions from the R–H value in the analysis of hypersonic flow over blunt bodies and past a flat plate at low Reynolds numbers. The analysis of hypersonic non-R–H shocks will be considered in detail below.

A. The Influence of Shock Curvature

Application of the conservation laws across a thin curved shock wave such as shown in Fig. 21 leads to the equations

$$\begin{aligned}
\rho_1 V_{1n} &= \rho_2 V_{2n}, && \text{(continuity)} \\
-\rho_1 V_{1n}\mathbf{V}_1 + \mathbf{p}_{n1} &= -\rho_2 V_{2n}\mathbf{V}_2 + \mathbf{p}_{n2}, && \text{(momentum)} \\
-\rho_1 V_{1n}[(V_1^2/2) + e_1] &+ \mathbf{p}_{n1}\cdot\mathbf{V}_1 + \lambda_1(\partial T_1/\partial n) && \\
&= -\rho_2 V_{2n}[(V_2^2/2) + e_2] + \mathbf{p}_{n2}\cdot\mathbf{V}_2 + \lambda_2(\partial T_2/\partial n). && \text{(energy)}
\end{aligned} \tag{6.1}$$

Here V_n is the component of velocity normal to the shock surface while $\mathbf{p}_n$ is the stress on an area element perpendicular to $\mathbf{n}$, the unit normal to the shock surface. If the flow upstream and downstream of the shock is uniform $\mathbf{p}_{n1}$ and $\mathbf{p}_{n2}$ are simply $-p_1\mathbf{n}$ and $-p_2\mathbf{n}$, the heat flux terms due to $\partial T/\partial n$ vanish, and the set (6.1) reduces to the equations needed to determine the R–H conditions across the shock.

However, Sedov *et al.* (1953) point out that gradients of velocity and temperature behind the detached shock in the supersonic flow past

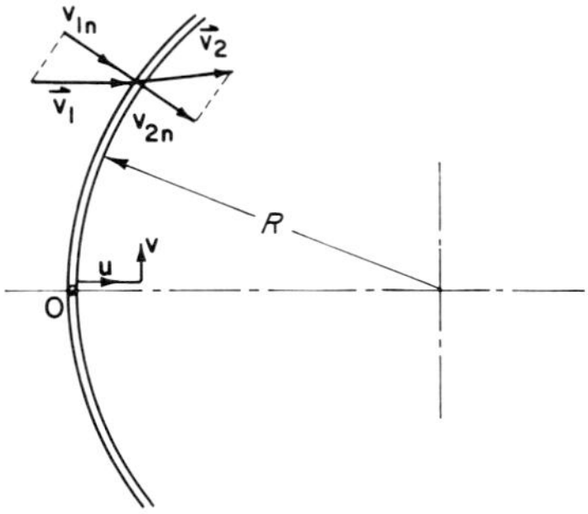

FIG. 21. Bow shock velocities.

bodies of small size may be sufficiently large to significantly influence the shock relations as computed from (6.1). Clearly the determination of shock jump conditions then will require a knowledge of the downstream flow if the set (6.1) is to be solved, in sharp contrast to the usual case when the set (6.1) plus an equation of state is sufficient to solve for $\mathbf{V}_2$, p_2, T_2, ρ_2, etc., in terms of the upstream flow $\mathbf{V}_1$, p_1, T_1, ρ_1, etc.

Sedov *et al.* (1953) determined approximate expressions for the corrections to the Hugoniot conditions due to shock curvature. The problem was specialized to the vicinity of the point 0 on the shock axis of symmetry (Fig. 21). Then the equations (6.1) assume the form

$$\rho_1 u_1 = \rho_2 u_2 ,$$

$$\rho_1 u_1^2 + p_1 = \rho_2 u_2^2 + p_2 - (\zeta_2 - \tfrac{2}{3}\mu_2)[(\partial u_2/\partial x) + (j+1)(\partial v_2/\partial y)] - 2\mu_2(\partial u_2/\partial x),$$

$$0 = \partial u_2/\partial y + \partial v_2/\partial x,$$

$$\begin{aligned} \rho_1 u_1 \left(\frac{u_1^2}{2} + e_1\right) + p_1 u_1 \\ = \rho_2 u_2 \left(\frac{u_2^2}{2} + e_2\right) + p_2 u_2 \\ - \left[(\zeta_2 - \tfrac{2}{3}\mu_2)\left(\frac{\partial u_2}{\partial x} + (j+1)\frac{\partial v_2}{\partial y}\right) + 2\mu_2 \frac{\partial u_2}{\partial x}\right] u_2 - k_2 \frac{\partial v_2}{\partial x} . \end{aligned} \tag{6.2}$$

In (6.2), ζ is the second coefficient of viscosity and $j = 0, 1$ for plane and axisymmetric flow, respectively. Two key assumptions are that terms of $O(\mathbf{R}^{-1})$, where $\mathbf{R} = (\rho_2 \mu_2 R)/\mu_2$, can be neglected in the flow behind the shock wave and that the tangential velocity is conserved across the shock. Then the flow behind the shock wave is simply governed by the inviscid Euler equations, and although the magnitude of the viscous stress and heat flux may be significant the gradients of these dissipative quantities are thus neglected. The Euler equations can be solved for the derivatives in (6.2) with the result that at O

$$\begin{aligned}
\frac{\partial v_2}{\partial y} &= -\frac{u_2}{R}\left(1-\frac{u_1}{u_2}\right),\\
\frac{\partial u_2}{\partial x} &= -\frac{1}{R}\frac{u_2}{(M_2^2-1)}(1+j)\left(1-\frac{u_1}{u_2}\right),\\
\frac{\partial T_2}{\partial x} &= \frac{1}{R}\frac{(u_2/C_P)}{M_2^2-1}(1+j)\left(1-\frac{u_1}{u_2}\right).
\end{aligned} \tag{6.3}$$

Now (6.2) and (6.3) can be solved for ρ_2, u_2, p_2, etc., in terms of the upstream flow parameters, the shock radius of curvature, and the transport coefficients. Thus, for example, neglecting the second viscosity,

$$\begin{aligned}
\frac{T_{02}}{T_{01}} &= 1-\frac{(1+j)}{\mathbf{R}}4\left(\frac{\gamma-1}{\gamma+1}\right)\left[\left(1-\frac{1}{2Pr}\right)\left(M_1^2-\frac{\gamma-1}{\gamma+1}\right)-\frac{2}{3(\gamma+1)}\right]\Big/M_1^2,\\
\frac{p_{02}}{p_{01}} &= \left(\frac{p_{02}}{p_{01}}\right)_{id}\left\{1-\frac{(1+j)}{\mathbf{R}}\frac{8\gamma}{(\gamma+1)^2}\left[\frac{\gamma+1}{2}-\frac{1}{3M_1^2}-\frac{\gamma-1}{3M_1^2}\left(1-\frac{1}{2Pr}\right)\right]\right\},
\end{aligned} \tag{6.4}$$

from (6.4) it follows that for $Pr \geqslant 3/4$ viscosity and heat flux behind the shock result in a decrease in T_{02}/T_{01} for all $M_1 \geqslant 1$, while p_{02}/p_{01} decreases for all values of M_1 and Pr. The corrections to the inviscid Hugoniot conditions are of $O(\mathbf{R}^{-1})$ and thus become significant only for very small values of **R**.

B. The Influence of Shock Thickness

The results of Sedov *et al.* (1953) are only valid for very thin shock waves, but fail to take into account the "shock thickness" effect which arises from the two-dimensional flow within shock waves when $\bar{\eta}/\bar{R}$ becomes sufficiently large. The conservation equations (6.1) can be derived by applying the conservation of mass, momentum, and energy to the control volume sketched in Fig. 22. However, in the derivation of (6.1) only the flux across the surfaces A^+ and A^- is considered while the flux across the lateral surface $\mathscr{L}$ is neglected. When $\bar{\eta}/\bar{R} \ll 1$ the net flux of mass, momentum, and energy across $\mathscr{L}$ will be negligible and then (6.1) will be valid, but when $\bar{\eta}/\bar{R}$ is sufficiently large the net flux across $\mathscr{L}$ will also influence the jump conditions across the shock wave, that is, there will be a "thickness effect."

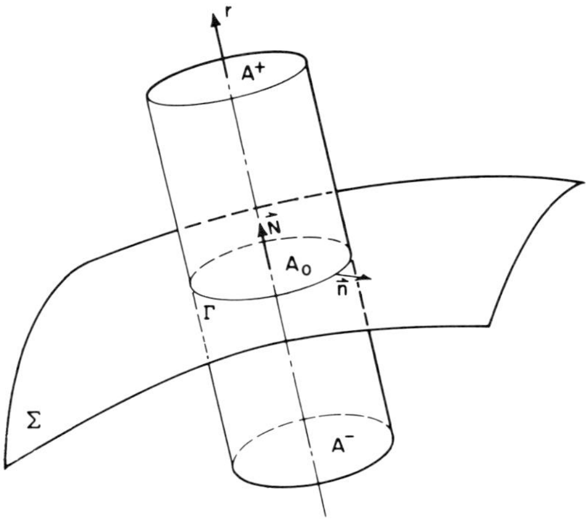

FIG. 22. Control volume across shock surface. (Reprinted from Pan and Probstein, 1966 with permission of Cornell University Press).

The "thickness effect" is taken into account in the quite similar analyses of Chow and Ting (1961) and Germain and Guiraud (1964). Chow and Ting (1961) divide the flow into a region upstream (region 1), a region downstream (region 2), and a region within the shock wave (Fig. 23). In regions 1 and 2 the variables are expanded in powers of $\epsilon = \mathbf{R}^{-1}$ so that

$$
\begin{gathered}
u_1 = u_1^{(0)} = U\cos\theta, \\
v_1 = v_1^{(0)} = U\sin\theta, \\
p_1 = p_\infty, \qquad \rho_1 = \rho_\infty, \qquad h_1 = h_{s\infty} - \tfrac{1}{2}U^2,
\end{gathered}
\tag{6.5}
$$

and

$$
\begin{aligned}
u_2(r,\theta,\epsilon) &= u_2^{(0)}(r,\theta) + \epsilon u_2^{(1)}(r,\theta) + \cdots, \\
v_2(r,\theta,\epsilon) &= v_2^{(0)}(r,\theta) + \epsilon v_2^{(1)}(r,\theta) + \cdots.
\end{aligned}
\tag{6.6}
$$

Within the shock layer the stretched inner variable

$$s = (r - R_0)/\epsilon$$

is introduced and the inner expansion is as follows:

$$
\begin{aligned}
u(s,\theta,\epsilon) &= u^{(0)} + \epsilon u^{(1)} + \cdots, \\
v(s,\theta,\epsilon) &= v^{(0)} + \epsilon v^{(1)} + \cdots.
\end{aligned}
\tag{6.7}
$$

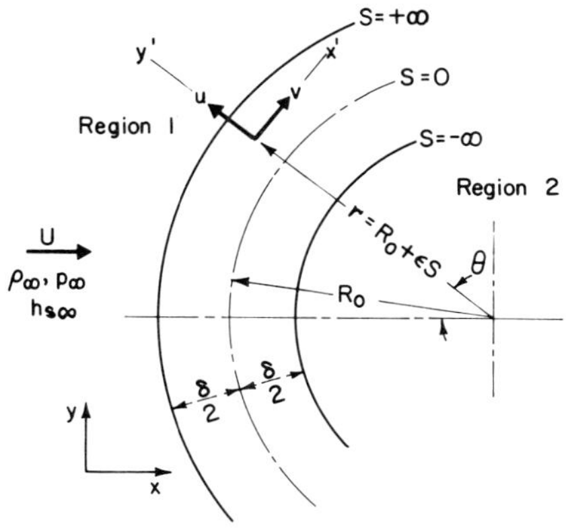

FIG. 23. Coordinates used in the analysis of the "thickness effect."

To match the inner and outer expansions the variation of $u_2(r, \theta, \epsilon)$ with r must be taken into account, at least for the first order terms. For example, for $u(r, \theta)$ the matching conditions are

$$\begin{aligned} u^{(0)}(+\infty, \theta) &= -U \cos \theta, \\ u^{(1)}(+\infty, \theta) &= 0, \end{aligned} \tag{6.8}$$

$$\begin{aligned} u^{(0)}(-\infty, \theta) &= u_2^{(0)}(R_0, \theta), \\ u^{(1)}(-\infty, \theta) &= u_2^{(1)}(R_0, \theta) + \lim_{s\to\infty} s \left[\frac{\partial}{\partial r} u_2^{(0)}\right]_{r=R_0}, \end{aligned} \tag{6.9}$$

where now $u_2^{(1)}(r, \theta)$ has been expanded in a Taylor series.

The zeroth order solution is taken as the one-dimensional R. Becker (1922) shock structure solution, and the expansions (6.7) are introduced in the conservation equations. Then, for example, the first order continuity equation in terms of the first order mass-flux density $m^{(1)} = \rho^{(0)}u^{(1)} + \rho^{(1)}u^{(0)}$ is

$$R_0\, \partial/\partial s(m^{(1)}) = -m^{(0)} - \partial/\partial\theta(\rho^{(0)}v^{(0)}). \tag{6.10}$$

Integration of (6.10) and application of the matching conditions (6.8) and (6.9) then leads to the following relation for $m_2^{(1)}$, the first order mass-flux density downstream of the shock:

$$\begin{aligned} m_2^{(1)}(R_0, \theta) = {} & \frac{1}{R_0}\int_0^\infty \left\{\frac{\partial}{\partial\theta}\left[(\rho^{(0)} - \rho_\infty)v_1\right]\right\} ds \\ & + \frac{1}{R_0}\int_{-\infty}^0 \left\{\frac{\partial}{\partial\theta}\left[(\rho^{(0)} - \rho_2^{(0)}|_{r=R_0})v_1\right]\right\} ds. \end{aligned} \tag{6.11}$$

For a normal or oblique shock, with $R_0 \to \infty$, (6.11) yields the expected result $m_2^{(1)} = 0$, that is the mass-flow density is invariant across the shock. The other conservation equations yield similar first order equations for the downstream quantities and this set can be solved for $u_2^{(1)}$, $v_2^{(1)}$, $p_2^{(1)}$, $\rho_2^{(1)}$, $h_2^{(1)}$, and $h_{s2}^{(1)}$, where the subscript s refers to the stagnation value. Since the Becker solution requires that $Pr = 3/4$, the results of Chow and Ting (1961) are only valid for this Prandtl number. For the mass flux, for example, Chow and Ting (1961) obtain

$$\frac{m_2^{(1)}}{\rho_\infty U} = \frac{16}{3}\left(\frac{\gamma}{\gamma+1}\right)\ln 2 \times \left\{\frac{\cos\theta(M_\infty{}^2\cos^2\theta-1)}{[(\gamma-1)M_\infty{}^2\cos^2\theta+2]} - \frac{(\gamma+1)M_\infty{}^2\sin\theta\sin 2\theta}{[(\gamma-1)M_\infty{}^2\cos^2\theta+2]^2}\right\}, \tag{6.12}$$

which, for $\theta = 0$ (corresponding to the case considered by Sedov *et al.*) becomes

$$\frac{m_2^{(1)}}{\rho_\infty U} = \frac{8}{3}\left(\frac{\gamma}{\gamma+1}\right)\ln 2\,\frac{M_\infty{}^2-1}{1+\frac{1}{2}(\gamma-1)M_\infty{}^2}\,. \tag{6.13}$$

Since $\rho_\infty u_\infty < 0$ in the coordinate system used here (Fig. 23), the result (6.13) indicates that the mass flow density decreases at least at the point of symmetry. Chow and Ting (1961), like Sedov *et al.* (1953), find that shock thickness and curvature result in a decrease in the stagnation temperature ratio T_{02}/T_{01}. The tangential velocity component is found to increase across the shock.

As can be seen from (6.11) the analysis of Chow and Ting (1961) requires the integration of discontinuous functions. Germain and Guiraud (1964) avoid this difficulty. Their analysis is again based on an expansion in $\mathbf{R}^{-1}$ and upon the fact that the nth order jump conditions across the shock depend on the $(n-1)$th order shock structure (Germain and Guiraud, 1964). The analysis starts from the flux conservation theorem

$$\iint_{A^+} \mathbf{A}\cdot\mathbf{N}\,dS - \iint_{A^-} \mathbf{A}\cdot\mathbf{N}\,dS + \int_\Gamma d\sigma\mathbf{n}\cdot\int_{-\delta/2}^{\delta/2}\mathbf{b}\,dy' = 0, \tag{6.14}$$

where $\mathbf{A}$ and $\mathbf{b}$ are the fluxes of physical quantities and $\mathbf{N}$ and $\mathbf{n}$ are unit vectors normal to the shock and to the curve Γ (Fig. 22). The areas A^+ and A^-, shown in Fig. 22, are located at $s = \pm\infty$. The flux conservation demanded by (6.14) will, by appropriate definition of $\mathbf{A}$, take downstream gradients into account, while the integral of the flux of $\mathbf{b}$ through the surface $\mathscr{L}$ accounts for the shock thickness effect.

A key contribution of Germain and Guiraud (1964) was to show that (6.14) can also be expressed in the form

$$\iint_{\mathbf{A}_0} [\mathbf{A}] \cdot \mathbf{N}\, dS + \epsilon \int_{\Gamma} \mathbf{b}^* \cdot \mathbf{n}\, d\sigma = 0, \tag{6.15}$$

where the bracket notation denotes

$$[\,] = (\,)^+ - (\,)^-,$$

and the asterisk has the meaning

$$F^*(x', \epsilon) = \int_{-\infty}^{\infty} [F(x', s, \epsilon) - \hat{F}(x', s, \epsilon)]\, dS, \tag{6.16}$$

with F indicating the nth order shock structure solution and $\hat{F}$ a special fitting function. $\hat{F}$ is symmetric with respect to the shock center and is asymptotic to the nth order values of F, i.e., F^+ and F^- as $S \to \infty$. Germain and Guiraud (1964) used the function

$$\hat{F} = \tfrac{1}{2}(F^+ + F^-) + \tfrac{1}{2}(F^+ - F^-)[\tanh S + (S/\cosh^2 S)].$$

For a two-dimensional wave it is convenient to use the local Cartesian coordinate system (x, y, z) as shown in Fig. 23, with z taken normal to the x–y plane. If $\mathbf{A}$ and $\mathbf{b}^*$ are vectors in the x–y plane such that

$$\mathbf{A} = A_x \mathbf{e}_x + A_y \mathbf{e}_y\,,$$

$$\mathbf{b}^* = b_x{}^* \mathbf{e}_x + b_y{}^* \mathbf{e}_y\,,$$

where $\mathbf{e}_x$ and $\mathbf{e}_y$ are unit vectors, then the flux conservation equation (6.15) becomes

$$\iint [A_y]\, dx\, dz + \epsilon \int_{\Gamma} b_x{}^*\, dz = 0. \tag{6.17}$$

Here $[A_y]$ is the change of the vector component normal to the shock and (6.17) shows that $[A_y]$ is equal to minus the net flux of $b_x{}^*$ out of the control volume, which in two dimensions is a rectangular paralellepiped. Differentiation of (6.17) with respect to x now yields the expression

$$[A_y] + \epsilon(\partial b_x{}^*/\partial x) = 0. \tag{6.18}$$

Since

$$A_y = A_y^{(0)} + \epsilon A_y^{(1)} + \cdots,$$

$$b_x{}^* = b_x^{*(0)} + \epsilon b_x^{*(1)},$$

it follows that

$$[A_y^{(1)}] + (db_x^{*(0)}/dx) = 0. \tag{6.19}$$

As an example (6.18) will be applied to the conservation of mass. Then

$$\mathbf{A} = \mathbf{b} = \rho v \mathbf{e}_x + \rho u \mathbf{e}_y\,,$$

and with $m = \rho u$(6.19) yields

$$m_2^{(1)} = (d/dx)(\rho v)^{*(0)}. \tag{6.20}$$

Equation (6.20) is identical to the result (6.11) obtained by Chow and Ting (1961), as is evident from a detailed examination of $(\rho v)^{*(0)}$. From (6.16) and the fact that $v^{(0)} = v_1 = v_2{}^{(0)}$ it follows that

$$(\rho v)^{*(0)} = \int_{-\infty}^{\infty} v_1(\rho^{(0)} - \hat{\rho}^{(0)})\, dS, \tag{6.21}$$

and a graphical interpretation of the integral (6.21) is given in Fig. 24. From this figure it is clear that

$$(\rho v)^{*(0)} = \int_0^{\infty} (\rho^{(0)} - \rho_\infty)\, v_1\, dS + \int_{-\infty}^{0} (\rho^{(0)} - \rho_2^{(0)})\, v_1\, dS, \tag{6.22}$$

which is the integral which occurs in Chow and Ting's expression for $m_2^{(1)}$. In the analysis of Germain and Guiraud (1964) the discontinuous integral (6.22) is replaced by a single integral of a continuous function. Further, rather than using the conservation equations in differential form, Germain and Guiraud (1964) start from the more general flux conser-

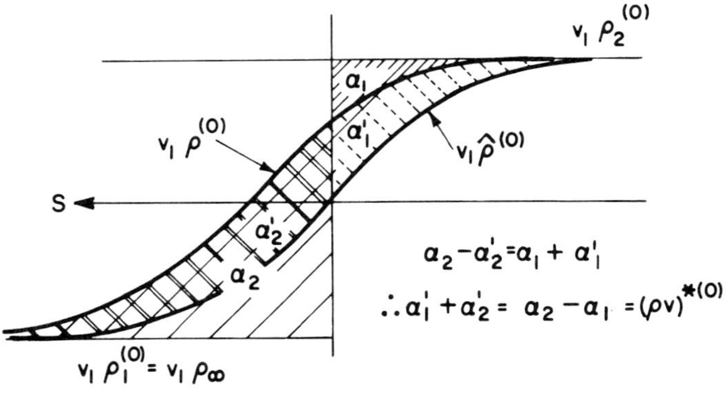

FIG. 24. Graphical interpretation of the integral $(\rho v)^{*(0)}$.

vation theorems. As in the analysis of Chow and Ting (1961) flux conservation theorems of momentum and energy lead to a set of simultaneous equations for the first order expansion coefficients downstream of the curved shock. Detailed development of the other conservation equations is presented both by Germain and Guiraud (1964), and by Pan and Probstein (1966).

The final results of the calculations of Pan and Probstein are the following expressions for the first order tangential and normal velocity, components v_s and u_s, and the stagnation enthalpy H_s downstream of the shock:

$$v_s = U\cos\theta_s - \frac{\mu_s}{\rho_\infty U\sin\theta_s}\left(\frac{\partial v}{\partial y}\right)_s - \frac{1}{\mathbf{R}}\frac{8}{3}\left(\frac{\gamma}{\gamma-1}\right)\frac{U}{K\sin\theta_s}$$
$$\times\left[\ln\frac{(\gamma+3)K-2(\gamma+1)\sqrt{K(K-1)}}{(\gamma+3)K+2(\gamma+1)\sqrt{K(K-1)}}+\frac{2\sqrt{K}}{\sqrt{K-1}}\right], \quad (6.23)$$

$$u_s = -\left(\frac{\gamma-1}{\gamma+1}\right)KU\sin\theta_s, \quad (6.24)$$

$$H_s = H_\infty - \frac{\mu_s}{\rho_\infty U\sin\theta_s}\left(\frac{\partial H}{\partial y}\right)_s - \frac{1}{\mathbf{R}\sin\theta_s}\frac{8}{3}\frac{\gamma}{\gamma-1}\frac{U^2}{K}$$
$$\times\left[\ln\frac{(\gamma+3)K-2(\gamma+1)\sqrt{K(K-1)}}{(\gamma+3)K+2(\gamma+1)\sqrt{K(K-1)}}+\frac{2\sqrt{K}}{\sqrt{K-1}}\right]\left(\frac{3}{4}-\frac{2K}{\gamma+1}\right), \quad (6.25)$$

with

$$K = 1 + [2/(\gamma-1)\,M_\infty^2\sin^2\theta_s].$$

In deriving (6.23)–(6.25) the R. Becker solution (1922) with $Pr = 3/4$ and $\mu \propto T^{1/2}$ was used for the zero order shock structure and it was further assumed that $\epsilon = (\gamma-1)/(\gamma+1) \ll 1$. The first terms of (6.23)–(6.25) are clearly the usual Rankine–Hugoniot conditions while the second terms of (6.23) and (6.25) are due to the shock curvature. The last terms of (6.23) and (6.25) represent the shock thickness effect. To the first order there is no change in u_s. Both curvature and thickness cause a reduction in the tangential velocity and stagnation enthalpy downstream of the shock wave, and results, in the hypersonic leading-edge problem, in reduced skin friction and heat transfer at the surface.

From the discussion above two main features of curved shock waves in low Reynolds number hypersonic flow are evident. Shock curvature changes the R–H conditions due to the large downstream gradients and the shock thickness effect represents the change in R–H conditions due to two-dimensional structure of thick shocks. All the theories described above essentially use expansions in powers of the inverse Reynolds number starting from the one-dimensional R–H shock as the zeroth order solution. A very careful study of the low **R** hypersonic shock structure, using the method of matched asymptotic expansions has also been made by Bush (1964, 1966, 1969) in connection with both the blunt body and the hypersonic leading-edge problems.

C. Comparison with Experiment

Direct comparison of the theories described above with experimental shock structure measurements is in general difficult because these theories are usually only a part of the overall analysis of the blunt body or flat plate flow. However, the semi quantitative surveys of blunt body flow by Broadwell and Rungalier (1967) and Ahouse and Harbour (1969), among others, and the experimental studies of leading-edge flow by McCrosky *et al.* (1966), Harbour and Lewis (1967), and M. Becker and Boylan (1967) clearly verify the existence of non-R–H shock waves at low Reynolds numbers.

The main verification of the non-R–H shock theories, above, comes from the fact that at low **R** experimentally determined flat plate and blunt body heat transfer and skin friction coefficients and pressures agree with theoretical values only when the deviation of the shock conditions from the R–H values is taken into account. In their analysis of blunt body flow Probstein and Kemp (1960) took both shock curvature and thickness into account although the exact zero order shock structure was replaced by linear variation of the flow parameters through the shock. With the inclusion of the non-R–H effects Probstein and Kemp found that the skin friction and heat transfer appeared to approach the free molecule values with decreasing **R**, as indeed they should. The results were in qualitative agreement with experiment. Cheng (1961, 1963) showed that for thin blunt body shock layers only the shock curvature effect is important so that the shock thickness effect can be neglected. Using a theory essentially the same as that due to Sedov *et al.* (1953) to compute the shock conditions and coupling this to a thin shock layer analysis, Cheng (1963) obtained values of stagnation point heat transfer in remarkable agreement with experiment over a wide range of **R**. In fact the theoretically computed heat transfer approaches the free molecule value as $\mathbf{R} \rightarrow 0$. A significant departure of Cheng's theory (1963) from those described above is that the deviations from the R–H conditions are not necessarily small. Subject to the thin shock layer assumption

Cheng (1963) also was able to compute the structure of the non-R–H shock wave. In at least one test case the shock structure computed by Cheng was almost identical to that determined by numerical integration of the complete Navier–Stokes equations by Levinsky and Yoshihara (1962).

Pan and Probstein (1966) coupled the modified shock conditions (6.23)–(6.25) with a boundary layer analysis of the viscous layer separating the shock and the flat plate surface to compute skin friction and heat transfer near the leading edge in low **R** hypersonic flow. The results were in qualitative agreement with experiment. The jump conditions (6.23)–(6.25) represent small deviations from the R–H conditions, and to first order the shock curvature does not affect the normal velocity and the density ratio across the shock. However, the flow field surveys of McCrosky *et al.* (1966), Harbour and Lewis (1967), and M. Becker and Boylan (1966) show that the shock is drastically different from a R–H shock. In particular the shock density ratio is much less than the R–H value in direct contradiction to the Pan–Probstein theory (1966).

Consequently, Shorenstein and Probstein (1968), following a procedure first used by Oguchi (1967), used an expansion scheme in which the R–H shock structure is not necessarily used as the zeroth order solution. The dimensionless variables and the stretching

$$\tilde{y} = \int_0^y \frac{\rho_\infty U}{\mu}\, dy, \qquad \tilde{u} = u/U, \qquad \tilde{v} = v/U, \qquad \tilde{p} = p/\rho_\infty U^2,$$
$$\tilde{\rho} = \rho/\rho_\infty, \qquad \tilde{\mu} = \mu/\mu_\infty, \qquad \tilde{H} = (H - \tfrac{1}{2}u^2)/U^2, \tag{6.26}$$

with u, v, and y as shown in Fig. 23, are used together with expansions of the form

$$\tilde{u} = \tilde{u}^{(0)} + \epsilon\tilde{u}^{(1)} + O(\epsilon^2) + \cdots,$$

where as before $\epsilon = \mu_\infty/(\rho_\infty U R_0)$. Then the zeroth order continuity, tangential and normal momentum, and energy equations become

$$\begin{aligned} &\partial(\tilde{\rho}^{(0)}\tilde{u}^{(0)})/\partial\tilde{y} = 0,\\ &\partial^2\tilde{v}^{(0)}/\partial\tilde{y}^2 - \tilde{\rho}^{(0)}\tilde{u}^{(0)}(\partial\tilde{v}^{(0)}/\partial\tilde{y}) = 0,\\ &\tilde{\rho}^{(0)}\tilde{u}^{(0)}\,\partial\tilde{u}^{(0)}/\partial\tilde{y} + \partial\tilde{p}^{(0)}/\partial\tilde{y} - \tfrac{4}{3}\,\partial^2\tilde{u}^{(0)}/\partial\tilde{y}^2 = 0,\\ &\tilde{\rho}^{(0)}\tilde{u}^{(0)}\,\partial\tilde{H}^{(0)}/\partial\tilde{y} - \tfrac{4}{3}\,\partial^2\tilde{H}^{(0)}/\partial\tilde{y}^2 + \tilde{\rho}^{(0)}\tilde{u}^{(0)}\tilde{v}^{(0)}\,\partial\tilde{v}^{(0)}/\partial\tilde{y}\\ &\qquad - \partial/\partial\tilde{y}[\tilde{v}^{(0)}\,\partial\tilde{v}^{(0)}/\partial\tilde{y}] = 0. \end{aligned} \tag{6.27}$$

Conditions downstream of the shock wave are determined by matching solutions of (6.27) with solutions of the boundary layer type for the flow between the shock and the plate surface. First order corrections were obtained by Shorenstein and Probstein (1968), using the curvature corrections of Chow and Ting (1961) rather than by solving the first order equations.

Theoretically computed and experimentally measured shock density ratios are compared in Fig. 25. Curve 1 corresponds to the R–H value corresponding to the theoretically computed shock angle. Zero order results obtained by solving (6.27) are shown by curve 2, while curves 3 and 4 include first order corrections for shock curvature and a correction for wall slip, respectively. It is obvious that appropriate considerations of the non-R–H shock conditions is a key element in the development of a theory which is consistent with experimental results.

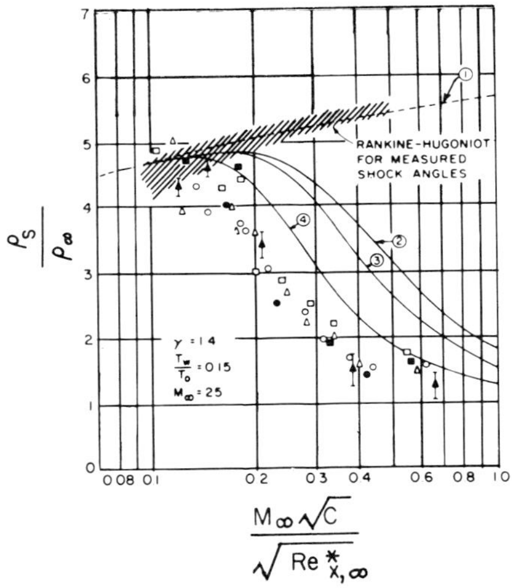

FIG. 25. Comparison of measured and computed density at the shock surface. Curve 1, Rankine–Hugoniot for theoretical shock angles; curve 2, zero-order theory; curve 3, + shock curvature; curve 4, + wall slip (from Shorenstein and Probstein, 1968, with permission).

Symbol	M_∞	$\mathbf{R}_\infty$/inch	T_w/T_0	γ	Reference
■	25.5	15,000	0.15	1.4	McCroskey *et al.* (1966)
▲	24.5	10,000	0.15	1.4	
●	23.3	7500	0.15	1.4	
□	25.5	15,000	0.15	1.4	Harbour and Lewis (1967)
△	24.5	10,000	0.15	1.4	
○	23.3	7500	0.15	1.4	

VII. Discussion

From Section III above it follows that transonic two-dimensional shock structure can be described by a single partial differential equation, the viscous-transonic (V-T) equation. However, as is evident from Sections IV and V, although mathematical properties and some solutions of the V-T equation have been obtained, satisfactory solutions for the central problems of V-T flow are still lacking. Further, there appears to be no experimental data with which the theoretical results could be compared. As is evident from Section VI, in the hypersonic case, analysis of two-dimensional shock structure cannot be reduced to the solution of a single equation. However, numerical solutions for the key problems of blunt body and flat plate flows, which are in substantial agreement with experimental measurements, have been obtained.

As indicated above, the mathematical properties of the V-T equation have been explored and the equation has been used to formulate similarity laws for V-T flow. V-T solutions describing nozzle flow and vortex source flow have been determined; however, the problem of finding V-T solutions for flow through a nozzle throat of arbitrary shape remains to be solved. Asymptotic solutions valid at large distances from two-dimensional and axisymmetric bodies have been obtained. However, the near field behavior, which will require an investigation of the transonic shock–boundary layer interaction problem, remains to be explored. The variation of the structure of the transonic bow shock with the shock radius of curvature has been determined; however, the relation between the shock curvature and the body generating the bow shock has not yet been established. Thus it is not yet clear whether in the limit $M_\infty \rightarrow 1$, the curvature actually becomes large enough to affect significantly the shock wave structure. There has been no progress in the analysis of the flow near the triple point in the Mach reflection of a weak shock wave.

Experimental data which could be used as chekpoints for the V-T theory do not appear to be available. It would ve very useful to have an experiment designed to test the interesting conclusion that at large distances viscous effects can be neglected only in two-dimensional flow. Detailed measurements describing the process of shock formation near a nozzle throat would also be desirable. The shock–boundary layer interaction problem is, perhaps, the problem of greatest practical importance. Here the early measurements of Ackeret *et al.* (1946) indicate that the shock structure immediately outside the boundary layer will be two-dimensional; however, further experimental studies of the detailed flow in the interaction region are needed.

In transonic flow changes across shock waves are very small; however, even in transonic flow the shock waves are extremely thin at ordinary densities. The experimental survey of transonic shock structure thus presents formidable difficulties. The use of transonic flow of a relaxing gas as an analogue for V-T flow presents a possible alternate approach to the direct measurement of transonic shock structure. In the flow of vibrationally relaxing gases a type of wave known as a fully dispersed wave can exist at velocities close to the equilibrium sonic speed. The structure of these waves is governed by a balance between the relaxation process and convection in contrast to the viscous shock whose structure results from a balance between convection and viscous and thermal dissipation. Such fully dispersed wave flows are, under appropriate conditions, completely analogous to V-T flow and satisfy the V-T equation with the shear viscosity replaced by a bulk viscosity (Yin, 1971). However, fully dispersed waves are thicker than viscous shocks by orders of magnitude. Interferometric observations of the structure of two-dimensional fully dispersed waves in CO_2 by Strehlow and Maxwell (1968) suggest that relaxing transonic flows may provide a powerful experimental tool for the sudy of two-dimensional shock structure.

The equations for the analysis of hypersonic two-dimensional shock structure are more complex than in the transonic case. However, the understanding of hypersonic non-Hugoniot shocks is more advanced. Asymptotic methods have been employed to obtain solutions for both the shock structure and the resultant flat plate and blunt body heat transfer which are in excellent agreement with experimental measurements. Because of the interest in re-entry prhenomena a large body of low Reynolds number hypersonic flow data is available. A detailed comparison between measured and theoretically calculated non-R–H shock structure, however, is still not available. Except for very thin shock layers (Cheng, 1963), determination of the non-R–H shock structure requires the solution of a system of nonlinear ordinary differential equations. It would be helpful if more simple solutions of these equations could be developed. While the blunt body and flat plate problems have been studied in considerable detail a comparable low-**R** hypersonic theory for slender body flow does not seem to be available.

Although the problem of non-Hugoniot shock structure is an old one going back to the early work of G. I. Taylor in 1930, many important questions still remain unresolved.

ACKNOWLEDGMENTS

The author is grateful to The University of Michigan for support in the preparation of this survey. Much of the work described herein was supported by the U.S. Army

Research Office in Durham, N.C. under contract DAHC 04-68-C-0008, as well as under earlier contracts. The author would like to thank Professor Robert W. Truitt for his interest and encouragement in connection with this review.

REFERENCES

Ackeret, J., Feldmann, F., and Rott, N. (1946). *Nat. Adv. Comm. Aeronaut., Tech. Notes* **1113**.

Ahouse, D. R., and Harbour, P. J. (1969). *In* "Rarefied Gas Dynamics" (L. Trilling and H. Y. Wachman, eds.), Vol. 1, p. 699. Academic Press, New York.

Axford, W. L., and Newman, R. C. (1967). *Astrophys. J.* **147**, 230.

Becker, M., and Boylan, D. E. (1967). *In* "Rarefied Gas Dynamics" (C. L. Brundin, ed.), Vol. 2, p. 993. Academic Press, New York.

Becker, R. (1922). *Z. Phys.* **8**, 321.

Block, H. (1912). *Ark. Mat., Astron. Fys.* **7**, Nos. 13 and 21.

Broadwell, J. E., and Rungalier, H. (1967). *In* "Rarefied Gas Dynamics" (C. L. Brundin, ed.), Vol. 2, p. 1145. Academic Press, New York.

Bush, W. B. (1964). *J. Fluid Mech.* **20**, Part 3, 353.

Bush, W. B. (1966). *J. Fluid Mech.* **25**, Part 1, 51.

Bush, W. B. (1969). *AIAA J.* **7**, 189.

Cheng, H. K. (1961). *Proc. Heat Transfer Fluid Mech. Inst.* p. 161.

Cheng, H. K. (1963). *Inst. Aeronaut. Sci., Pap.* **63-92**.

Chow, R. R., and Ting, L. (1961). *J. Aerosp. Sci.* **28**, 428; see also *Polytech. Inst. Brooklyn, Dep. Aerosp. Eng. Appl. Mech., PIBAL Rep.* **609** (1960) (also under ARL Tech. Note 60-142).

Cole, J. D. (1968). "Perturbation Methods in Applied Mathematics," Chapter 2. Ginn (Blaisdell), Boston, Massachusetts.

Dezin, A. A. (1958). *Dokl. Akad. Nauk SSSR* **123**, No. 4, 595–598.

Dezin, A. A. (1959). *Usp. Mat. Nauk* **14**, No. 3, 21–73.

Diesperov, V. N., and Ryzhov, O. S. (1967). *Prikl. Mat. Mekh.* **31**, 783.

Emmons, H. W. (1946). *Nat. Adv. Comm. Aeronaut., Tech. Notes* **1003**.

Emmons, H. W. (1948). *Nat. Adv. Comm. Aeronaut., Tech. Notes* **1746**.

Euvrard, D. (1967). *J. Mec.* **6**, 547.

Fal'kovitch, S. V., and Chernov, I. A. (1966). *Prikl. Mat. Mekh.* **30**, 848.

Ferrari, C., and Tricomi, F. G. (1968). "Transomic Aerodynamics." Academic Press, New York.

Frankl', F. I. (1947). *Dokl. Akad. Nauk SSSR* **57**, No. 7.

Gadd, G. E. (1960). *Z. Angew. Math. Phys.* **11**, 51.

Germain, P. (1964). *In* "Symposium Transsonicum" (K. Oswatitsh, ed.), p. 24. Springer, Berlin.

Germain, P., and Guiraud, J. P. (1964). *J. Math. Pures Appl.* **45**, 311.

Görtler, H. (1939). *Z. Angew. Math. Mech.* **19**, 325.

Guderley, K. G. (1947). Rep. F-TR-2168-ND. Wright Field.

Guderley, K. G. (1962). "The Theory of Transonic Flow." Addison-Wesley, Reading, Massachusetts.

Guderley, K. G., and Yoshihara, H. (1951). *Quart. Appl. Math.* **8**, No. 4, 333.

Hall, I. M., and Sutton, E. P. (1964). *In* "Symposium Transsonicum" (K. Oswatitsch, ed.), p. 325. Springer, Berlin.

Harbour, P. J., and Lewis, J. H. (1967). *In* "Rarefied Gas Dynamics" (C. L. Brundin, ed.), Vol. 2, p. 1031. Academic Press, New York.

HARTUNIAN, R. A. (1961). *Phys. Fluids* **4**, 1059.
HAYES, W. D. (1958). *In* "Fundamentals of Gas Dynamics" (H. W. Emmons, ed.), Sect. D, p. 416. Princeton Univ. Press, Princeton, New Jersey.
HAYES, W. D., and PROBSTEIN, R. F. (1959). "Hypersonic Flow Theory," 1st ed., Chapter IX. Academic Press, New York.
HOLDER, D. W. (1964). *J. Roy. Aeronaut. Soc.* **68**, 501.
HOSOKAWA, I. (1960a). *J. Phys. Soc. Jap.* **15**, 149.
HOSOKAWA, I. (1960b). *J. Phys. Soc. Jap.* **15**, 2080.
HUBERT, J. (1968). *C. R. Acad. Sci., Ser. A* **267**, 846.
KOPYSTYŃSKI, J., and SZANIAWSKI, A. (1965). *Arch. Mech. Stosowanej* **17**, 453.
LEVEY, H. C. (1954). *Quart. Appl. Math.* **12**, 25.
LEVEY, H. C. (1959). *Quart. Appl. Math.* **17**, 77.
LEVINSKY, E. S., and YOSHIHARA, H. (1962). *Progr. Astronaut. Rocketry* **7**, 81.
LIEPMANN, H. W., ASHKENAS, H., and COLE, J. D. (1948). *U.S. Air Force, Tech. Rep.* **5667**; see also HILTON, W. J. (1952). "High Speed Aerodynamics." Longmans, Green, New York.
LIGHTHILL, M. J. (1950). *Quart. J. Mech.* **3**, 303.
LIGHTHILL, M. J. (1956). *In* "Surveys in Mechanics" (G. K. Batchelor and R. M. Davies, eds.), p. 250. Cambridge Univ. Press, London and New York.
LIN, C. C., and RUBINOV, S. I. (1948). *J. Math. Phys.* **27**, 105.
LUDFORD, G. S. S. (1961). *J. Fluid Mech.* **10**, Part 1, 141.
MCCROSKEY, W. J., BOGDONOFF, S. M., and MCDOUGALL, J. G. (1966). *AIAA J.* **4**, 1580.
MCCROSKEY, W. J., BOGDONOFF, S. M., and GENCHI, A. P. (1967). *In* "Rarefied Gas Dynamics" (C. L. Brundin, ed.), Vol. 2, p. 1047. Academic Press, New York.
MCLACHLAN, N. W. (1947). "Theory and Application of Mathieu Functions." Oxford Univ. Press (Clarendon), London and New York.
MEYER, T. (1908). Ph.D. Dissertation, Göttingen; see also CARRIER, G. (1951). "Foundations of High Speed Aerodynamics." Dover, New York.
OGUCHI, H. (1967). Rep. No. 418. Inst. Space Aero. Sci., Univ. of Tokyo, Tokyo, Japan.
OSWATITSCH, K. (1956). "Gas Dynamics." Academic Press, New York.
OSWATITSCH, K., AND KEUNE, F. (1955). *Proc. Conf. High Speed Aeronaut.*, A. Ferri, N. J. Hoff, and P. A. Libby, eds., Polytech. Inst. of Brooklyn, p. 113.
OSWATITSCH, K., and ZIEREP, J. (1960). *Z. Angew. Math. Mech.* **40T**, Special Report 143.
PAN, Y. S., and PROBSTEIN, R. F. (1966). *In* "Fundamental Phenomena in Hypersonic Flow" (J. G. Hall, ed.), p. 259. Cornell Univ. Press, Ithaca, New York.
PARKER, E. N. (1965). *Space Sci. Rev.* **4**, 666.
PEARCEY, H. H. (1964). *In* "Symposium Transsonicum" (K. Oswatitsch, ed.), p. 264. Springer, Berlin.
PROBSTEIN, R. F., and KEMP, N. H. (1960). *J. Aerosp. Sci.* **27**, 174.
RAE, W. J. (1960). Rep. AFOSR-TN-60-409. Cornell University, Ithaca, New York.
RYZHOV, O. S. (1963). *Prikl. Mat. Mekh.* **27**, 309.
RYZHOV, O. S. (1965). *Prikl. Mat. Mekh.* **29**, 1004.
RYZHOV, O. S. (1968). *Zh. Vychisl. Mat. Mat. Fiz.* **8**, 472.
RYZHOV, O. S. (1969). *Fluid Dyn. Trans.* **4**, 379.
RYZHOV, O. S., and SHEFTER, G. M. (1964). *Prikl. Mat. Mekh.* **28**, 996.
RYZHOV, O. S., and TERENT'EV, E. D. (1967). *Prikl. Mat. Mekh.* **31**, 1035.
SAKURAI, A. (1958). *Quart. J. Mech.* **11**, 274.
SEDOV, L. I., MICHAILOVA, M. P., and CHERNYI, G. G. (1953). *Vestn. Mosk. Univ. Ser. III, Fiz. Astron.*, pp. 95–100 (translated by R. F. Probstein as WADC TN No. 59-349. Div. Eng., Brown University, Providence, Rhode Island, 1959).

Shorenstein, M. L., and Probstein, R. F. (1968). *AIAA J.* **6**, 1898.
Sichel, M. (1960). *J. Aerosp. Sci.* **27**, 635.
Sichel, M. (1961). Rep. No. 541. Dept. Aero. Eng., Princeton University, Princeton, New Jersey; see also Ph.D. Thesis, 1961.
Sichel, M. (1962). *Phys. Fluids* **5**, 1168.
Sichel, M. (1963). *Phys. Fluids* **6**, 653.
Sichel, M. (1966). *J. Fluid Mech.* **25**, 769.
Sichel, M. (1970). Interim Tech. Rep. 01361-1-T. Dept. Aerospace Eng., University of Michigan, Ann Arbor, Michigan.
Sichel, M., and Yin, Y. K. (1967). *J. Fluid Mech.* **28**, 513.
Sichel, M., and Yin, Y. K. (1969a). *Appl. Sci. Res.* **20**, 357.
Sichel, M., and Yin, Y. K. (1969b). *Fluid Dyn. Trans.* **4**, 403.
Sichel, M., Yin, Y. K., and David, T. S. (1968). ORA Rep. 07146-3-F. University of Michigan, Ann Arbor, Michigan.
Sinnott, C. S. (1960). *J. Aerosp. Sci.* **27**, 767.
Spreiter, J. R., and Alksne, A. Y. (1958). *Nat. Adv. Comm. Aeronaut.*, *Rep.* **1359**; see also 44th Ann. Rep., 1958.
Sternberg, J. (1959). *Phys. Fluids* **2**, 179.
Strehlow, R. A., and Maxwell, K. R., (1968). *AIAA J.* **6**, 2431.
Szaniawski, A. (1962). *Arch. Mech. Stosowanej* **14**, 905.
Szaniawski, A. (1963). *Arch. Mech. Stosowanej* **15**, 904.
Szaniawski, A. (1964a). *Arch. Mech. Stosowanej* **16**, 643.
Szaniawski, A. (1964b). "Importance des effects de dissipation en écoulement transsonique," Paper No. 64-587. Intern. Council Aeronaut. Sci. Congr., Paris.
Szaniawski, A. (1966). *Archi. Mech. Stosowanej* **18**, 127.
Szaniawski, A. (1967). *Z. Angew. Math. Mech.* **47**, 343.
Szaniawski, A. (1968). *Acta Mech.* **5**, 189.
Szaniawski, A. (1969). *Fluid Dyn. Trans.* **4**, 415.
Taylor, G. I. (1910). *Proc. Roy. Soc.*, *Ser. A* **84**, 371.
Taylor, G. I. (1930). Aeronaut. Res. Council. Repts. and Memoranda 1381 and 1382; see also *in* "The Scientific Papers of G. I. Taylor" (G. K. Batchelor, ed.), Vol. 3, pp. 128 and 142. Cambridge Univ. Press, London and New York.
Tomotika, S., and Tamada, K. (1950). *Quart. Appl. Math.* **7**, 381.
Tomotika, S., and Hasimoto, Z. (1950). *J. Math. Phys.* **29**, 105.
Tricomi, F. G. *et al.* (1954). "Tables of Integral Transforms," Vol. I, Appendix. McGraw Hill, New York.
Tsien, H. S. (1947). *J. Math. Phys.* **26**, 69.
Vas, I. E., Iacavazzi, C., Garlomagno, G., and Bogdonoff, S. M. (1969). *In* "Rarefied Gas Dynamics" (L. Trilling and H. Y. Wachman, eds.), Vol. 1, p. 501. Academic Press, New York.
Vasileva, A. B. (1963a). *Usp. Mat. Nauk* **18**, 15.
Vasileva, A. B. (1963b). *Th. Vychisl. Mat. Mat. Fiz.* **3**, 611.
Vidal, R. J., and Bartz, J. A. (1969). *AIAA J.* **7**, 1099.
von Mises, R. (1958). "Mathematical Theory of Compressible Fluid Flow," p. 73. Academic Press, New York.
Wu, T. Y. (1955). *Quart. Appl. Math.* **13**, 393.
Yin, Y. K., Ph.D. Thesis, The University of Michigan, 1971.
Zierep, J. (1958). *Z. Angew. Math. Phys.* **9b**, 764.

Nonlinear Theory of Random Vibrations

T. K. CAUGHEY

California Institute of Technology, Pasadena, California

I. Introduction

Nonlinear Theory of Random Vibrations

The general problem of random excitation of physical systems was first investigated theoretically by Einstein (1905) and was generalized and extended by von Smoluchowski (1916) in the context of the theory of Brownian motion. In 1931, Kolmogorov derived a precise mathematical formulation of the equations governing the probability densities satisfied by such processes. Contributions of major importance were also made by Fokker, Planck, Burger, Furth, Ornstein, Uhlenbeck,

Chandrasekhar, Kramers, and others, while Wiener, Khintchine, Feller, Levy, Doob, Ito, and others made important contributions to the purely mathematical aspects of the problem. A number of the important papers from this era have been reproduced in the book by Wax (1954).

The early studies were confined to the effects of additive noise on linear systems. The solution of linear problems of this type can be obtained by a number of standard techniques, which are adequately covered in current textbooks. In the case of nonlinear systems, however, considerable difficulties occur and the solution of such problems has not yet been completely resolved.

The earliest work on the problem of random excitation of nonlinear systems was that of Andronov *et al.* (1933), who used the Kolmogorov–Fokker–Planck equations (Kolmogorov, 1931) to study the motion of a general dynamic system subject to random disturbances. Kramers (1940) used this technique to study chemical reaction rates. More recently, Caughey and Dienes (1961), Lyon (1960a, 1961), Payne (1967), Wong (1964), Atkinson (1970), Ariaratnam (1960, 1962), Crandall (1962, 1964a,b), Klein (1964), and Herbert (1964, 1965) have used the Fokker–Planck equation to study the response of nonlinear systems to white noise excitation. Chuang and Kazda (1959), McFadden (1959), Wishner (1960), Barrett (1961), Pugachev (1957), Merklinger (1963), Stratonovich (1963), and Sawaragi *et al.* (1958-1961) have applied the technique to solve nonlinear control problems.

In almost all these investigations only first order statistical properties were obtained. While first order statistics are important parameters in the description of random processes, there are numerous applications where additional statistical information is required. For example, the spectral density of a random process requires a knowledge of the second order statistics of the process.

A number of approximate techniques have been developed to obtain second order statistics for the response of nonlinear systems to random excitation. Booton (1954) and Caughey (1959a) independently developed the method of equivalent linearization, which is simply a statistical extension of the well known equivalent linearization technique of Krylov and Bogoliubov (1937). Crandall (1961) developed a perturbational method based on calssical perturbation theory. Payne (1967), Wong (1964), and Atkinson (1970) have developed approximate techniques based on eigenfunction expansions and variational techniques.

Random vibration analysis of mechanical systems has become an important subject in recent years, principally because of advances in high speed flight. In order to design structures and equipment which will survive the randomly fluctuating loads caused by the flow of turbulent air or the efflux of jet or rocket engines, it has become necessary to develop a theory capable of analyzing the effect of such fluctuating loads on structures and equipment. Hight speed flight is, of course, only one example of a field where an understanding of random vibration theory is essential to the successful design of structures and equipment. Other examples include the response of structures to such forces as earthquakes, turbulence in air or water, storm waves, random seas, and the forces experienced by a vehicle traversing rough terrain.

A common feature of such problems is that the excitation is often so complex that is can be described only statistically. In addition, most physical systems behave in a linear manner only for a limited range of excitation, and since under random excitation large responses can be expected, at least occasionally, one is therefore forced to undertake a study of the response of nonlinear systems to random excitation.

Many of the techniques developed for the analysis of random excitation of nonlinear control systems are applicable to the analysis of nonlinear random vibrations, and conversely many of the techniques developed in the theory of nonlinear random vibrations are equally applicable to problems in communication theory and electronics. For this reason no distinction will be made amongst problems originating in different physical fields, rather the underleying unity of the subject will be stressed.

The literature on nonlinear random vibrations is quite extensive, and while an attempt has been made to include all the major references, the list of references cannot be claimed to be exhaustive.

II. Modeling in Nonlinear Random Vibrations by Markov Processes

The response of physical systems to external disturbances is often modeled by ordinary differential equations. The response is then represented by the solution of the ordinary differential equation, for example,

$$(d^2x/dt^2) + f[x, (dx/dt)] = g(t) \tag{2.1}$$

where $g(t)$ represents the external disturbance.

In some applications, the external disturbances cannot be successfully modeled by a single time function, because only the statistical nature of the disturbance is known. One possible approach to this situation is to determine the solution of the differential equation for each member of a family of functions $\{g(t)\}$ which represent the disturbance. It is clear that the response of the system can no longer be regarded as a single solution of an ordinary differential equation. The response would be more appropriately regarded as a family of solutions $\{x(t)\}$ characterized by some suitable statistics; for example, their mean and covariance.

In this way, one is led to consider the problem of determining the statistical nature of the solutions of differential equations in which the time functions representing the external disturbances are known only statistically. Of course, for a specific choice of the disturbance, say $g_k(t)$, there will be a specific solution $x_k(t)$. Such specific realizations are called sample functions since they are samples drawn from a large collection of disturbing functions and their corresponding responses. The entire collection of possible disturbing functions is called the stochastic process of the excitation. The entire collection of possible solutions to such excitation is called the stochastic process of the response.

The simplest situation to model is that in which the spectral density of the excitation is constant up to some frequency, f_c, which is very large compared with the characteristic frequencies of the dynamic system. In this case the excitation can be modeled by a white, Gaussian noise process which, formally, has the properties of the derivative of a Wiener process $W(t)$. In other situations the excitation process cannot be modeled by white noise, but can be modeled by the response of a secondary system to white noise. For example, $g(t)$, as defined by

$$\sum_{i=0}^{N} a_i \frac{d^i g}{dt^i} = n(t) \tag{2.2}$$

with $n(t)$ white noise, may serve as a model for the excitation of the dynamical system represented by Eq. (2.1). The desired response can then be thought of as the response represented by the enlarged system of differential equations, for example, (2.1) and (2.2) excited by white noise. The procedure of enlarging the system of differential equations expands the possible range of applicability of the method to a wider class of problems.

The chief reason for adopting the idealized model of a system of differential equations excited by white noise is, of course, that the computations are much simpler in this case. In fact, choosing white noise excitation allows the development of a rather complete theory, since it can be shown that the response in this case is a Markov process, for which there exists a large body of theory.

One of the difficulties involved in modeling nonlinear random vibrations by Markov processes is that, in order to guarantee the existence of a unique solution to a nonlinear stochastic differential equation, one is restricted to quasi-linear systems. At first sight this may seem to be a severe limitation on the class of problems which may be treated. It should be remembered, however, that all physical excitations are bounded in nature, and hence most physical systems exhibit bounded response. Thus, if a physical nonlinearity is modeled by a quasi-linear mathematical nonlinearity in such a way that the two coincide over the range of physical response, the modeling of random vibrations by Markov processes will usually be satisfactory, except for the prediction of extreme statistics.

In the subsequent development of the theory no distinction will be made between the physical nonlinearity and the mathematical model of that nonlinearity.

III. Basic Theory of Stochastic Processes

A. The Wiener Process

A Wiener process, $w(t)$, is characterized by the following properties:

(i) If $t_1 < t \leqslant \cdots < t_n$, the differences

$$[w(t_2) - w(t_1)], [w(t_3) - w(t_2)],..., [w(t_n) - w(t_{n-1})]$$

are mutually independent.

(ii) $w(t) - w(s)$ is normally distributed (Gaussian), with*

$$E[w(t) - w(s)] = 0$$
$$E[(w(t) - w(s))^2] = \sigma^2 | t - s |.$$

Formally one can express these properties in terms of differentials:

(i) $dw(t_i)$, $i = 1,..., n$, are mutually independent.

(ii) $E[dw(t)] = 0$.

$$E[(dw(t))^2] = \sigma^2 \, dt.$$

It is also useful to note that formal differentiation leads to

$$E[dw(t) \, dw(s)] = \sigma^2 \, \delta(t - s) \, dt \, ds,$$

where $\delta(t - s)$ is the Dirac delta function.

The formal properties of white noise, defined by

$$n(t) = dw/dt,$$

* $E[\cdot]$ denotes expectation.

are then:

(i) $n(t_i)$, $i = 1, 2, \ldots, n$ are mutually independent.
(ii) $n(t)$ is normally distributed (Gaussian) with

$$E[n(t)] = 0$$
$$E[n(t)n(s)] = \sigma^2 \, \delta(t - s).$$

Many problems in mechanics and related fields, involving the response of dynamical systems to stochastic excitation can be modeled as systems of first order differential equations of the form

$$\frac{d\mathbf{x}}{dt} = \mathbf{a}(t, \mathbf{x}) + \sum_{k=1}^{m} \mathbf{b}_k(t, \mathbf{x}) \frac{dw_k(t)}{dt}, \qquad \mathbf{x}(t_0) = \mathbf{y}, \tag{3.1}$$

where $\mathbf{x}$, $\mathbf{a}$, $\mathbf{b}_k$, $k = 1, 2, \ldots, m$ are m vectors and the $w_k(t)$ for $k = 1, 2, \ldots, m$ are independent processes of Brownian motion. The vectors $\mathbf{a}(t, \mathbf{x})$, $\mathbf{b}_k(t, \mathbf{x})$ are defined for $t \in [t_0, T]$ and $\mathbf{x} \in R^m$ [the m-dimensional Euclidean space] and their range is in R^m. Some authors prefer to write (3.1) in differential form, since in many cases $d\mathbf{x}/dt$ exists almost nowhere, thus

$$d\mathbf{x} = \mathbf{a}(t, \mathbf{x})\, dt + \sum_{k=1}^{m} \mathbf{b}_k(t, \mathbf{x})\, dw_k(t), \qquad \mathbf{x}(t_0) = \mathbf{y}. \tag{3.2}$$

At this point, we require an interpretation of the relation (3.2) between stochastic differentials which is in accord with the physical phenomena being studied. If $\mathbf{b}_k(t, \mathbf{x})$ is independent of $\mathbf{x}$ and t, there is general agreement on the interpretation. First, if there were no stochastic disturbances $\mathbf{b}_k \equiv 0$, the system would evolve deterministically, and in that case

$$\lim_{\Delta t \to 0} \{(\Delta t)^{-1}[x_i(t + \Delta t) - x_i(t)]\} = a_i(t, \mathbf{x}).$$

Often, the effect of the stochastic disturbances is to cause the system to wander randomly about the deterministic path in such a way that the average behavior is the deterministic path. This can be interpreted by the statement

$$\lim_{\Delta t \to 0} \{(\Delta t)^{-1} E[x_i(t + \Delta t) - x_i(t)]\} = a_i(t, \mathbf{x}), \tag{3.3}$$

differing from the previous statement concerning the deterministic behavior only in that the expectation, $E(\cdot)$, has been introduced. The effect of the stochastic excitation can be seen by considering the higher order incremental moments. Equation (3.3) is a statement concerning the first incremental moments. Using the properties of the Wiener process, it is seen that

$$\begin{aligned}\lim_{\Delta t\to 0}&\{(\Delta t)^{-1}E[(x_i(t+\Delta t)-x_i(t))(x_j(t+\Delta t)-x_j(t))]\}\\ &=\sum_{k=1}^{m} b_{ik}(t,\mathbf{x})\, b_{jk}(t,\mathbf{x})\\ &=B_{ij}(t,\mathbf{x}).\end{aligned} \tag{3.4}$$

Due to the Gaussian nature of the Wiener process, it is easily seen that

$$\lim_{\Delta t\to 0}\{(\Delta t)^{-1}E[(x_i(t+\Delta t)-x_i(t))(x_j(t+\Delta t)-x_j(t))(x_k(t+\Delta t)-x_k(t))]\}=0 \tag{3.5}$$

for all triple and higher order incremental moments. Equation (3.4) has the interpretation that in a small time Δt, the sample paths spread out around the deterministic paths by an amount characterized by $B_{ij}\,\Delta t$.

When $\mathbf{b}_k(t, x)$ depends on $\mathbf{x}(t)$, there is some question as to whether the sole effect of the stochastic disturbance is to diffuse the sample paths; the stochastic disturbances may also effect changes in the average behavior as expressed by the first incremental moments. The resolution of this particular point depends ultimately upon a proper consideration of the physical process being studied. A discussion of this point has been taken up by Caughey and Gray (1965). Ito (1946) has chosen the interpretation expressed by (3.3), (3.4), and (3.5) even when $B_{ij}(t, \mathbf{x})$ depends on $\mathbf{x}(t)$. Stratonovich (1963) has offered an alternate interpretation in which the stochastic disturbances do affect the first incremental moments.

In many physical systems with additive stochastic excitation, the $\mathbf{b}_k(t, \mathbf{x})$ are constants, independent of t and $\mathbf{x}(t)$. In this case, the stochastic excitation does not affect the first incremental moments, even in the interpretation adopted by Stratonovich.

Problems of existence and unicity are basic to the study of systems of ordinary differential equations. In the study of stochastic differential equations there is the analogous problem of establishing existence and unicity of solutions. Ito (1946) designed a stochastic calculus which allows one, with appropriate conditions on the incremental moments, to do just that. The main theorems will be stated here without proof; the details of the proofs can be found in Doob (1953) or Gikhman and Skorokhod (1965).

B. Existence and Unicity of Solutions of Stochastic Differential Equations

As stated previously, many problems in mechanics and related fields involving the response of dynamical systems to stochastic excitations can be modeled by stochastic differential equations of the form

$$\begin{aligned} d\mathbf{x}(t) &= \mathbf{a}[t, \mathbf{x}(t)]\, dt + \sum_{k=1}^{m} \mathbf{b}_k[t, \mathbf{x}(t)]\, dw_k(t), \\ \mathbf{x}(t_0) &= \mathbf{y}, \end{aligned} \tag{3.6}$$

where $\mathbf{x}$, $\mathbf{a}$, $\mathbf{b}_k$, $k = 1, 2, \ldots, m$, are m vectors and the $w_k(t)$, for $k = 1, 2, \ldots m$ are independent Wiener processes. The vector functions $\mathbf{a}(t, \mathbf{x})$, $\mathbf{b}_k(t, \mathbf{x})$ are defined for $t \in [t_0, T]$ and $\mathbf{x} \in R^m$ and their ranges are in R^m.

Equation (3.6) is equivalent to the integral equation

$$\mathbf{x}(t) = \mathbf{y} + \int_{t_0}^{t} \mathbf{a}[s, \mathbf{x}(s)]\, ds + \sum_{k=1}^{m} \int_{t_0}^{t} \mathbf{b}_k[s, \mathbf{x}(s)]\, dw_k(s). \tag{3.7}$$

This equation is solved for a given $\mathbf{y}$, which is assumed independent of the Wiener processes $W_k(t)$, $k = 1, 2, \ldots, m$.

Theorem 1. [Ito, 1951]. *Let* $\mathbf{a}(t, \mathbf{x})$, $\mathbf{b}_1(t, \mathbf{x})$, $\mathbf{b}_2(t, \mathbf{x}), \ldots, \mathbf{b}_m(t, \mathbf{x})$ *denote Borel functions defined for* $t \in [t_0, T]$ *and* $\mathbf{x} \in R^m$ *with ranges in* R^m. *If there exists a constant* K, *such that*

$$\begin{aligned} |\mathbf{a}(t, \mathbf{x})|^2 + \sum_{k=1}^{m} |\mathbf{b}_k(t, \mathbf{x})|^2 &\leqslant K^2(1 + |\mathbf{x}|^2) \\ |\mathbf{a}(t, \mathbf{x}) - \mathbf{a}(t, \mathbf{y})| + \sum_{k=1}^{m} |\mathbf{b}_k(t, \mathbf{x}) - \mathbf{b}_k(t, \mathbf{y})| &\leqslant K|\mathbf{x} - \mathbf{y}| \end{aligned} \tag{3.8}$$

for every $\mathbf{x}, \mathbf{y} \in R^m$, *then* (3.7) *has a solution* $\mathbf{x}(t)$ *which is unique up to a stochastic equivalence and continuous with probability one. This solution*

$\mathbf{x}(t)$ is a Markov process whose transition probabilities $P(A, t \mid \mathbf{y}, s)$ for $s < t$ are given by the relation

$$P(A, t \mid \mathbf{y}, s) = P\{\underset{s,y}{\mathbf{x}(t)} \in A\}$$

where $\underset{s,y}{\mathbf{x}(t)}$ is the solution of the integral equation

$$\underset{s,y}{\mathbf{x}(t)} = \mathbf{y} + \int_s^t \mathbf{a}[\xi, \underset{s,y}{\mathbf{x}(\xi)}] \, d\xi + \sum_{k=1}^{m} \int_s^t \mathbf{b}_k[\xi, \underset{s,y}{\mathbf{x}(\xi)}] \, dw_k(\xi). \tag{3.9}$$

If the functions $\mathbf{a}(t, \mathbf{x})$ and $\mathbf{b}_k(t, \mathbf{x})$, $k = 1, 2, \ldots, m$ are continuous with respect to t, then the process $\mathbf{x}(t)$ is a diffusion process with transfer vector $\mathbf{a}(t, \mathbf{x})$ and diffusion operator $B(t, \mathbf{x})$ satisfying the equation

$$[B(t, \mathbf{x})\mathbf{z}, \mathbf{z}] = \sum_{k=1}^{m} [\mathbf{b}_k(t, \mathbf{x}), \mathbf{z}]^2. \tag{3.10}$$

If in (3.6) the vector functions $\mathbf{a}(t, \mathbf{x})$ and $\mathbf{b}_k(t, \mathbf{x})$ do not depend explicity on t, that is, if the stochastic differential equation is of the form

$$d\mathbf{x} = \mathbf{a}(\mathbf{x}) \, dt + \sum_{k=1}^{m} \mathbf{b}_k(\mathbf{x}) \, dw_k(t) \tag{3.11}$$

and $\mathbf{a}(\mathbf{x})$ and $\mathbf{b}_k(\mathbf{x})$ satisfy the conditions of Theorem 1, then the solution $\mathbf{x}(t)$ is a homogeneous Markov process; that is, the transistion probability $P(A, t + \tau \mid \mathbf{y}, t)$ is independent of t.

The Markovian property of these solutions has the interpretation that the statistics of the future states depend only on the present state and the Wiener process $w_k(t)$, $k = 1, 2, \ldots, m$. This is analogous to the situation in dynamical systems.

The difficulty in extending the method employed by Ito to establish existence and unicity for other than quasi-linear systems is related to the distinction between establishing local and global existence for ordinary differential equations. In the case of ordinary differential equations, local or global existence depends on the existence of a local or a global Lipschitz condition on the nonlinearities. It is inherent in the nature of the stochastic processes being considered that the system variables have a finite probability of exceeding any fixed bound, so that only global existence has any meaning.

It is important to remember that the conditions of Theorem 1 are sufficient, not necessary and sufficient, for the existence of a unique solution to the system of stochastic differential equations. Thus the system of stochastic differential equations may possess a unique solution even though the conditions of Theorem 1 are violated.

C. The Statistical Description of a Continuous Parameter Markov Process

As already noted, for each sample function representing the external disturbance there is a corresponding sample function representing the response of the system. For a specified collection of sample functions representing the external disturbances there is a corresponding collection of sample functions representing the response of the system. Questions of general interest would involve statements about the values of sample functions at discrete instants of time, or statements about the values of the sample functions over prescribed time intervals. What is desired then is a statistical description of the response process which will provide answers to these two general types of questions. To this end it will be necessary to introduce the concept of an m-component stochastic process.

1. *Definition*

An m-component stochastic process takes on values in an m-dimensional space, called the phase space, R^m. Sets in phase space shall be denoted by Γ. The conditional probability that the system occupy sets $\Gamma_1, \Gamma_2, \Gamma_3, \ldots, \Gamma_n$ at successive times $t_1, t_2, \ldots, t_n$, given that the system occupied states $\mathbf{y}_1, \mathbf{y}_2, \ldots, \mathbf{y}_p$, at successive times $s_1, s_2, \ldots, s_p$ shall be denoted by

$$P(\Gamma_1, t_1; \Gamma_2, t_2; \cdots; \Gamma_n, t_n \mid \mathbf{y}_1, s_1; \mathbf{y}_2, s_2; \cdots; \mathbf{y}_p, s_p).$$

If P has a density such that

$$P(\Gamma_1, t_1; \Gamma_2, t_2; \cdots; \Gamma_n, t_n \mid \mathbf{y}_1, s_1; \cdots; \mathbf{y}_p, s_p)$$

$$= \int_{\Gamma_1} d\mathbf{x}_1 \cdots \int_{\Gamma_n} d\mathbf{x}_n p(\mathbf{x}_1, t_1; \mathbf{x}_2, t_2; \cdots; \mathbf{x}_n, t_n \mid \mathbf{y}_1, s_1; \cdots; \mathbf{y}_p, s_p),$$

then $p(\mathbf{x}_1, t_1; \cdots \mid \mathbf{y}_1, s_1; \cdots)$ shall be called the conditional probability density function. It will be assumed that P has a density, and much of the subsequent theory will be developed in terms of the probability density function. If a consistent set of joint probability density functions $p_1(\mathbf{x}_1, t_1)$, $p_2(\mathbf{x}_1, t_1; \mathbf{x}_2, t_2)$, etc., could be formulated, then general questions of the first type could be answered.

Kolmogorov (1950) has shown that necessary and sufficient conditions for the determination of a valid joint probability density function are:

(i) $p_n(\mathbf{x}_1, t_1; \mathbf{x}_2, t_2; \cdots; x_n, t_n) \geqslant 0, \forall n = 1, 2, \ldots.$

(ii) $p_n(\mathbf{x}_1, t_1; \mathbf{x}_2, t_2; \cdots; \mathbf{x}_n, t_n)$ is a symmetric function in the n sets of variables $\mathbf{x}_1 t_1; \mathbf{x}_2, t_2; \cdots; \mathbf{x}_n, t_n$.

(iii) $p_k(\mathbf{x}_1, t_1; \cdots; \mathbf{x}_k, t_k) = \int d\mathbf{x}_{k+1} \underset{R^{(n-k)}}{\cdots} \int d\mathbf{x}_n \, p_n(\mathbf{x}_1, t; \cdots; \mathbf{x}_n, t_n)$.

Furthermore, Kolmogorov has shown by means of his Extension Theorem that the hierarchy of joint probability density functions can be extended uniquely to assign probabilities to events such as $|x(t)| \leqslant M$, $0 \leqslant t \leqslant T$. Thus the distribution function can be used to answer questions of the two general types posed above. In this paper we shall be concerned only with questions of the first types.

2. *The Transition Probability Density Function*

A stochastic process is Markovian if, for $t_1 < t_2 < \cdots < t_n < \cdots < t_{m+n}$,

$$p(\mathbf{x}_{n+1}, t_{n+1}; \cdots; \mathbf{x}_{n+m}, t_{n+m} \mid \mathbf{x}_1, t_1; \cdots; \mathbf{x}_n, t_n)$$
$$= p(\mathbf{x}_{n+1}, t_{n+1}; \cdots; \mathbf{x}_{n+m}, t_{n+m} \mid \mathbf{x}_n, t_n).$$

This is not the only definition of a Markov process, but it is a convenient one which will serve our present purposes. It is generally true of probability density functions that

$$\begin{aligned} p(\mathbf{x}, t_1; \cdots; \mathbf{x}_n, t_n) &= p(\mathbf{x}_2, t_2; \cdots; \mathbf{x}_n, t_n \mid \mathbf{x}_1, t_1) p(\mathbf{x}_1, t_1) \\ &= p(\mathbf{x}_3, t_3; \cdots; \mathbf{x}_n, t_n \mid \mathbf{x}_1, t_1; \mathbf{x}_2, t_2) p(\mathbf{x}_2, t_2 \mid \mathbf{x}_1, t_1) p(\mathbf{x}_1, t_1) \\ &= p(\mathbf{x}_n, t_n \mid \mathbf{x}_1, t_1; \mathbf{x}_2, t_2; \cdots; \mathbf{x}_{n-1}, t_{n-1}) \cdots p(\mathbf{x}_2, t_2 \mid \mathbf{x}_1, t_1) p(\mathbf{x}_1, t_1). \end{aligned}$$

If $t_1 < t_2 < \cdots < t_n$, the Markovian property of $\mathbf{x}(t)$ allows considerable simplification:

$$p(\mathbf{x}_1, t_1; \mathbf{x}_2, t_2; \cdots; \mathbf{x}_n, t_n)$$
$$= p(\mathbf{x}_n, t_n \mid \mathbf{x}_{n-1}, t_{n-1}) \cdots p(\mathbf{x}_2, t_2 \mid \mathbf{x}_1, t_1) p(\mathbf{x}_1, t_1).$$

The conditional probability density function $p(\mathbf{x}_i, t_i \mid \mathbf{x}_{i-1}, t_{i-1})$ is, for the Markov process, called the transition probability density function. The transition probability density function gives the density of probability of a transition from one point in phase space, $\mathbf{x}_{i-1}$, at time t_{i-1} to a point in phase space, $\mathbf{x}_i$, at time t_i, where $t_i > t_{i-1}$.

3. *The Chapman–Kolmogorov Equation*

Using the Markov property; if $t_1 < t_2 < t_3$,

$$p(\mathbf{x}_2, t_2; \mathbf{x}_3, t_3 \mid \mathbf{x}_1, t_1) = p(\mathbf{x}_3, t_3 \mid \mathbf{x}_2, t_2) p(\mathbf{x}_2, t_2 \mid \mathbf{x}_1, t_1).$$

Integrating over $\mathbf{x}_2$, this becomes

$$p(\mathbf{x}_3, t_3 \mid \mathbf{x}_1, t_1) = \int_{R^m} p(\mathbf{x}_3, t_3 \mid \mathbf{x}_2, t_2) p(\mathbf{x}_2, t_2 \mid \mathbf{x}_1, t_1)\, d\mathbf{x}_2, \tag{3.12}$$

the reduction on the left hand side being a consequence of the necessity of Kolmogorov's consistency condition. This equation is known as the Chapman–Kolmogorov or Smoluchowski equation.

If the stochastic process $\mathbf{x}(t)$ is a homogeneous Markov process, then

$$p(\mathbf{x}_2, t_2 \mid \mathbf{x}_1, t_1) = p(\mathbf{x}_2, t \mid \mathbf{x}_1, 0) \qquad t_2 > t_1$$

where

$$t = (t_2 - t_1).$$

In this case, the transition probability density function is simply written $p(\mathbf{x}_2, t \mid \mathbf{x}_1)$. The Chapman–Kolmogorov equation becomes

$$p(\mathbf{x}_3, t + \tau \mid \mathbf{x}_1) = \int_{R^m} p(\mathbf{x}_3, t \mid \mathbf{x}_2) p(\mathbf{x}_2, \tau \mid \mathbf{x}_1)\, d\mathbf{x}_2. \tag{3.13}$$

The joint probability density functions in this case are then functions only of the time differences

$$p(\mathbf{x}_1, t_1; \mathbf{x}_2, t_2; \cdots; \mathbf{x}_n, t_n) = p(\mathbf{x}_1; \mathbf{x}_2, \tau_2; \cdots; \mathbf{x}_n, \tau_n)$$

where

$$\tau_j = t_j - t_1.$$

If the limit

$$\lim_{t\to\infty} p(\mathbf{x}, t \mid \mathbf{y}) = p_s(\mathbf{x})$$

exists, independent of t and $\mathbf{y}$, $p_s(\mathbf{x})$ is called the steady-state probability density function.

4. *Moments, Covariance Matrix, and Spectral Density Function*

It is clear that the set of joint probability density functions generated by the transition probability density and an initial probability density suffices to answer any questions which may be described in terms of the values taken by the stochastic process at discrete instants of time. A few important statistics of this type will now be considered.

Of obvious interest are the moments of the process, defined by

$$m_{j_1,j_2,\ldots,j_n}(t \mid \mathbf{y}) = \int_{R^m} x_1^{j_1} x_2^{j_2} \cdots x_n^{j_n} p(\mathbf{x}, t \mid \mathbf{y})\, d\mathbf{x}, \tag{3.14}$$

where

$$n \leqslant m.$$

In particular the first moments are

$$m_k(t \mid \mathbf{y}) = \int_{R^m} x_k p(\mathbf{x}, t \mid \mathbf{y})\, d\mathbf{x} \qquad k = 1, 2, \ldots, m. \tag{3.15}$$

For a dynamical system governed by a set of linear stochastic equations, the transition probability density functions is normal (Gaussian). This follows from the fact that the Wiener process is normal and a linear transformation of a normal process is itself a normal process.

The transition probability density function of a normal process is completely characterized by its first and second moments. The first moments, $m_k(t \mid \mathbf{y})$, as defined above, are the components of the mean vector $\mathbf{m}(t \mid \mathbf{y})$. The second moment matrix, with elements

$$K_{jk}(t \mid \mathbf{y}) = \int_{R^m} [x_j - m_j(t \mid \mathbf{y})][x_k - m_k(t \mid \mathbf{y})]\, p(\mathbf{x}, t \mid \mathbf{y})\, d\mathbf{x}, \tag{3.16}$$

is called the correlation matrix.

Thus the transition probability density function $p(\mathbf{x}, t \mid \mathbf{y})$ is given by

$$p(\mathbf{x}, t \mid \mathbf{y}) = (1/(2\pi)^{m/2})(1/|K|^{1/2}) \exp\{-\tfrac{1}{2}[(\mathbf{x} - \mathbf{m}), K^{-1}(\mathbf{x} - \mathbf{m})]\}, \tag{3.17}$$

where K is the matrix with elements $K_{ij}(t \mid \mathbf{y})$. $|K|$ is the determinant of the correlation matrix K.

Also of particular interest for linear dynamical systems is the matrix of second order moments for the second joint probability density function:

$$\begin{aligned} R_{jk}(t, t+\tau \mid \mathbf{y}) &= \iint_{R^{2m}} [z_k - m_k(t+\tau \mid \mathbf{y})][x_j - m_j(t \mid \mathbf{y})] \cdot p(\mathbf{x}, t; \mathbf{z}, t+\tau)\, d\mathbf{x}\, d\mathbf{z} \\ &= \iint_{R^{2m}} [z_k - m_k(t+\tau \mid \mathbf{y})][x_j - m_j(t \mid \mathbf{y})] \cdot p(\mathbf{z}, \tau \mid \mathbf{x}) p(\mathbf{x}, t \mid \mathbf{y})\, d\mathbf{x}\, d\mathbf{z}, \end{aligned} \tag{3.18}$$

where, in obtaining the last relationship, use has been made of the Markov property and stationarity.

If the linear dynamical system is asymptotically stable in the absence of external disturbances, then one has the following results:

(i) $\lim_{t\to\infty} p(\mathbf{x}, t \mid \mathbf{y}) = p_s(\mathbf{x})$ independent of t and $\mathbf{y}$.
(ii) $\lim_{t\to\infty} m_j(t \mid \mathbf{y}) = 0$.

Hence, as $t \to \infty$, the elements of the covariance matrix are given by

$$\lim_{t\to\infty} R_{jk}(t, t+\tau \mid \mathbf{y}) = R_{jk}(\tau) = \iint_{R^{2m}} x_j z_k p(\mathbf{z}, \tau \mid \mathbf{x}) p_s(\mathbf{x})\, d\mathbf{x}\, d\mathbf{z}. \tag{3.19}$$

The elements, $\Phi_{jk}(\omega)$, of the spectral density matrix, $\Phi(\omega)$, are related to the elements of the elements, $R_{jk}(\tau)$, of the covariance matrix, $R(\tau)$, by the Wiener–Khintchine relations (Khintchine, 1934)

$$\Phi_{jk}(\omega) = \frac{1}{\pi}\int_0^\infty [R_{jk}(\tau) + R_{kj}(\tau)] \cos \omega t\, dt$$
$$- \frac{i}{\pi}\int_0^\infty [R_{jk}(\tau) - R_{kj}(\tau)] \sin \omega\tau\, d\tau, \tag{3.20}$$

$$R_{jk}(\tau) = \frac{1}{2}\int_0^\infty [\Phi_{jk}(\omega) + \Phi_{kj}(\omega)] \cos \omega\tau\, d\omega$$
$$+ \frac{i}{2}\int_0^\infty [\Phi_{jk}(\omega) - \Phi_{kj}(\omega)] \sin \omega\tau\, d\omega. \tag{3.21}$$

From the first of these relations it is seen that the cross-spectral density $\Phi_{jk}(\omega)$ is Hermitian.

In particular, if $j = k$, we have

$$\Phi_{jj}(\omega) = \frac{2}{\pi}\int_0^\infty R_{jj}(\tau) \cos \omega\tau\, d\tau, \tag{3.22}$$

$$R_{jj}(\tau) = \int_0^\infty \Phi_{jj}(\omega) \cos \omega\tau\, d\omega. \tag{3.23}$$

The spectral density function $\Phi_{jj}(\omega)$ and the autocorrelation function $R_{jj}(\tau)$ play an important role in harmonic analysis and prediction theory for linear systems.

In nonlinear stochastic differential equations the structure of the transition probability density function is usually much more complex than that for linear systems, and cannot be obtained in as direct a fashion. The most common method of obtaining the transition probability density function for nonlinear stochastic differential equations is through the use of the Kolmogorov–Fokker–Planck equations.

D. Diffusion Processes and the Kolmogorov Equations

A stochastic process $\mathbf{x}(t)$ is said to be a diffusion process if the following conditions are satisfied.

(a) For every $\mathbf{y}$ and every $\epsilon > 0$, the transition probability density function $p(\mathbf{x}, t \mid \mathbf{y}, s)$ satisfies the condition

$$\int_{|\mathbf{x}-\mathbf{y}|>\epsilon} p(\mathbf{x}, t \mid \mathbf{y}, s)\, d\mathbf{x} = O(t-s)$$

uniformly over $t > s$ and $\mathbf{x}, \mathbf{y} \in R^m$.

(b) There exist functions $a_i(t, \mathbf{x})$, $B_{ij}(s, \mathbf{y})$ such that for every $\mathbf{y} \in R^m$ and every $\epsilon > 0$

$$\text{(i)} \quad \int_{|x-\mathbf{y}|\leqslant\epsilon} (x_i - y_i)\, p(\mathbf{x}, t \mid \mathbf{y}, s)\, d\mathbf{x} = a_i(t, \mathbf{y})(t-s) + O(t-s) \qquad i = 1, 2,\ldots, m$$

$$\text{(ii)} \quad \int_{|x-\mathbf{y}|\leqslant\epsilon} (x_i - y_i)(x_j - y_j)\, p(\mathbf{x}, t \mid \mathbf{y}, s)\, d\mathbf{x} = B_{ij}(s, \mathbf{y})(t-s) + O(t-s) \qquad i, j = 1, 2,\ldots, m$$

uniformly for $t \geqslant s$ and $\mathbf{x}, \mathbf{y} \in R^m$.

Conditions (a) and (b) are not sufficient for a unique determination of the transition probability densities of the process. If, in addition to conditions (a) and (b), certain assumptions are made about the differentiability of the probability density functions, $p(\mathbf{x}, t \mid \mathbf{y}, s)$, it can be shown that $p(\mathbf{x}, t \mid \mathbf{y}, s)$ is completely determined by the coefficients $a_i(t, \mathbf{x})$, $B_{ij}(t, \mathbf{x})$ provided the Cauchy problem for the Fokker–Planck–Kolmogorov equations possesses a unique solution.

The Fokker–Planck–Kolmogorov Equation

THEOREM 2 [Fokker–Planck–Kolmogorov]. (i) *If the transition Probability $P(A, t \mid \mathbf{y}, s)$ has a density $p(\mathbf{x}, t \mid \mathbf{y}, s)$ so that*

$$P(A, t \mid \mathbf{y}, s) = \int_A p(\mathbf{x}, t \mid \mathbf{y}, s)\, d\mathbf{x},$$

and

(ii) *if conditions (a) and (b) of Section III, D are satisfied uniformly with respect to* $\mathbf{y}$, *and if there exist continuous derivatives*

$$(\partial/\partial t)[p(\mathbf{x}, t \mid \mathbf{y}, s)], \qquad (\partial/\partial x_i)[a_i(t, \mathbf{x})\, p(\mathbf{x}, t \mid \mathbf{y}, s)]$$

and

$$(\partial^2/\partial x_i\, \partial x_j)[B_{ij}(t, \mathbf{x})\, p(\mathbf{x}, t \mid \mathbf{y}, s)] \qquad i, j = 1, 2,\ldots, m,$$

then $p(\mathbf{x}, t \mid \mathbf{y}, s)$ *for* $\mathbf{x}, \mathbf{y} \in R^m$ *and* $t \in [s, T]$ *satisfies the equation*

$$\frac{\partial p}{\partial t} = -\sum_{i=1}^{m} \frac{\partial}{\partial x_i}[a_i(t, \mathbf{x})p] + \frac{1}{2}\sum_{i,j=1}^{m} \frac{\partial^2}{\partial x_i\, \partial x_j}[B_{ij}(t, \mathbf{x})p]$$

with initial codition

$$\lim_{t \downarrow s} p(\mathbf{x}, t \mid \mathbf{y}, s) = \delta(\mathbf{x} - \mathbf{y}). \tag{3.24}$$

This equation is known as the Fokker–Planck equation (Fokker, 1914) or the forward Kolmogorov equation (Kolmogorov, 1931). The proof of this theorem is given in most modern texts on stochastic processes, for example, Gikhman and Skorokhod (1965).

THEOREM 3 [Kolmogorov]. (i) *If the transition probability density* $P(A, t \mid \mathbf{y}, s)$ *has a density* $p(\mathbf{x}, t \mid \mathbf{y}, s)$, *and* (ii) *if conditions* (*a*) *and* (*b*) *above are satisfied, and if* $a_i(s, \mathbf{y})$, $B_{ij}(s, \mathbf{y})$ *are continuous and if there exist continuous derivatives*

$$(\partial/\partial s)[p(\mathbf{x}, t \mid \mathbf{y}, s)], \qquad (\partial/\partial y_i)[p(\mathbf{x}, t \mid \mathbf{y}, s)], \qquad (\partial^2/\partial y_i\, \partial y_j)[p(\mathbf{x}, t \mid \mathbf{y}, s)],$$

then $p(\mathbf{x}, t \mid \mathbf{y}, s)$ *for* $\mathbf{y} \in R^m$ *and* $s \in [0, t]$ *satisfies the equation*

$$-\frac{\partial p}{\partial s} = \sum_{i=1}^{m} a_i(s, \mathbf{y}) \frac{\partial p}{\partial y_i} + \frac{1}{2}\sum_{i,j=1}^{m} B_{ij}(s, \mathbf{y}) \frac{\partial^2 p}{\partial y_i\, \partial y_i}$$

with initial condition

$$\lim_{s \uparrow t} p(\mathbf{x}, t \mid \mathbf{y}, s) = \delta(\mathbf{x} - \mathbf{y}). \tag{3.25}$$

This equation is known as the backward or converse Kolmogorov equation. The proof of this theorem is given in Gikhman and Skorokhod (1965).

If we denote the spatial operator of the forward equation by $L_\mathbf{x}$, the forward and backward Kolmogorov equation may be written

$$\partial p/\partial t = L_\mathbf{x} p \qquad \text{and} \qquad \partial p/\partial s = -L_\mathbf{y}{}^* p$$

where L^* is the formal adjoint of L.

If the incremental moments are independent of time, the transition probability density function is stationary, and the backward and forward Kolmogorov equations for $p(\mathbf{x}, t \mid \mathbf{y})$ are

$$\partial p/\partial t = L_\mathbf{y}{}^* p$$

with

$$\lim_{t\downarrow 0} p(\mathbf{x}, t \mid \mathbf{y}) = \delta(\mathbf{x} - \mathbf{y}), \tag{3.26}$$

$$\partial p/\partial t = L_{\mathbf{x}} p$$

with

$$\lim_{t\downarrow 0} p(\mathbf{x}, t \mid \mathbf{y}) = \delta(\mathbf{x} - \mathbf{y}). \tag{3.27}$$

A solution, $p(\mathbf{x}, t \mid \mathbf{y})$, must satisfy both (3.26) and (3.27) and, in order to be a transition probability density function, the following conditions are satisfied:

$$p \geqslant 0 \qquad \mathbf{x}, \mathbf{y} \in R^m, \qquad \forall t$$

$$\int_{R^m} p(\mathbf{x}, t \mid \mathbf{y})\, d\mathbf{x} = 1$$

$$\lim_{t\downarrow 0} \int_{R^m} p(\mathbf{x}, t \mid \mathbf{y})\, \varphi(\mathbf{y})\, d\mathbf{y} = \varphi(\mathbf{x})$$

for any initial probability density, $\varphi(\mathbf{y})$.

E. Existence and Unicity of Solutions of the Fokker–Planck–Kolmogorov Equations

The existence and unicity of solutions of the Fokker–Planck equations in the case where the operator $L_{\mathbf{y}}^*$ is a degenerate elliptic operator was first established by Ilin and Khasminskii (1964) and has recently been extended by Kushner (1969). Kushner's results will be stated without proof.

Theorem 4 [Kushner]. *Let the pair (A, B) be controllable,* and let the vector function $\mathbf{g}(\mathbf{y})$ and its first derivatives be bounded. Then, for $t > 0$ there exists a unique continuous Green's function $G(\mathbf{x}, \mathbf{y}, t)$ for the equation*

$$u_t = L_y{}^*u + f(\mathbf{y}, t) = [L_0{}^* + (Bg)'\text{grad}_{\mathbf{y}}]\, u + f(\mathbf{y}, t). \tag{3.28}$$

Furthermore the function $G(\mathbf{x}, \mathbf{y}, t)$ is the transition probability density function $p(\mathbf{x}, t \mid \mathbf{y})$ of the continuous Markov process which satisfies the stochastic differential equation

$$\begin{aligned} d\mathbf{x} &= A\mathbf{x}\, dt + B\mathbf{g}\, dt + B\, d\mathbf{w}(t), \\ \mathbf{x}(0) &= \mathbf{y}, \end{aligned} \tag{3.29}$$

* In Kushner's paper the pair (A, B) is required to satisfy other mild restrictions which do not appear to be strictly necessary.

where $\mathbf{w}(t)$ *is a vector of independent Markov processes and* $\mathbf{x}(t)$ *is a strong Feller process.*

F. Invariant Measure and Approach to Steady State

A number of authors, Feller (1952, 1954), Gray (1964), and Benes (1968), have studied the problem of invariant measure and approach to the steady state. Only a few of the more important results will be stated here.

Khasminskii (1960) has shown, under Conditions (C1)–(C5), that there exists a unique σ-finite invariant probability $P(U, t \mid \mathbf{y})$.

(C1) For any ϵ neighborhood U_ϵ of $\mathbf{y}$, $1 - P((U, t \mid \mathbf{y}) = O(t)$ uniformly in $\mathbf{y}$, for $\mathbf{y}$ in any compact set.

(C2) $\mathbf{x}(t)$ is a strong Markov process and a strong Feller process.

(C3) $P(U, t \mid \mathbf{y}) > 0$ for all open set and $t > 0$.

(C4) The paths of $\mathbf{x}(t)$ are continuous with probability one.

(C5) $\mathbf{x}(t)$ is recurrent; i.e., for some compact set K, there exists with probability one, a random time $\tau < \infty$, such that $\mathbf{x}(\tau)$ belongs to K with probability one.

In the case that $\mathbf{x}(t)$ is a diffusion process with elliptic differential generator L^*, Khasminskii proved that a necessary and sufficient condition for (C5) is the existence of a unique bounded solution to the exterior Dirichlet problem $L^*u = 0$ in $R^m - G$, where G is bounded and open with a smooth boundary, and arbitrary continuous boundary data are assigned on ∂G. Wonham (1966) showed that Khasminskii's criterion is valid if there exists a continuous Liapunov function $V(\mathbf{y}) \geqslant 0$ in $R^m - G$, which satisfies the condition $L_y{}^*V(y) \leqslant 0$ in $R^m - G$. Recently Kushner (1969) has extended these results to the case of degenerate elliptic generators L^*. Kushner's results are stated below without proof.

Theorem 5 [Kushner]. *Let* $\mathbf{x}(t)$ *be a right continuous strong Markov process and a strong Feller process. Define the stopped process* $\mathbf{x}_m(t)$ *by*

$$\mathbf{x}_m(t) = \mathbf{x}(t) \qquad 0 \leqslant t \leqslant \tau_m\,,$$

$$\mathbf{x}_m(t) = \mathbf{x}(\tau_m) \qquad t \geqslant \tau_m\,.$$

Let $\tilde{A}$ *and* $\tilde{A}_m$ *be the weak infinitesimal operators of the processes* $\mathbf{x}(t)$ *and* $\mathbf{x}_m(t)$, *respectively, with domains* $\mathscr{D}(\tilde{A})$ *and* $\mathscr{D}(\tilde{A}_m)$ *,respectively.*

Let $V(\mathbf{y}) \geqslant 0$ *be a continuous function; suppose that* $V(y)$ *is scaled so that* $\{\mathbf{y} : V(y) \leqslant 1\} \equiv K$ *has a nonempty interior, and let* $V(\mathbf{y}) \to \infty$ *as*

$|\mathbf{y}| \to \infty$. Define $Q_m = \{\mathbf{y} : m > V(\mathbf{y}) > 1\}$. Let $V(\mathbf{y}) \in \mathscr{D}(\tilde{A}_m)$ and $\tilde{A}_{m\mathbf{y}} V(\mathbf{y}) \leqslant 0$ for each $m > 1$. Suppose that either (C3) *or the weaker condition: for each* $\mathbf{y}$, *there exists a $t > 0$ such that*

$$P_{\mathbf{y}}\{1 < |V(\mathbf{y}(t))| < m\} \leqslant 1 - \alpha(\mathbf{y}) < 1$$

holds, then the process $\mathbf{x}(t)$ is recurrent.

THEOREM 6 [Kushner]. *Suppose that conditions* (C1)–(C4) *hold. Let $V(\mathbf{y})$ satisfy the properties of Theorem 5 except that $\tilde{A}_{m\mathbf{y}} V(\mathbf{y}) \leqslant -h < 0$ in each Q_m, for some constant $h > 0$. Then the process $\mathbf{x}(t)$ has a finite invariant measure.*

COROLLARY 1. *Suppose that conditions* (C1)–(C4) *hold. Let $\mathbf{x}(t)$ be a diffusion process, the terms of whose differential generator satisfy a local Lipshitz condition. Let there exist a continuous function $V(\mathbf{y}) \geqslant 0$, which is twice continuously differentiable in its variables, and $V(\mathbf{y}) \to \infty$ $|\mathbf{y}| \to \infty$, and $L_{\mathbf{y}}^* V(\mathbf{y}) \leqslant -h < 0$ in $R^m - G$, for some bounded set G. Then $\mathbf{x}(t)$ has a finite invariant measure.*

COROLLARY 2. *Let $L_{\mathbf{y}}^*$ be the differential generator of the process*

$$d\mathbf{x} = A\mathbf{x}\, dt + B\mathbf{g}\, dt + B\, d\mathbf{w}(t),$$

$$\mathbf{x}(0) = \mathbf{y}.$$

Then, if there exists a function $V(\mathbf{y})$ satisfying the conditions of Corollary 1, *there exists a finite invariant measure. The measure has a continuous density $p_s(\mathbf{x})$*

$$p_s(\mathbf{x}) = \int \mu(d\mathbf{y})\, G(\mathbf{y}, \mathbf{x}, t)$$

and

$$G(\mathbf{y}, \mathbf{x}, t) \to p_s(\mathbf{x}) \qquad \text{as} \quad t \to \infty.$$

COROLLARY 3. *If the eigenvalues of A have negative real parts, then $\mathbf{x}(t)$ has an invariant measure for any nonlinearity $\mathbf{g}(\mathbf{x})$ which is bounded and has bounded first derivatives.*

IV. Applications and Solution Techniques

The previous section has been devoted to the development of the basic theory of stochastic differential equations which led to the Fokker–Planck–Kolmogorov equations satisfied by the transition probability density function for the stochastic process. This section is concerned with the application of the basic theory to systems of stochastic differential equations.

Exact solutions of the Fokker–Planck–Kolmogorov equations have been found for two types of stochastic differential equations: (1) systems of linear equations, and (2) certain first order nonlinear equations. The steady-state probability density can always be obtained for first order nonlinear systems, and has also been found for a certain class of coupled nonlinear oscillator problems.

A. Exact Solutions

1. *Linear Systems*

As previously remarked, the transition probability density function for a linear stochastic system is Gaussian due to the Gaussian nature of the Wiener process, so that the transition probability density function is completely characterized by the first and second moments processes. Even without this fact, the Fokker–Planck equation is easily solved by Fourier transform techniques. Uhlenbeck and Wang (1945) used this technique to solve the Fokker–Planck equation for a system of linear oscillators excited by white noise. To illustrate some of these ideas, consider the following system of linear stochastic equations

$$\begin{aligned} d\mathbf{x} &= A\mathbf{x}\,dt + B\,d\mathbf{w}(t), \\ \mathbf{x}(0) &= \mathbf{y}, \end{aligned} \tag{4.1}$$

where A and B are $m \times m$ matrices and $\mathbf{x}$, $\mathbf{y}$, $\mathbf{w}(t)$, m vectors. The mean value vector $\mathbf{m}(t \mid \mathbf{y})$ is given by

$$\mathbf{m}(t \mid \mathbf{y}) = e^{At}\mathbf{y}. \tag{4.2}$$

The correlation matrix $K(t \mid \mathbf{y})$ is given by

$$K(t \mid \mathbf{y}) = \int_0^t e^{As}BB'\,e^{A's}\,ds. \tag{4.3}$$

The transition probability density function, $p(\mathbf{x}, t \mid \mathbf{y})$, is given by

$$p(\mathbf{x}, t \mid y) = [1/(2\pi)^{m/2}|\,K(t \mid \mathbf{y})|^{1/2}]\exp\{-\tfrac{1}{2}[(\mathbf{x} - e^{At}\mathbf{y}), K^{-1}(t \mid \mathbf{y})(\mathbf{x} - e^{At}\mathbf{y})]\}, \tag{4.4}$$

where $|\,K()|$ is the determinant of K.

If A is a stability matrix, then as $t \to \infty$

$$p_s(\mathbf{x}) = \lim_{t\to\infty} p(\mathbf{x}, t \mid \mathbf{y}) = [1/(2\pi)^{m/2}|\,K\,|^{1/2}]\exp\{-\tfrac{1}{2}[\mathbf{x}, K^{-1}\mathbf{x}]\} \tag{4.5}$$

independent of t and $\mathbf{y}$ where

$$K(t \mid \mathbf{y}) = \int_0^\infty e^{As} BB' \, e^{A's} \, ds. \tag{4.6}$$

In this case the first order moments are all zero as time increases without bound. The second moments are given by (4.6).

Caughey and Dienes (1962) and Bogdanoff and Kozin (1962) have shown that the moments of the process $\mathbf{x}(t)$ can be obtained from the Fokker–Planck equation without first solving for the transition probability density function $p(\mathbf{x}, t \mid \mathbf{y})$. The Fokker–Planck equation for the system (4.1) is

$$\begin{gathered} \frac{\partial p}{\partial t} = -\sum_{i=1}^{m} \frac{\partial}{\partial x_i} \sum_{j=1}^{m} (A_{ij} x_j p) + \frac{1}{2} \sum_{i,j=1}^{m} \frac{\partial^2}{\partial x_i \, \partial x_j} [D_{ij} p], \\ \lim_{t \to 0} p(\mathbf{x}, t \mid \mathbf{y}) = \delta(\mathbf{x} - \mathbf{y}) \qquad \mathbf{x}, \mathbf{y} \in R^m, \end{gathered} \tag{4.7}$$

where

$$D_{ij} = (BB')_{ij} \, .$$

If the sample paths of $\mathbf{x}(t)$ are not to vanish upon reaching the boundary ∂R^m of R^m, p and its first and second partial derivates must vanish as $|\mathbf{x}| \to \infty$. If Eq. (4.7) is multiplied by x_j and the resulting equation integrated over R^m, then

$$E_t[x_l] = \sum_{j=1}^{m} A_{lj} E[x_j] \qquad l = 1, 2, \ldots, m. \tag{4.8}$$

Alternatively,

$$\begin{gathered} \frac{d}{dt} [E[\mathbf{x}]] = AE[\mathbf{x}], \\ E[\mathbf{x}(0)] = \mathbf{y}. \end{gathered} \tag{4.9}$$

Therefore,

$$E[\mathbf{x}(t)] = \mathbf{m}(t \mid \mathbf{y}) = e^{At} \mathbf{y}. \tag{4.10}$$

If Eq. (4.7) is multiplied by $(x_l - E[x_l])(x_k - E[x_k])$ and the resulting equation integrated over R^m, then

$$(d/dt) \, Y = AY + YA' + D \tag{4.11}$$

where

$$Y = (\mathbf{x} - E[\mathbf{x}])(\mathbf{x} - E[\mathbf{x}])',$$

$$Y(0) = [0].$$

Thus

$$Y(t) = K(t \mid \mathbf{y}) = \int_0^t e^{As} D\, e^{A's}\, ds. \tag{4.12}$$

Equations (4.10) and (4.12) are exactly the same as (4.2) and (4.3), respectively. This is, of course, exactly what one would expect. It should be noted, however, that if A is a stability matrix, then

$$\mathbf{m}(t, \mathbf{y}) \to 0, \qquad dK(t \mid \mathbf{y})/dt \to 0, \qquad \text{as} \quad t \to \infty.$$

Let X be the matrix with elements $X_{ij} = E[x_i x_j]$. In the steady state, Eq. (4.11) may be written

$$0 = AX + XA' + D. \tag{4.13}$$

Recognizing that, since X is symmetric, there are only $N = m(m + 1)/2$ independent quantities. Thus (4.13) may be rewritten

$$G\mathbf{v} = -\mathbf{d}$$

where

$$\mathbf{v} = \begin{Bmatrix} X_{11} \\ \vdots \\ X_{1m} \\ X_{22} \\ \vdots \\ X_{2m} \\ \vdots \\ X_{mm} \end{Bmatrix} \qquad \text{and} \qquad \mathbf{d} = \begin{Bmatrix} D_{11} \\ \vdots \\ D_{1m} \\ D_{22} \\ \vdots \\ D_{2m} \\ \vdots \\ D_{mm} \end{Bmatrix} \tag{4.14}$$

are $N = m(m + 1)/2$ vectors and G is a $N \times N$ matrix. Thus

$$\mathbf{v} = -G^{-1}\mathbf{d}. \tag{4.15}$$

Equation (4.15) is often more convenient than (4.12) for computing the second order moments.

2. *Nonlinear Systems*

a. Exact Solutions of the Fokker–Planck Equation. The transition probability density function has been found for certain first order systems. Caughey and Dienes (1961), and later Atkinson and Caughey (1968), obtained the transition probability density function for a class of piecewise linear systems excited by white noise.

In Caughey and Dienes (1961) the transition probability density function for the process, $\mathbf{x}(t)$, governed by the stochastic differential equation

$$dx = -k \operatorname{sgn} x \, dt + dw(t); \quad x(0) = y, \qquad k > 0$$

was obtained by Laplace transforming the associated Fokker–Planck equation with respect to t and piecing together the solutions of the resulting ordinary differential equations. Inversion of the solution gave the following results.

$$p(x, t \mid y) = \frac{2k}{\sqrt{\pi}} e^{-2k|x|} \int_{(y+|x|)2\sqrt{t}}^{\infty} e^{-u^2} \, du + \frac{1}{\sqrt{2\pi t}} \exp\left[k(x - |x|) - \frac{(x - y - kt)^2}{2t}\right]$$

As $t \to \infty$, $p(x, t \mid y \mid) \to p_s(x)$, where $p_s(x)$ is given by

$$p_s(x) = ke^{-2k|x|}.$$

The spectral density function for the process is

$$\Phi(\omega) = \frac{1}{2\pi} \left\{ \frac{k^2}{2\omega^2} + \frac{2k^4}{\omega^4} \left[1 - \left(\frac{1 + 4\omega^2/k^4}{2}\right)^{1/2}\right] \right\}.$$

In Atkinson and Caughey (1968) the techniques used in Caughey and Dienes (1961) were extended to the class of problems governed by the stochastic differential equation

$$dx = -f(x) \, dt + dw(t), \qquad x(0) = y$$

where $f(x)$ is a piecewise linear function of x.

Wong (1964) obtained eigenfunction expansions for the transition probability density function satisfying Fokker–Planck equations of the form

$$\frac{\partial p}{\partial t} = \frac{\partial}{\partial x} \left\{ (ax + b) p + \frac{\partial}{\partial x} [(cx^2 + dx + e) p] \right\}$$

where a, b, c, d, and e are constants. Bluman (1967) and Bluman and Cole (1969) have recently used group theoretic methods to construct exact solutions for a class of one dimensional Fokker–Planck equations.

3. *Exact Steady-State Solutions of the Fokker–Planck Equation*

a. First-Order Systems. The exact steady-state probability density (if it exists) for any first order nonlinear system excited by white noise can readily be determined by direct integration. The Fokker–Planck equation for a stochastic differential equation of the type

$$dx = -f(x)\,dt + dw(t), \qquad x(0) = y$$

is given by

$$\frac{\partial p}{\partial t} = \frac{\partial}{\partial x}\left[f(x)p + D\frac{\partial p}{\partial x}\right]$$
$$p(x, 0 \mid y) = \delta(x - y) \qquad D \text{ a positive constant.} \tag{4.16}$$

If $f(x)$ satisfies the axioms of Theorems 2, 4, and 6 then there exists a unique steady-state probability density function $p_s(x)$ satisfying

$$0 = (\partial/\partial x)[f(x)p + D\,\partial p/\partial x].$$

Direct integration yields

$$p_s(x) = C^{-1} \exp\left[-\int_0^x f(\xi)/D\,d\xi\right], \tag{4.17}$$

where C is the normalizing constant

$$C = \int_{-\infty}^{\infty} \exp\left[-\int_0^x f(\xi)/D\,d\xi\right] dx. \tag{4.18}$$

b. Second Order Equations and Systems of Second Order Equations. Exact solutions for nonlinear equations of second order excited by white noise have been found only for the steady-state probability density function for equations of the form

$$\ddot{x} + f(H)\,\dot{x} + g(x) = \dot{w}(t); \qquad x(0) = y, \quad \dot{x}(0) = \dot{y},$$
$$E[dw(t)^2] = 2D\,dt, \tag{4.19}$$

where

$$H = \tfrac{1}{2}\dot{x}^2 + \int_0^x g(\eta)\,d\eta.$$

The associated Fokker–Planck equation is easily shown to be

$$\frac{\partial p}{\partial t} = -\dot{x}\frac{\partial p}{\partial x} + \frac{\partial}{\partial \dot{x}}[(g(x) + f(H)\dot{x})p] + D\frac{\partial^2 p}{\partial \dot{x}^2}$$
$$p(x, \dot{x}; 0 \mid y, \dot{y}) = \delta(x - y)\,\delta(\dot{x} - \dot{y}) \qquad \mathbf{x} \in R^2 \tag{4.20}$$

If $g(x)$ and $f(H)\,\dot{x}$ satisfy the axiom of Theorems 2, 4, and 6, then the existence of a unique solution of (4.20) is guaranteed. Furthermore there exists a unique steady-state density $p_s(x, \dot{x})$ independent of y, $\dot{y}$, and t. $p_s(x, \dot{x})$ satisfies the equation

$$0 = \left[g(x)\frac{\partial p}{\partial \dot{x}} - \dot{x}\frac{\partial p}{\partial x}\right] + \frac{\partial}{\partial \dot{x}}\left[f(H)\,\dot{x}p + D\frac{\partial p}{\partial \dot{x}}\right]. \tag{4.21}$$

Caughey (1964) observed that this equation can be solved readily if the following separation is used.

$$\begin{aligned} 0 &= g(x)\,\partial p/\partial x - \dot{x}(\partial p/\partial x), \\ 0 &= (\partial/\partial \dot{x})[f(H)\,\dot{x}p + D\,\partial p/\partial \dot{x}]. \end{aligned} \tag{4.22}$$

Equation (4.22) is easily integrated to give

$$p_s(x, \dot{x}) = \frac{\exp[-(1/D)\int_0^H f(\eta)\,d\eta]}{\int_{-\infty}^{\infty}\int \exp[-(1/D)\int_0^H f(\eta)\,d\eta]\,dx\,d\dot{x}}. \tag{4.23}$$

The same technique can be applied to the system of coupled nonlinear equations

$$\begin{aligned} \ddot{x}_i + \beta_i f(H)x_i + \partial V(x)/\partial x_i &= w_i(t), \qquad i = 1, 2,\dots, n \\ x_i(0) = y_i\,, \qquad \dot{x}_i(0) &= \dot{y}_i \\ E[dw_i(t)\,dw_j(t)] &= 2D_i\,\delta_{ij}\,dt, \qquad \mathbf{x} \in R^n \\ D_i/\beta_i &= \text{constant} \\ H = \left[\frac{1}{2}\sum_{i=1}^{n} x_i^2 + V(\mathbf{x})\right]&. \end{aligned} \tag{4.24}$$

It is readily shown that the steady-state density for this system is

$$p_s(\mathbf{x}, \dot{\mathbf{x}}) = \frac{\exp[-(\beta_i/D_i)\int_0^H f(\eta)\,d\eta]}{\int_{-\infty}^{\infty} dx_1 \int_{-\infty}^{\infty} d\dot{x}_1 \cdots \int_{-\infty}^{\infty} dx_n \int_{-\infty}^{\infty} d\dot{x}_n \exp[-(\beta_i/D_i)\int_0^H f(\eta)\,d\eta]}. \tag{4.25}$$

Unfortunately this solution technique does not appear to work for other equations or systems of equations.

B. Approximate Techniques

Since, in general, it is not possible to obtain exact statistics for the response of a nonlinear system excited by white noise, a number of techniques have been developed to obtain approximate solutions.

1. *Techniques Based on the Use of the Fokker–Planck Equation*

a. Iterative Solution of the Fokker–Planck Equation. The technique used by Ilin and Khasminskii (1964), and which was generalized by Kushner (1969) to establish the existence and unicity of solutions of the Fokker–Planck–Kolmogorov equations, is a constructive technique and hence can be used, in principle, to construct approximate solutions of the Fokker–Planck–Kolmogorov equations. To illustrate the ideas, consider the system below:

$$\begin{gathered} \ddot{x} + \beta\dot{x} + x + \epsilon g(x, \dot{x}) = \dot{w}(t), \qquad \beta > 0 \\ E[dw(t)^2] = 2D\, dt, \qquad |\epsilon| < 1 \\ x(0) = y, \qquad \dot{x}(0) = \dot{y}. \end{gathered} \tag{4.26}$$

The Fokker–Planck equation associated with (4.26) is

$$\begin{gathered} \frac{\partial p}{\partial t} = -\dot{x}\frac{\partial p}{\partial x} + \frac{\partial}{\partial \dot{x}}[(\beta\dot{x} + x + \epsilon g(x, \dot{x}))] + D\frac{\partial^2 p}{\partial \dot{x}^2} \\ p(\mathbf{x}, 0 \mid \mathbf{y}) = \delta(x - y)\,\delta(\dot{x} - \dot{y}), \qquad \mathbf{x} \in R^2. \end{gathered} \tag{4.27}$$

Let L_0 be the operator

$$L_0 u = -\dot{x}\frac{\partial u}{\partial x} + \frac{\partial}{\partial \dot{x}}[(\beta\dot{x} + x)u] + D\frac{\partial^2 u}{\partial \dot{x}^2}. \tag{4.28}$$

Thus (4.27) may be rewritten in the form

$$\frac{\partial p}{\partial t} = L_0 p + \epsilon\frac{\partial}{\partial \dot{x}}[g(x, \dot{x})p]. \tag{4.29}$$

If $g(x, \dot{x})$, $g_x(x, \dot{x})$, $g_{\dot{x}}(x, \dot{x})$ are bounded for all $\mathbf{x} \in R^2$, (4.29) satisfies the axioms of Theorems 2, 4, and 6, thus the existence of a unique solution is guaranteed.

The fundamental solution of the equation

$$\begin{gathered} \partial G/\partial t = L_0 G \\ G(\mathbf{x}, 0 \mid \mathbf{y}) = \delta(x - y)\,\delta(\dot{x} - \dot{y}) \end{gathered} \tag{4.30}$$

is

$$G(\mathbf{x}, t \mid y) = p_0(\mathbf{x}, t \mid \mathbf{y}), \tag{4.31}$$

where $p_0(\mathbf{x}, t \mid \mathbf{y})$ is the transition probability density function for the linearized system (4.26) with ϵ set to zero.

Equation (4.29) may be rewritten in the form of the integral equation (4.32)

$$p(\mathbf{x}, t \mid \mathbf{y}) = p_0(\mathbf{x}, t \mid \mathbf{y}) + \epsilon \int_0^t \int_{R^2} p_0(\mathbf{x}, t - \tau \mid \boldsymbol{\xi}) \frac{\partial}{\partial \xi_2} (g \mid \boldsymbol{\xi}) p(\boldsymbol{\xi}, \tau \mid \mathbf{y})\, d\boldsymbol{\xi}\, d\tau. \tag{4.32}$$

Partial integration of the second term in (4.32) with respect to ξ_2 yields

$$p(\mathbf{x}, t \mid \mathbf{y}) = p_0(\mathbf{x}, t \mid \mathbf{y}) - \epsilon \int_0^t \int_{R^2} g(\boldsymbol{\xi}) p(\boldsymbol{\xi}, \tau \mid \mathbf{y}) \frac{\partial}{\partial \xi_2} [p_0(\mathbf{x}, t - \tau \mid \xi)]\, d\boldsymbol{\xi}\, d\tau. \tag{4.33}$$

Equation (4.33) may be solved, in principle, by the following iterative scheme

$$p_n(\mathbf{x}, t \mid \mathbf{y}) = p_0(\mathbf{x}, t \mid \mathbf{y}) - \epsilon \int_0^t \int_{R^2} g(\boldsymbol{\xi}) p_{n-1}(\boldsymbol{\xi}, \tau \mid \mathbf{y}) \frac{\partial}{\partial \xi_2} \times [p_0(\mathbf{x}, t - \tau \mid \boldsymbol{\xi})]\, d\boldsymbol{\xi}\, d\tau \qquad n = 1, 2, \ldots . \tag{4.34}$$

Now $p_0(\mathbf{x}, t \mid \mathbf{y})$ is given by

$$p_0(\mathbf{x}, t \mid \mathbf{y}) = (1/2\pi |K|^{1/2}) \exp[-\tfrac{1}{2}((\mathbf{x} - e^{At}\mathbf{y}), K^{-1}(t)(\mathbf{x} - e^{At}\mathbf{y}))],$$

where

$$K(t) = \int_0^t e^{As} \mathscr{D} e^{A's}\, ds, \tag{4.35}$$

$$A = \begin{bmatrix} 0 & 1 \\ -1 & -\beta \end{bmatrix}, \qquad \text{and} \qquad \mathscr{D} = \begin{bmatrix} 0 & 0 \\ 0 & D \end{bmatrix}.$$

It is readily shown, using (4.35), that

$$\left| \frac{\partial p_0}{\partial \xi_2} (\mathbf{x}, t \mid \mathbf{y}) \right| \leqslant \frac{K e^{-\gamma t}}{(1 - e^{-2\gamma t})^{1/2}} p_0(\alpha \mathbf{x}, t \mid \alpha \mathbf{y}), \qquad 1 < \alpha < 0 \quad \gamma > 0 \tag{4.36}$$

where K is a bounded constant. Thus if $|g(\boldsymbol{\xi})| \leqslant M$, $\forall \boldsymbol{\xi} \in R^2$,

$$| p_1(\mathbf{x}, t \mid \mathbf{y}) - p_0(\mathbf{x}, t \mid \mathbf{y})|$$

$$\leqslant | \epsilon | KM \int_0^t \frac{e^{-\gamma(t-\tau)}}{(1 - e^{-2\gamma(t-\tau)})^{1/2}} \int_{R^2} p_0(\alpha \mathbf{x}, t - \tau \mid \alpha \boldsymbol{\xi}) p_0(\boldsymbol{\xi}, \tau \mid \mathbf{y})\, d\boldsymbol{\xi}\, d\tau,$$

$$\leqslant | \epsilon | KM \int_0^t \frac{e^{-\gamma(t-\tau)}}{(1 - e^{-2\gamma(t-\tau)})^{1/2}} \int_{R^2} p_0(\alpha \mathbf{x}, t - \tau \mid \alpha \boldsymbol{\xi}) p_0(\alpha \boldsymbol{\xi}, \tau \mid \alpha \mathbf{y})\, d\boldsymbol{\xi}\, d\tau.$$

Making use of the Chapman–Kolmogorov equation,

$$p_1(\mathbf{x}, t \mid \mathbf{y}) - p_0(\mathbf{x}, t \mid \mathbf{y})| \leqslant | \epsilon | KM p_0(\alpha \mathbf{x}, t \mid \alpha \mathbf{y}) \int_0^t \frac{e^{-\gamma \eta}}{(1 - e^{-2\gamma \eta})^{1/2}}\, d\eta. \tag{4.37}$$

Elementary integration yields

$$| p_1(\mathbf{x}, t \mid \mathbf{y}) - p_0(\mathbf{x}, t \mid y | \leqslant (| \epsilon |/\gamma) KM \cos^{-1} e^{-\gamma t} p_0(\alpha\mathbf{x}, t \mid \alpha\mathbf{y}). \qquad (4.38)$$

Similarily

$$| p_n(\mathbf{x}, t \mid \mathbf{y}) - p_{n-1}(\mathbf{x}, t \mid \mathbf{y})| \leqslant [(| \epsilon |KM \cos^{-1} e^{-\gamma t})/\gamma]^n p_0(\alpha\mathbf{x}, t \mid \alpha\mathbf{y}). \qquad (4.39)$$

If $| \epsilon |$ is sufficiently small, so that

$$r = (| \epsilon |KM\pi/2\gamma) < 1, \qquad (4.40)$$

then it is readily seen that the series

$$p(\mathbf{x}, t \mid \mathbf{y}) = p_0(\mathbf{x}, t \mid \mathbf{y}) + \sum_{n=0}^{\infty} [p_{n+1}(\mathbf{x}, t \mid \mathbf{y}) - p_n(\mathbf{x}, t \mid \mathbf{y})] \qquad (4.41)$$

converges uniformly and satisfies the integral equation (4.33). Thus, an approximate solution of (4.33) is given by

$$p(\mathbf{x}, t \mid \mathbf{y}) \sim p_n(\mathbf{x}, t \mid \mathbf{y}),$$

where

$$| p(\mathbf{x}, t \mid \mathbf{y}) - p_n(\mathbf{x}, t \mid \mathbf{y})| \leqslant [r^{n+1}/(1 - r)] p_0(\alpha\mathbf{x}, t \mid \alpha\mathbf{y}). \qquad (4.42)$$

Hence, for any ϵ for which (4.40) holds,

$$\lim_{n\to\infty} | p(\mathbf{x}, t \mid \mathbf{y}) - p_n(\mathbf{x}, t \mid \mathbf{y})| = 0.$$

In addition, for any n, we have

$$\lim_{|\epsilon|\to 0} | p(\mathbf{x}, t \mid \mathbf{y}) - p_n(\mathbf{x}, t \mid \mathbf{y})| = 0.$$

If the system of (4.26) belongs to the class of problems for which the steady-state probability density can be readily obtained, then the autocorrelation function, and hence the spectral density, can be readily computed. Since

$$R_{ij}(\tau) = \int_{R^4} y_i x_j p(\mathbf{x}, \tau \mid \mathbf{y}) p_s(\mathbf{y}) \, d\mathbf{x} \, d\mathbf{y},$$

let

$$R_{ij}^{(n)}(\tau) = \int_{R^4} y_i x_j p_n(\mathbf{x}, \tau \mid \mathbf{y}) p_s(\mathbf{y}) \, d\mathbf{x} \, d\mathbf{y}.$$

From the properties of p_n, it follows that for $|\epsilon|$ sufficiently small

(i) $\lim_{n\to\infty} | R_{ij}(\tau) - R_{ij}^{(n)}(\tau)| = 0 \qquad \forall\tau \geqslant 0$

(ii) $\lim_{|\epsilon|\to 0} | R_{ij}(\tau) - R_{ij}^{(n)}(\tau)| = 0 \qquad \forall n, \forall\tau \geqslant 0.$

b. The Technique of Eigenfunction Expansion. The technique of eigenfunction expansions has been used by Wong (1964) and Payne (1967, 1968) for first order systems, in which case Sturm–Liouville theory applies. For many higher order systems the Fokker–Planck–Kolmogorov equations are of degenerate form and cannot be rendered self-adjacent. Recently Atkinson (1970) has used eigenfunction expansion techniques for second order systems excited by white noise.

Using the technique of separation of variables, a solution of the Fokker–Planck–Kolmogorov equations is sought in the form

$$p(\mathbf{x}, t \mid \mathbf{y}) = \sum_{i=1}^{\infty} c_i u_i(\mathbf{x}) v_i(\mathbf{y}) T_i(t).$$

Substituting into (3.26) and (3.27) leads to

$$\begin{aligned} T_i(t) &= e^{-\lambda_i t}, \\ L_{\mathbf{x}} u_i + \lambda_i u_i &= 0, \\ L_{\mathbf{y}}{}^* v_i + \lambda_i v_i &= 0. \end{aligned} \tag{4.43}$$

Thus u_i and v_i are the eigenfunctions of L and L^*, respectively, if they exist, while the λ_i are the corresponding eigenvalues. Since L and L^* are adjoint in the sense that

$$\int_{R^m} (uL^*v - vLu)\, d\mathbf{x} = 0,$$

then L and L^* have the same eigenvalues. If the problem under consideration is such that the steady-state density $p_s(\mathbf{x})$ can be calculated, then it will be convenient to use, in place of L and its eigenfunctions u_i, the operator G and its eigenfunctions w_i, where

$$G(w_i) = p_s^{-1} L(p_s w_i), \qquad w_i = p_s^{-1} u_i\,. \tag{4.44}$$

The operators G and L^* are adjoint operators with respect to the inner product

$$\langle v, w \rangle = \int_{R^m} p_s(\mathbf{x}) v(\mathbf{x}) w(\mathbf{x})\, d\mathbf{x} = E[vw], \tag{4.45}$$

the expectation of vw.

If the eigenvalues λ_i are discrete and distinct then the eigenfunctions are bi-ortho normal and

$$\langle v_i , w_j \rangle = \delta_{ij} . \tag{4.46}$$

The solution of the Fokker–Planck–Kolmogorov equations may be written

$$p(\mathbf{x}, t \mid \mathbf{y}) = p_s(\mathbf{x}) \sum_{i=0}^{\infty} e^{-\lambda_i t} w_i(\mathbf{x}) v_i(\mathbf{y}). \tag{4.47}$$

This expression assumes that the spectrum of G is discrete. The existence of $p_s(\mathbf{x})$ implies that there always exists at least one discrete eigenvalue, namely, $\lambda_0 = 0$, with corresponding eigenfunctions $v_0 = w_0 = 1$.

However, there may be a continuous range of eigenvalues R_λ, say, distinct from λ_0. This leads to the additional term in (4.47) given by

$$p_s(\mathbf{x}) \int_{R^m} e^{-\lambda t} \frac{w_\lambda(\mathbf{x}) v_\lambda(\mathbf{y})}{\langle w_\lambda , v_\lambda \rangle} \, d\lambda, \tag{4.48}$$

where w_λ and v_λ are the eigenfunctions of G and L^*, respectively, corresponding to λ. The auto- and cross-correlation functions and the corresponding spectral densities are readily expressed in terms of the eigenfunctions

$$R_{jk}(\tau) = E[x_j(t) x_k(t + \tau)] = \int_{R^m} d\mathbf{x} \int_{R^m} d\mathbf{y} y_j x_k p_s(\mathbf{y}) p(\mathbf{x}, \tau \mid \mathbf{y}), \tag{4.49}$$

$$\Phi_{jk}(\omega) = \frac{1}{\pi} \int_0^\infty [R_{jk}(\tau) + R_{kj}(\tau)] \cos \omega\tau \, d\tau - \frac{i}{\pi} \int_0^\infty [R_{jk}(\tau) - R_{kj}(\tau)] \sin \omega\tau \, d\tau. \tag{4.50}$$

Using (4.47) and (4.48) the above expressions reduce to

$$R_{jk}(\tau) = \sum_{l=0}^{\infty} e^{-\lambda_l |\tau|} \alpha_{jl} \beta_{kl} + \int_{R\lambda} e^{-\lambda |\tau|} \alpha_j(\lambda) \beta_k(\lambda) \, d\lambda, \tag{4.51}$$

$$\begin{aligned} \Phi_{jk}(\omega) = {} & \frac{1}{\pi} \sum_{l=0}^{\infty} \frac{\lambda_l}{\lambda_l^2 + \omega^2} [\alpha_{jl} \beta_{kl} + \alpha_{kl} \beta_{jl}] \\ & + \frac{1}{\pi} \int_{R\lambda} \frac{\lambda}{\lambda^2 + \omega^2} [\alpha_j(\lambda) \beta_k(\lambda) + \alpha_k(\lambda) \beta_j(\lambda)] \, d\lambda \\ & - \frac{i}{\pi} \sum_{l=0}^{\infty} \frac{\omega}{\lambda_l^2 + \omega^2} [\alpha_{jl} \beta_{kl} - \alpha_{kl} \beta_{jl}] \\ & - \frac{i}{\pi} \int_{R\lambda} \frac{\omega}{\lambda^2 + \omega^2} [\alpha_j(\lambda) \beta_k(\lambda) - \alpha_k(\lambda) \beta_j(\lambda)] \, d\lambda, \end{aligned} \tag{4.52}$$

where

$$\begin{aligned}\alpha_{ij} &= \langle x_i\,, w_j\rangle,\\ \beta_{ij} &= \langle x_i\,, v_j\rangle,\\ \alpha_i(\lambda) &= \langle x_i\,, w(\lambda)\rangle/\langle w(\lambda), v(\lambda)\rangle^{1/2},\\ \beta_i &= \langle x_i\,, v(\lambda)\rangle/\langle w(\lambda), v(\lambda)\rangle^{1/2}.\end{aligned} \tag{4.53}$$

The existence of expansions (4.47), (4.51), and (4.52) for arbitrary systems has not been rigorously established to date. This would require a careful investigation of the nature of the operator G and its spectrum. In general for a system of higher order than the first, the coefficient matrix of the second derivative terms is singular, thus the extensive theory available for positive definite operators does not apply.

Application to second order systems. If the above analysis is applied to second order systems of the type

$$\begin{gathered}\ddot{x} + f(H)\dot{x} + g(x) = \dot{w}(t),\\ x(0) = y, \qquad \dot{x}(0) = \dot{y},\\ E[dw^2] = 2D\,dt,\\ H = \tfrac{1}{2}\dot{x}^2 + \int_0^x g(\eta)\,d\eta,\end{gathered} \tag{4.54}$$

the operators L^* and G take the form

$$\begin{aligned}L^*(v) &= D\frac{\partial^2 v}{\partial \dot{x}^2} - [\dot{x}f(H) + g(x)]\frac{\partial v}{\partial \dot{x}} + \dot{x}\frac{\partial v}{\partial x},\\ G(w) &= D\frac{\partial^2 w}{\partial \dot{x}^2} - [\dot{x}f(H) - g(x)]\frac{\partial w}{\partial \dot{x}} - \dot{x}\frac{\partial w}{\partial x}, \qquad \mathbf{x} \in R^2.\end{aligned} \tag{4.55}$$

Thus L^* is obtained from G by replacing $\dot{x}$ by $-\dot{x}$ throughout. The same result also holds for the eigenfunctions, thus

$$w_i(x, \dot{x}) = v_i(x, -\dot{x}).$$

Since G is not in general self-adjoint, except for first order systems, it may have complex eigenvalues. Since the operator is real the eigenvalues, if complex, will occur in complex conjugate pairs. The same will be true also for the corresponding normalized eigenfunctions and hence for the coefficients α_{jk} and β_{jk} in (4.51) and (452). This may give rise to oscillatory correlation functions and spectral densities with peaks, a situation which is known to occur in linear systems of second and higher order.

Perturbation techniques applied to the eigenvalue problem. In many problems, the nonlinearity is associated with a small parameter, for example,

$$\ddot{x} + \beta\dot{x} + x + \epsilon g(x) = \dot{w}(t).$$

It is natural then to look for a perturbation expansion of the associated eigenequations

$$\begin{aligned} Gw_i + \lambda_i w_i &= 0, \\ L^* v_i + \lambda_i v_i &= 0. \end{aligned} \tag{4.56}$$

Suppose that the eigenequations of (4.56) are rewritten in the form

$$\begin{aligned} (G_0 + \epsilon G_1) w_i + \lambda_i w_i &= 0, \\ (L_0{}^* + \epsilon L_1{}^*) v_i + \lambda_i v_i &= 0. \end{aligned} \tag{4.57}$$

Assuming expansions of the form

$$\begin{aligned} v_i &= v_{i0} + \epsilon v_{i1} + \epsilon^2 v_{i2} + \cdots, \\ w_i &= w_{i0} + \epsilon w_{i1} + \epsilon^2 w_{i2} + \cdots, \\ \lambda_i &= \lambda_{i0} + \epsilon \lambda_{i1} + \epsilon^2 \lambda_{i2} + \cdots, \qquad i = 1, 2, \ldots, \end{aligned} \tag{4.58}$$

substituting (4.58) into (4.57), and equating the coefficients of like powers of ϵ to zero, the following equations are obtained

$$\epsilon^0 \quad \begin{aligned} G_0 w_{i0} + \lambda_{i0} w_{i0} &= 0, \\ L_0{}^* v_{i0} + \lambda_{i0} v_{i0} &= 0, \end{aligned} \tag{4.59}$$

$$\epsilon^1 \quad \begin{aligned} G_0 w_{i1} + \lambda_{i0} w_{i1} &= -G_1 w_{i0} - \lambda_{i1} w_{i0}, \\ L_0{}^* v_{i1} + \lambda_{i0} v_{i1} &= -L_1{}^* v_{i0} - \lambda_{i1} v_{i0}, \end{aligned} \tag{4.60}$$

$$\epsilon^2 \quad \begin{aligned} G_0 w_{i2} + \lambda_{i0} w_{i2} &= -G_1 w_{i1} - \lambda_{i1} w_{i1} - \lambda_{i2} w_{i0}, \\ L_0{}^* v_{i2} + \lambda_{i0} v_{i2} &= -L_1{}^* v_{i1} - \lambda_{i1} v_{i1} - \lambda_{i2} v_{i0}, \end{aligned} \tag{4.61}$$

... etc. ... .

Suppose that the eigenvalues of (4.59) are discrete and distinct. Then

$$\begin{aligned} &\text{(i)} \quad \int_{R^m} p_s(\mathbf{x}) v_{i0}(\mathbf{x}) w_{j0}(\mathbf{x})\, d\mathbf{x} = \delta_{ij}, \\ &\text{(ii)} \quad \int_{R^m} p_s(\mathbf{x}) v_{i0}(\mathbf{x}) G w_{j0}(\mathbf{x})\, d\mathbf{x} = -\lambda_{i0}\, \delta_{ij}, \\ &\text{(iii)} \quad \int_{R^m} p_s(\mathbf{x}) w_{j0}(\mathbf{x}) L^* v_{i0}(\mathbf{x})\, d\mathbf{x} = -\lambda_{i0}\, \delta_{ij}. \end{aligned} \tag{4.62}$$

If it is assumed that the bi-orthonormal set of functions $\{v_{i0}, w_{i0}\}$; $i = 1, 2, \ldots$, is complete, then the function v_{i1}, v_{i2}, etc., can be expanded in terms of the function v_{i0}, w_{i0}. That is,

$$w_{i1} = \sum_{j=1}^{\infty} a_{ij} w_{j0} . \tag{4.63}$$

Substituting (4.63) into (4.60), taking the inner product with respect to $p_s(\mathbf{x})\, v_{k0}(\mathbf{x})$, and making use of (4.62), yields

$$(\lambda_i - \lambda_k) a_{ik} = -\langle v_{k0}, G_1 w_{i0} \rangle - \lambda_{i1}\, \delta_{ik} . \tag{4.64}$$

Thus if $i \neq k$,

$$a_{ik} = -\langle v_{k0}, G_1 w_{i0} \rangle / \lambda_i - \lambda_k . \tag{4.65}$$

If $i = k$, then (4.64) gives

$$\lambda_{i1} = \langle v_{i0}, G_1 w_{i0} \rangle = \langle w_{i0}, L_1^* v_{i0} \rangle. \tag{4.66}$$

In a similar manner, if v_{i1} is expanded in terms of v_{j0}, then

$$v_{i1} = \sum_{j=1}^{\infty} b_{ij} v_{j0} ,$$

where

$$b_{ik} = -\langle w_{k0}, L_1{}^* v_{i0} \rangle / \lambda_i - \lambda_k \qquad i \neq k.$$

The coefficients a_{ii} and b_{ii} can be obtained by using the normalization condition that

$$\int_{R^m} p_s(\mathbf{x}, \epsilon) w_i(\mathbf{x}) v_i(\mathbf{x})\, d\mathbf{x} = 1.$$

Expanding the steady-state density $p_s(\mathbf{x}, \epsilon)$ in a series in ϵ,

$$p_s(\mathbf{x}, \epsilon) = p_s(\mathbf{x})[1 + \epsilon s_1(\mathbf{x}) + \epsilon^2 s_2(\mathbf{x}) \cdots],$$

gives

$$\int_{R^m} p_s(\mathbf{x})[v_{i0} w_{i0} + \epsilon[s_1(\mathbf{x}) v_{i0} w_{i0} + (v_{i1} w_{i0} + v_{i0} w_{i1})] + O(\epsilon^2)]\, d\mathbf{x} = 1.$$

Equating coefficients of like powers of ϵ to zero gives, for the first power in ϵ,

$$\langle s_1 v_{i0}, w_{i0} \rangle + (a_{ii} + b_{ii}) = 0. \tag{4.67}$$

It will be noted that neither a_{ii} nor b_{ii} can be obtained separately; however, it is readily shown that in expanding an arbitrary function in terms of v_{i0} or w_{i0}, only the sum $(a_{ii} + b_{ii})$ is required.

The second and higher order terms in the perturbation series can be obtained in a similar manner. From (4.51), the cross-correlation function is given by

$$R_{jk}(\tau) = \sum_{l=0}^{\infty} \exp(-\lambda_l |\tau|)\alpha_{jl}\beta_{kl} \,. \tag{4.68}$$

Since the spectrum of G is assumed discrete, $R\lambda \equiv 0$. The coefficients α_{jl} and β_{kl} are given by (4.53):

$$\alpha_{jl} = \langle x_j \,, w_l \rangle = \int_{R^m} x_j w_l(\mathbf{x})\, p_s(\mathbf{x}, \epsilon)\, d\mathbf{x},$$

$$\beta_{kl} = \langle x_k \,, v_l \rangle = \int_{R^m} x_k v_l(\mathbf{x})\, p_s(\mathbf{x}, \epsilon)\, d\mathbf{x}.$$

Expanding $p_s(\mathbf{x}, \epsilon)$, $w_l(\mathbf{x})$, and $v_l(\mathbf{x})$ in a power series in ϵ gives

$$\begin{aligned}\alpha_{jl} = &\int_{R^2} p_s(\mathbf{x}) x_j v_{l0}(\mathbf{x})\, d\mathbf{x} + \epsilon \\ &\times \left[\iint_{R^2} p_s(\mathbf{x}) x_j \left[s_1(\mathbf{x}) v_{l0} + \sum_{k=0}^{\infty} a_{lk} v_{k0}(\mathbf{x})\right] d\mathbf{x}\right] + O(\epsilon^2),\end{aligned} \tag{4.69}$$

$$\begin{aligned}\beta_{kl} = &\int_{R^2} p_s(\mathbf{x}) x_k w_{l0}(\mathbf{x})\, d\mathbf{x} + \epsilon \\ &\times \left[\iint_{R^2} p_s(\mathbf{x}) x_k \left[s_1(\mathbf{x}) w_{l0} + \sum_{n=0}^{\infty} a_{ln} w_{n0}(\mathbf{x})\right] d\mathbf{x}\right] + O(\epsilon^2),\end{aligned} \tag{4.70}$$

$$\lambda_l = \lambda_{l0} + \epsilon\lambda_{l1} + O(\epsilon^2). \tag{4.71}$$

The chief advantage of this method is that it yields uniformly valid asymptotic expansions in τ at each stage of the computation.

The variational method. It has been shown above that the correlation functions and spectral densities of nonlinear systems excited by white noise may be found by eigenfunction expansion of the Fokker–Planck–Kolmogorov operators. Unfortunately for many nonlinear systems of second and higher order, the eigenvalue problem cannot be solved exactly. In some cases, perturbation techniques may be used to extend the class of systems which may be analyzed by this method. This requires that the eigenvalues and eigenfunctions of an associated Fokker–Planck operator be known a priori, a situation which unfortunately occurs rather infrequently. The Rayleigh–Ritz method (Mikhlin, 1964) has been widely used to approximate the eigenvalues and eigenfunctions of self-adjacent linear differential operators. The method is readily generalized to nonself-adjacent operators such as G.

The functional $\mu(w, v)$, the Rayleigh quotient, is defined in this case as

$$\mu(w, v) = -\langle Gw, v\rangle/\langle v, w\rangle = -\langle w, L^*v\rangle/\langle v, w\rangle, \tag{4.72}$$

where w and v are arbitrary functions in the domains of G and L^*, respectively, and hence satisfy the necessary boundary conditions. The calculus of variation shows that $\mu(w, v)$ is stationary if

$$\begin{aligned} Gw + \mu w &= 0, \\ L^*v + \mu v &= 0, \end{aligned} \tag{4.73}$$

that is, the stationary values of $\mu(w, v)$ are the eigenvalues of G and L^*, the corresponding functions w and v being the eigenfunctions of G and L^*, respectively. To obtain an approximate solution to the eigenvalue problem (4.73) the following method is used. Let $\{w_i(\mathbf{x}), i = 1, 2,..., N\}$ and $\{v_i(\mathbf{x}), i = 1, 2,..., N\}$ be two sets of bi-ortho normal functions satisfying the necessary boundary conditions on G and L^*. Trial solutions for w and v are assumed in the form

$$w^* = \sum_{i=1}^{N} e_i w_i \qquad \text{and} \qquad v^* = \sum_{i=1}^{N} d_i v_i ,$$

where the coefficients e_i and d_i are to be determined. The $\mu(w^*v^*)$ will be stationary if

$$\partial\mu/\partial c_i = \partial\mu/\partial d_i = 0 \qquad \forall i = 1, 2,..., N.$$

This leads to

$$\langle [G + \mu]w^*, v_i\rangle = \langle w_i , [L^* + \mu]v^*\rangle = 0 \qquad i = 1, 2,..., N. \tag{4.74}$$

If M denotes the $N \times N$ matrix, whose ijth element is $\langle Gw_j , v_i\rangle$, and $\mathbf{e}$ and $\mathbf{d}$ denote the N vectors $\{\mathbf{e}_i\}$ and $\{d_i\}$, then (4.74) can be rewritten in the following way:

$$\begin{aligned} M\mathbf{e} + \mu\mathbf{e} &= 0, \\ M^T\mathbf{d} + \mu\mathbf{d} &= 0. \end{aligned} \tag{4.75}$$

The eigenvalues μ_k of these two eigenvalue problems will be the same, and will approximate N of the eigenvalues λ_k of the original problem. The functions $w_k{}^*$ and $v_k{}^*$ corresponding to the eigenvectors $\mathbf{e}_k$ and $\mathbf{d}_k$ will approximate the corresponding eigenfunctions for the original problem. If the matrix M is symmetric or if it is known that $w_i(x, \dot{x}) = v_i(x, -\dot{x})$, then it is necessary to solve only one of the matrix equations (4.75).

As noted previously, if G is the operator corresponding to a first order nonlinear system excited by white noise, then G is self-adjacent and the two equations of (4.75) are identical. Payne (1967) has used this technique for first order systems, while Atkinson (1970) has used the technique for both first and second order systems.

c. Other Analytical Methods. The methods described above are applicable to a wide class of problems. For some problems, however, it may be advantageous to develop methods which take account of special features inherent in the problem. The problem of self-excited oscillators driven by white noise is such a problem, and a variety of special methods have been developed by Caughey and Payne (1967), Stratanovich (1963), Blaquiere and Grivet (1963a,b), Bernstein (1950), and Zakai (1963). Since these methods are not of general use they will not be discussed here.

d. Numerical Methods. In recent years the capacity of high speed digital computers has increased to the point where it is now practicable to solve the Fokker–Planck–Kolmogorov equations associated with first and second order systems numerically. The Doctoral Theses of Wolaver (1964) and Cumming (1967) give good accounts of the use of numerical methods to obtain response statistics for a variety of first and second order nonlinear systems driven by white noise.

2. *Approximate Techniques Applicable to Nonlinear Stochastic Differential Equations*

In addition to the approximate techniques developed for use with the Fokker–Planck–Kolmogorov equations a number of approximate techniques have been developed which are directly applicable to nonlinear stochastic differential equations.

a. The Method of Equivalent Linearization. The method of equivalent linearization, sometimes called the method of stochastic linearization, was developed independently by Booton (1954) and Caughey (1959a, b). This method is a statistical extension of the method of equivalent linearization of Krylov and Bogoliubov (1937). To illustrate the ideas of method of stochastic linearization consider a second order nonlinear oscillator driven by white noise

$$\ddot{x} + \beta_0 \dot{x} + K_0 + \epsilon f(x, \dot{x}) = \dot{w}(t), \tag{4.76}$$

where $E[dw^2] = 2D\,dt$, and ϵ is a small parameter. If $\epsilon f(x, \dot{x})$ satisfies the axioms of Theorem 1, the existence of a unique continuous solution is guaranteed. If K_0, β_0 are positive, Theorem 6, Corollary 3 guarantees the existence of a steady-state density $p_s(x, \dot{x})$. The basic idea of the method is to write (4.76) in the form

$$\ddot{x} + \beta_e \dot{x} + K_e x + \mathscr{E}(x, \dot{x}) = \dot{w}(t), \tag{4.77}$$

where the equivalent damping and stiffness parameters β_e and K_e are chosen in such a way as to make the equation deficiency term, $\mathscr{E}(x, \dot{x})$, as small as possible. To this end β_e and K_e are chosen in such a way that the mean square, $E[\mathscr{E}^2]$, of the equation deficiency is minimized. Now

$$E[\mathscr{E}^2] = E[(\beta_0 \dot{x} + K_0 x + \epsilon f(x, \dot{x}) - \beta_e \dot{x} - K_e x)^2] \tag{4.78}$$

To minimize $E[\mathscr{E}^2]$, set $\partial E[\mathscr{E}^2]/\partial \beta_e$ and $\partial E[\mathscr{E}^2]/\partial K_e$ equal to zero. Then

$$\beta_e = \beta_0 + \epsilon \frac{E[\dot{x} f(x, \dot{x}) + (K_0 - K_e) x \dot{x}]}{E[\dot{x}^2]}, \tag{4.79}$$

$$K_e = K_0 + \epsilon \frac{E[x f(x, \dot{x}) + (\beta_0 - \beta_e) x \dot{x}]}{E[x^2]}. \tag{4.80}$$

For a differentiable, stationary stochastic process, the expected value of $x\dot{x}$ is zero. Thus

$$\beta_e = \beta_0 + \epsilon \frac{E[\dot{x} f(x, \dot{x})]}{E[\dot{x}^2]}, \tag{4.81}$$

$$K_e = K_0 + \epsilon \frac{E[x f(x, \dot{x})]}{E[x^2]}. \tag{4.82}$$

A simple calculation shows that these values of β_e and K_e do indeed minimize $E[\mathscr{E}^2]$. An approximate solution of (4.77) is obtained by dropping the deficiency term $\mathscr{E}(x, \dot{x})$. Thus

$$x(t) \doteq z(t)$$

where

$$\ddot{z} + \beta_e \dot{z} + K_e z = \dot{w}(t). \tag{4.83}$$

Since (4.83) is a linear stochastic equation with $\dot{w}(t)$ Gaussian, $z(t)$ is also Gaussian. The first and second order statistics are easily calculated by standard techniques thus:

$$E[x^2] \doteq E[z^2] = D/\beta_e K_e, \tag{4.84}$$

$$E[\dot{x}^2] = E[\dot{z}^2] = D/\beta_e. \tag{4.85}$$

Hence

$$p_s(x, \dot{x}) \simeq 1/(2\pi D/\beta_e K_e)^{1/2} \exp(-\beta_e K_e x^2/2D)\, 1/(2\pi D/\beta_e)^{1/2} \exp(-\beta_e \dot{x}^2/2D) \qquad (4.86)$$

Using this approximate expression for $p_s(x, \dot{x})$ in (4.81) and (4.82), two coupled nonlinear equations are obtained, from which β_e and K_e can be obtained. Once β_e and K_e are known the standard methods for linear stochastic differential equations may be used to obtain any desired statistic. This method has been used by several authors including Lyon *et al.* (1961) and Smith (1962).

If the system under study belongs to the class of problems for which the exact steady-state density $p_s(x, \dot{x})$ can be obtained, then the statistical linearization can be carried out using $p_s(x, \dot{x})$. In particular if the nonlinearity $f(x, \dot{x})$ depends on x alone then it can be shown Caughey (1963a,b) that (4.84) and (4.85) yield the exact mean squared displacement and velocity, respectively. Though the mean squared displacement and velocity are correctly predicted in this case, the autocorrelation and spectral density are predicted only approximately. The technique of statistical linearization has been extended to cover the response of nonlinear systems, exhibiting hystertic behavior, excited by white noise (Caughey 1960b).

Caughey (1960a, 1963b), Foster (1968), and Yang (1970) have extended the equivalent linearization technique to systems of nonlinear stochastic differential equations. Yang (1970) used the technique of (4.14) and (4.15) to obtain the second moments of the equivalent linear systems. Caughey (1959a) has also used equivalent linearization to obtain approximate solutions for a nonlinear partial differential equation excited by spatially uncorrelated white noise. It should be pointed out that the technique of statistical linearization applies equally well to nonlinear differential equations excited by weakly stationary random Gaussian processes, the only difference being that (4.83) is replaced by the equation

$$\ddot{z} + \beta_e \dot{z} + K_e z = p(t). \qquad (4.87)$$

If $\Phi_{pp}(\omega)$ is the spectral density of the weakly stationary random Gaussian process $p(t)$, then the mean square statistics are given by

$$E[x^2] \doteq \int_0^\infty \frac{\Phi_{pp}(\omega)\, d\omega}{(K_e - \omega^2)^2 + (\omega\beta_e)^2}, \qquad (4.88)$$

$$E[\dot{x}^2] \doteq \int_0^\infty \frac{\omega^2 \Phi_{pp}(\omega)\, d\omega}{(K_e - \omega^2)^2 + (\omega\beta_e)^2}. \qquad (4.89)$$

b. Perturbation Method. A perturbation method, based on classical perturbation theory, has been developed by Crandall (1963) to obtain approximate solutions to nonlinear systems, containing a small parameter, excited by weakly stationary random Gaussian processes. To illustrate the ideas involved in Crandall's perturbation method, consider the following nonlinear system

$$\ddot{x} + \beta_0 \dot{x} + K_0 x + \epsilon f(x, \dot{x}) = p(t), \tag{4.90}$$

where ϵ is a small parameter and $p(t)$ is a weakly stationary random Gaussian noise excitation. It is assumed that the solution of (4.90) can be expanded in a power series ϵ, for ϵ sufficiently small, that is

$$x(t) = x_0(t) + \epsilon x_1(t) + \epsilon^2 x_2(t) + \cdots. \tag{4.91}$$

A rigorous proof of the convergence of the series (4.91) is lacking at the present time. Substituting (4.91) into (4.90) and sitting the coefficients of like powers of ϵ to zero, leads to a set of linear equations for x_0, x_1, x_2 etc.

$$\begin{aligned} L_0 x_0 &= p(t), \\ L_0 x_1 &= -f(x_0, \dot{x}_0), \\ L_0 x_2 &= -f_{x_0}(x_0, \dot{x}_0)\, x_1 - f_{\dot{x}_0}(x_0, \dot{x}_0)\dot{x}_1, \\ &\vdots \end{aligned} \tag{4.92}$$

where

$$L_0 u = \ddot{u} + \beta_0 \dot{u} + K_0 u.$$

The steady-state solutions of these equations are readily constructed.

$$\begin{aligned} x_0(t) &= \int_0^\infty h(\xi)\, p(t - \xi)\, d\xi, \\ x_0(t) &= -\int_0^\infty h(\xi) f(x_0(t - \xi), \dot{x}_0(t - \xi))\, d\xi, \\ &\vdots \end{aligned} \tag{4.93}$$

where $h(\xi)$ is the impulse response of the linear operator L_0.

Equations (4.93) can be used to compute the various statistics of the response. The expectation of $x(t)$ is given by

$$E[x(t)] = E[x_0(t)] + \epsilon E[x_1(t)] + \epsilon^2 [x_2(t)] + \cdots. \tag{4.94}$$

Using (4.93) in (4.94) yields

$$E[x(t)] = \int_0^\infty h(\xi)\, E[p(t - \xi)]\, d\xi - \epsilon \int_0^\infty h(\xi)\, E[f(x_0(t - \xi), \dot{x}_0(t - \xi)]\, d\xi + \cdots. \tag{4.95}$$

To evaluate terms like $E[f(x_0, \dot{x}_0)]$, use is made of the fact that since $p(t)$ is Gaussian, so also is $x_0(t)$; thus

$$E[f(x_0, \dot{x}_0)] = \int_{-\infty}^{\infty} \int f(x_0, \dot{x}_0)\, p_s(x_0, \dot{x}_0)\, dx_0\, d\dot{x}_0, \tag{4.96}$$

where $p_s(x_0, \dot{x}_0)$ is the steady-state density for the process $\{x_0, \dot{x}_0\}$.

Higher order terms can be evaluated in a similar manner; however, the computational difficulties increase very rapidly with the order of the terms included. Frequently, only the first order terms in the perturbation method are evaluated. The autocorrelation of the response, computed to first order in ϵ, is

$$\begin{aligned} E[x(t+\tau)\, x(t)] &= E[x_0(t+\tau)\, x_0(t)] \\ &\quad + \epsilon[E[x_0(t+\tau)\, x_1(t)] + E[x_0(t)\, x_1(t+\tau)]] \\ &\quad + \text{higher order terms}. \end{aligned} \tag{4.97}$$

From symmetry,

$$E[x_0(t)\, x_1(t+\tau)] = E[x_0(t-\tau) x_1(t)]; \tag{4.98}$$

thus (4.97) may be rewritten in the following way:

$$R_{xx}(\tau) = R_{x_0x_0}(\tau) + \epsilon[R_{x_0x_1}(\tau) + R_{x_0x_1}(-\tau)] + \text{higher order terms}, \tag{4.99}$$

where

$$\begin{aligned} R_{x_0x_0} &= E[x_0(t+\tau)\, x_0(t)] \\ &= \int_{-\infty}^{\infty} \int E[p(t+\tau-\xi_1)\, p(t-\xi_2)]\, h(\xi_1)\, h(\xi_2)\, d\xi_1\, d\xi_2, \end{aligned} \tag{4.100}$$

$$\begin{aligned} R_{x_0x_1}(\tau) &= E[x_0(t+\tau)\, x_1(t)] \\ &= \int_{-\infty}^{\infty} E[x_0(t+\tau) f(x_0(t-\xi), \dot{x}_0(t-\xi))]\, h(\xi)\, d\xi, \end{aligned} \tag{4.101}$$

To evaluate the expectation $E(\cdot)$ occurring in (4.10), use is again made of the fact that $x_0(t)$ is Gaussian; thus,

$$E[x_0(t_1)\, f(x_0(t_2), \dot{x}_0(t_2))] = \int_{-\infty}^{\infty} \int y_1 f(x_0, \dot{x}_0)\, p(\mathbf{y}, t_1; \mathbf{x}_0, t_2)\, d\mathbf{x}_0\, d\mathbf{y}, \tag{4.102}$$

where $p(\mathbf{y}, t_1; \mathbf{x}_0, t_2)$ is the joint probability density. The spectral density of the process is obtained by applying the Wiener–Khintchine relations to (4.100) and (4.101) giving

$$\Phi_{xx}(\omega) = \Phi_{x_0x_0}(\omega) + 2\epsilon \operatorname{Re}(\Phi_{x_0x_1}(\omega)) + O(\epsilon^2), \tag{4.103}$$

where Re = real part and use has been made of the fact that the cross-spectral density of two jointly weak stationary processes is Hermitian.

Crandall (1963) and Crandall *et al.* (1964) have used this technique extensively in the study of nonlinear random vibrations. In principle, the perturbation method can be extended to systems of coupled nonlinear oscillators in which the nonlinearities contain a small parameter ϵ. Tung, Penzien, and Horonjeff (1964) used this technique to study the dynamic response of a nonlinear two degrees of freedom system to random excitation. Lyon (1960b) used the perturbation method to study the responses of a nonlinear string to random excitation.

c. Other Perturbation Techniques. Dienes (1961) in his Ph.D. thesis combined the method of statistical linearization with Crandall's perturbation method to examine the response of Duffing's equation to white noise excitation. The method of statistical linearization, using the exact steady-state probability density, predicts the correct mean squared response; however, it predicts that the spectral density has only a single peak close to the linear natural frequency. By combining Crandall's perturbation method with the method of equivalent linearization, Dienes was able to show that to first order in ϵ the spectral density exhibited a secondary peak close to three times the linear natural frequency. This result is not surprising and can easily be verified in the laboratory.

Kraichnan (1959, 1961) and Wyld (1961) have developed the method of consolidated expansions in connection with the study of turbulence. Morton (1967) has applied this method to obtain a second order approximation for Duffing's equation driven by white noise.

d. Comparison of Approximate Methods. Payne (1967, 1968) has made a comparison of the accuracy of a number of approximate methods. He found that equivalent linearization, Crandall's perturbation method, and eigenfunction expansion techniques all gave the same result to first order in ϵ. The method of equivalent linearization was found to be in error in the second order terms in ϵ; however, the relative error remained bounded as ϵ tended to infinity. Crandall's perturbation method could, with enough effort, yield results correct to second order in ϵ; however, the autocorrelation function so predicted was not uniformly valid in τ. The perturbation method, as applied to the eigenfunction expansion techniques, was found to yield the best overall results for a given level of effort. In addition this method yields uniformly valid asymptotic expansions for the autocorrelation function at each step.

ACKNOWLEDGMENTS

I gratefully acknowledge the many helpful suggestions of my former students Drs. J. K. Dienes, A. H. Gray Jr., H. J. Payne, T. W. MacDowell, and J. D. Atkinson.

REFERENCES

Andronov, A., Pontryagin, L., and Witt, A. (1933). On the statistical investigation of dynamical systems. *Zh. Eksp. Teor. Fiz.* **3**, 165–180.

Ariaratnam, S. T. (1960). Random vibrations of nonlinear suspensions. *J. Mech. Eng. Sci.* **2**, 195–201.

Ariaratnam, S. T. (1962). Response of a loaded nonlinear string to random excitation. *J. Appl. Mech.* **29**, 483–485.

Atkinson, J. D. (1970). A variational method for randomly excited systems. Tech. Note D-1. Dept. Mech. Eng., University of Sydney.

Atkinson, J. D., and Caughey, T. K. (1968). Spectral density of piecewise linear first order systems excited by white noise. *Int. J. Nonlinear Mech.* **3**, 137–156.

Barrett, J. F. (1961). Application of Kolmogorov's equations to randomly disturbed automatic control systems. *Proc. IFAC Congr., 1st*, 1960 pp. 724–729.

Benes, V. E. (1968). Finite regular invariant measures for Feller processes. *J. Appl. Prob.* **5**.

Bernstein, I. (1950). Amplitude and phase fluctuations of valve oscillators. *Bull. Acad. Sci. USSR, Phys. Ser.* **14**, 187.

Blaquiere, A., and Grivet, P. (1963a). Masers and classical oscillators. *Proc. Symp. Opt. Masers*, 1963, pp. 69–93.

Blaquiere, A., and Grivet, P. (1963b). Nonlinear effects of noise in electronic clocks. *Proc. IEEE* **15**, No. 11, 1606–1614.

Bluman, G. W. (1967). Construction of solutions to partial differential equations by use of transformation groups. Ph.D. Thesis, California Institute of Technology.

Bluman, G. W., and Cole, J. D. (1969). The general similarity solution of the heat equation. *J. Math. Mech.* **18**, 1025–1042.

Bogdanoff, J. L., and Kozin, F. (1962). Moments of the output of linear random systems. *J. Acoust. Soc. Amer.* **34**, No. 8, 1063–1066.

Booton, R. C. (1954). The analysis of nonlinear control systems with random inputs. *IRE Trans. Circuit Theory* **1**, 32–34.

Caughey, T. K. (1959a). Response of a nonlinear string to random loading. *J. Appl. Mech.* **26**, 341–344.

Caughey, T. K. (1959b). Response of Van der Pol's oscillator to random excitation. *J. Appl. Mech.* **26**, 345–348.

Caughey, T. K. (1960a). Random excitation of a loaded nonlinear string. *J. Appl. Mech.* **27**, 575–578.

Caughey, T. K. (1960b). Random excitation of a system with bilinear hysterisis. *J. Appl. Mech.* **27**, 649–652.

CAUGHEY, T. K. (1963a). Derivation and application of the Fokker–Planck equation to discrete nonlinear dynamic systems subjected to white random excitation. *J. Acoust. Soc. Amer.* **35**, No. 11, 1683–1692.

CAUGHEY, T. K. (1963b). Equivalent linearization techniques. *J. Acoust. Soc. Amer.* **35**, No. 11, 1706–1711.

CAUGHEY, T. K. (1964). On the response of a class of nonlinear oscillators to stochastic excitation. *Colloq. Int. Cent. Nat. Rech. Sci.* **148**, 393–402.

CAUGHEY, T. K., and DIENES, J. K. (1961). Analysis of first order system with white noise input. *J. Appl. Phys.* **32**, 2476–2479.

CAUGHEY, T. K., and DIENES, J. K. (1962). The behavior of linear systems with random parametric excitation. *J. Math. Phys.* **41**, No. 4, 300–318.

CAUGHEY, T. K., and GRAY, A. H., JR. (1965). A controversey in problems involving random parametric excitation. *J. Math. Phys.* **44**, No. 3, 288–296.

CAUGHEY, T. K., and PAYNE, H. J. (1967). On the response of a class of self excited oscillators to stochastic excitation. *Int. J. Nonlinear Mech.* **2**, 125–151.

CHUANG, K., and KAZDA, L. F. (1959). Study of nonlinear systems with random inputs. *Trans. AIEE* **78**, Part II, 100–105.

CRANDALL, S. H. (1961). Random vibrations of systems with nonlinear with nonlinear restoring forces. *Proc. Int. Symp. Nonlinear Vibrations*, No. 1, pp. 306–314.

CRANDALL, S. H. (1962). Random vibrations of a nonlinear system with a set-up spring. *J. Appl. Mech.* **29**, 477–482, pp. 306–314.

CRANDALL, S. H. (1963). Perturbation techniques for random vibrations of nonlinear systems. *J. Acoust. Soc. Amer.* **35**, No. 11, 1700–1705.

CRANDALL, S. H. (1964a). The envelope of random vibration of a lightly damped nonlinear oscillator. *Zagadnienia Drgan Nielinowych* **5**, 120–130.

CRANDALL, S. H. (1964b). The spectrum of random vibration of a nonlinear oscillator. *Proc. Int. Congr. Appl. Mech., 11th*, 1964, No. 1, pp. 239-244.

CRANDALL, S. H. (1965). Random forcing of nonlinear systems. *Colloq. Int. Cent. Nat. Rech. Sci.* **148**, 57–68.

CRANDALL, S. H., KHABBAZ, G. R., and Manning, J. E. (1964). Random vibration of an oscillator with nonlinear damping. *J. Acoust. Soc. Amer.* **36**, No. 7, 1330–1334.

CUMMING, I. G. (1967). Computing aspects of problems in nonlinear prediction and filtering. Ph.D. Thesis, Imperial College, University of London.

DIENES, J. K. (1961). Some applications of the theory of continuous Markoff processes to random oscillation problems. Ph.D. Thesis, California Institute of Technology.

DOOB, J. L. (1953). "Stochastic Processes." Wiley, New York.

EINSTEIN, A. (1905). Über dievan der molekular-Kineteschen Theorii der Warme gefordert Bewegung von in ruhenden Flüssigkeiten. *Ann. Phys. (Leipzig)* [4] **17**, 549.

FELLER, W. (1952). The parabolic differential equation and the associated semigroups of transformations. *Ann. Math.* **55**, No. 3, 468–519.

FELLER, W. (1954). Diffusion processes in one dimension. *Trans. Amer. Math. Soc.* **77**, No. 1, 1–31.

FOKKER, A. P. (1914). Die mittlere Energie rotierender electricher Dipole im Strahlungsfeld. *Ann. Phys. (Leipzig)* [4] **42**, 810.

FOSTER, E. T. (1968). Semilinear random vibrations in discrete systems. *J. Appl. Mech.* **35**, No. 3, 560–564.

GIKHMAN, I. I., and SKOROKHOD, A. V. (1965). "Introduction to the Theory of Random Processes." Saunders, Philadelphia, Pennsylvania.

GRAY, A. H., JR. (1964). Stability and related problems in randomly excited systems. Ph.D. Thesis, California Institute of Technology.

HERBERT, R. E. (1964). Random vibrations of a nonlinear elastic beam. *J. Acoust. Soc. Amer.* **36**, No. 11, 2090–2094.
HERBERT, R. E. (1965). Random vibrations of plates with large amplitudes. *J. Appl. Mech.* **32**, 547–552.
ILIN, A. M., and KHASMINSKII, R. Z. (1964). On equations of Brownian Motion. *Theory Probab. Appl. (USSR)* **9**, 421–444.
ITO, K. (1946). On a stochastic integral equation. *Proc. Jap. Acad.* Vol. **22**, No. 2, 32–35.
ITO, K. (1951). On stochastic differential equations. *Mem. Amer. Math. Soc.* No. 4.
KHASMINSKII, R. Z. (1960). Ergodic properties of recurrent diffusion processes and stabilization of the solution of the Cauchy problem for parabolic equations. *Theory Probab. Appl. (USSR)* **5**, 179–196.
KHAZEN, E. M. (1961). Evaluation of the one dimensional probability densities and moments of a random process in the output of an essentially nonlinear system. *Theory Probab. Appl. (USSR)* **6**, pp. 117–123.
KHINTCHINE, A. (1934). Korrelations Theorie der Stationaren Stochastischen Prozesse. *Math. Ann.* **109**, 604–615.
KLEIN, G. H. (1964). Random excitation of a nonlinear system with tangent elasticity characteristics. *J. Acoust. Soc. Amer.* **36**, No. 11, 2095–2105.
KOLMOGOROV, A. (1931). Über die Analytischen Methoden in Wahrsheinlichkeits-rechnung. *Math. Ann.* **104**, 415–458.
KOLMOGOROV, A. (1950). "Foundations of Probability Theory." Chelsea, New York.
KRAICHNAN, R. N. (1959). The theory of isotropic turbulence at very high Reynolds numbers. *J. Fluid Mech.* **5**, 497–543.
KRAICHNAN, R. N. (1961). Dynamics of nonlinear stochastic systems. *J. Math. Phys.* **2**, 124–148.
KRAMERS, H. A. (1940). Brownian motion in a field of force and the diffusion model of chemical reactions. *Physica* **7**, 284–304.
KRYLOV, N., and BOGOLIUBOV, N. (1937). "Introduction to Nonlinear Mechanics." Kiev. Translation by Princeton University Press, Princeton, New Jersey, 1943.
KUSHNER, H. J. (1969). The Cauchy problem for a class of degenerate parabolic equations and asymptotic properties of the related diffusion processes. *J. Diff. Equations* **6**, No. 2, 209–231.
LYON, R. (1960a). On the vibration statistics of a randomly excited hard-spring oscillator. *J. Acoust. Soc. Amer.* **32**, 716–719.
LYON, R. (1960b). Response of a nonlinear string to random excitation. *J. Acoust. Soc. Amer.* **32**, No. 8, 953–960.
LYON, R. (1961). On the vibration statistics of a randomly excited hard-spring oscillator II. *J. Acoust. Soc. Amer.* **33**, No. 10, 1395–1403.
LYON, R., HECKL, M., and HAZELGROVE, C. B. (1961). Narrow band excitation of the hard spring oscillator. *J. Acoust. Soc. Amer.* **33**, 1404–1411.
MCFADDEN, J. A. (1959). The probability density of the output of a filter when the input is a random telegraphic signal. *IRE Trans. Circuit Theory* **6**, 228–233.
MERKLINGER, K. L. (1963). Numerical analysis of nonlinear control systems using the Fokker–Planck–Kolmogorov equations. *Proc. IFAC Congr., 2nd*, 1963, pp. 81–89.
MIKHLIN, S. G. (1964). "Variational Method in Mathematical Physics," Chapter 5. Pergamon Press, Oxford.
MORTON, J. B. (1967). The use of consolidated expansions in solving for the statistical properties of a nonlinear oscillator. Ph.D. Thesis, Johns Hopkins University.
PAYNE, H. J. (1967). The response of nonlinear systems to stochastic excitation. Ph.D. Thesis, California Institute of Technology.

PAYNE, H. J. (1968). An approximate method for nearly linear, first order stochastic differential equations. *Int. J. Contr.* **7**, No. 5, 451–463.

PLANCK, M. (1917). Über einen Satz der statistischen Dynamik und seine Erweiterung in der Quanten-theorie. *Sitzungsber. Berlin Akad. Wiss.* pp. 324–341.

PUGACHEV, V. S. (1957). "The Theory of Random Functions and Its Application to Problems in Automatic Control." Moscow.

SAWARAGI, Y., *et al.* (1968–1961). Reports, Vol. VIII, No. 10; Vol. IX, Nos. 57, 60, 61, 68, and 79. Eng. Res. Inst., Kyoto University.

SMITH, P. W., JR. (1962). Response of nonlinear structures to random excitation. *J. Acoust. Soc. Amer.* **34**, 827–835.

STRATONOVICH, R. L. (1963). "Topics in the Theory of Random Noise," Vols. 1 and 2. Gordon & Breach, New York.

TUNG, C. C., PENZIEN, J., and HORONJEFF, R. (1967). The effects of runway unevenness on the dynamic respones of supersonic transports. NASA CR-119, University of California, Berckeley, California.

UHLENBECK, G. E., and WANG, M. C. (1945). On the theory of Brownian motion II. *Rev. Mod. Phys.* **17**, Nos. 2-3, 323–342.

VON SMOLUCHOWSKI, M. (1916). Drei Vortage uber Diffusion Brownsche Bewegung und Koagulation von Kolloidteilchen. *Phys. Z.* **17**, 557.

WAX, N., ed. (1954). "Noise and Stochastic Processes" (Select Papers). Dover, New York.

WIENER, N. (1930). Generalized harmonic analysis. *Acta Math.* **55**, 117–174.

WIENER, N. (1950). "Extrapolation, Interpolation, and Smoothing of Stationary Time Series." Wiley, New York.

WISHNER, R. P. (1960). On Markov processes in control systems. Rep. R-110. Corod. Sci. Lab., University of Illinois, Urbana, Illinois.

WOLAVER, L. E. (1964). Second order properties of nonlinear systems driven by random noise. Ph.D. Thesis, University of Michigan, Ann Arbor, Michigan.

WONG, E. (1964). The construction of a class of Markov processes. *Proc. Symp. Appl. Math.* **16**, 264–276.

WONHAM, W. M. (1966). A Liapunov method for the estimation of statistical averages. *J. Diff. Equations* **2**, 365–377.

WYLD, H. W. (1961). Formulation of the theory of turbulence in an incompressible fluid. *Ann. Phys. (Leipzig)* [7] **14**, 143–165.

YANG, I. M. (1970). Stationary random response of multidegree of freedom systems. Ph.D. Thesis, California Institute of Technology.

ZAKAI, M. (1963). On the first-order probability distribution of the Van der Pol type oscillator. *J. Electron. Contr.* **14**, No. 4, 381–388.

Physical Theory of Plasticity

T. H. LIN

Mechanics and Structures Department,
University of California, Los Angeles, California

I. Introduction

The stress–strain relation in the plactic range of ductile materials has been studied by many investigators. Their studies may be deivided into two classes: one is known as the mathematical theory of plasticity and the other, the physical theory of plasticity. A mathematical theory is mainly a representation of experimental data as a necessary extension of elasticity theory to furnish more realistic estimates of the load-carrying capacities. Mathematical simplicity is essential to this representation so as to be readily applicable to design and analysis. As pointed out by Drucker (1962), this type of plasticity theory is only a formalization of known experimental results and does not inquire deeply into the physical and chemical basis. This type of theory is generally started from hypotheses and assumptions of a phenomenological character based on certain exeperimental observations. The assumed phenomenological laws cannot claim generality and are apt to give a reliable approach only to a relatively limited class of real processes, as pointed out by Ilyushin (1955) and Olszak *et al.* (1963). On the other hand, a physical theory does attempt to explain why things happen the way they do, but may not embody mathematical simplicity. The plastic stress–strain relationship of polycrystalline solids is derived from that of single crystals or subcrystals. The present review is mainly concerned with the development of the physical theory, although its correlations with the basic hypotheses and assumptions of the mathematical theory are discussed.

II. Dislocation and Plastic Deformation of Single Crystals

A. Crystalline State of Metals

Metals are crystalline solids which consist of atoms arranged in a pattern that is repeated periodically in three dimensions. The regular arrangement of atoms in three dimensions is a space lattice. The smallest lattice representing a lattice type is called a unit cell. The lattice space is built with unit cells. The geometrical pattern of the atoms creates differences in the physical properties, such as elastic modulus and thermal coefficient in different directions. The crystalline substance is thus intrinsically anisotropic. An idealized simple lattice is cubic with one atom at each corner of a cube. The most common space lattices in metals for engineering use are face-centered cubic, body-centered cubic, and closely packed hexagon lattices. The unit cell of a face-centered cubic crystal (which is hereafter written as fcc crystal) has one atom located at each corner and one atom at the center of each face of the cubic, as shown in Fig. 1. In the present paper, discussions will be mainly confined to fcc crystals.

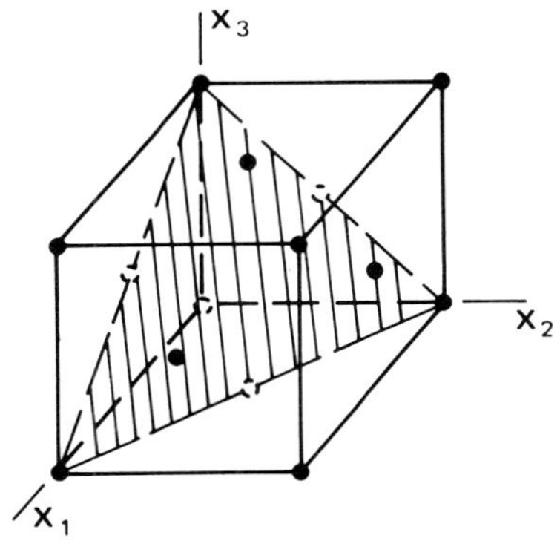

FIG. 1. Face-centered cubic lattice.

B. Shear Strength of Perfect Crystals

The shearing of two rows of atoms in a homogeneously strained crystal is shown in Fig. 2. The spacing between the rows is a, that between two adjacent atoms in a row is b, and the shear displacement of the upper row over the lower one is denoted by x. Clearly, the shear stress is zero when $x = 0, b$ since these are the normal lattice sites. At $x = b/2$, the shear stress is also zero by symmetry. Each atom of the top row is attracted towards its nearest lattice site as defined by the atoms of the lower row, so that the shearing stress τ must be a periodic function of x with period b. Assume this function to be represented by

$$\tau = k \sin(2\pi x/b), \tag{2.1}$$

where k is the maximum shear stress that the crystal can take and is generally called the shear strength. When x/b is small, we can write (2.1) as

$$\tau = k(2\pi a/b)(x/a). \tag{2.2}$$

Since x/a is the shear strain, $k(2\pi a/b)$ must be equal to the shear modulus G. This gives

$$k = (G/2\pi)(b/a). \tag{2.3}$$

The spacing of the atomic planes a is maximum when the density of atoms in the plane is maximum, corresponding to a minimum b. This gives the direction and plane for weakest shear strength to be those of highest atomic density. For a fcc crystal, this $b/a = \sqrt{3}/\sqrt{2}$, giving a shear strength of approximately $G/2\pi$. A refined calculation of this theoretical shear strength by Frankel (1926) gives a value of $G/30$, approximately. However, the observed shear strength of annealed crystals of various metals including fcc crystals are of the order of 10^{-6} to $10^{-4}G$. This large disparity has been ingeniously explained by Taylor (1934), Orowan (1934), and Polanyi (1934) by the concept of dislocations.

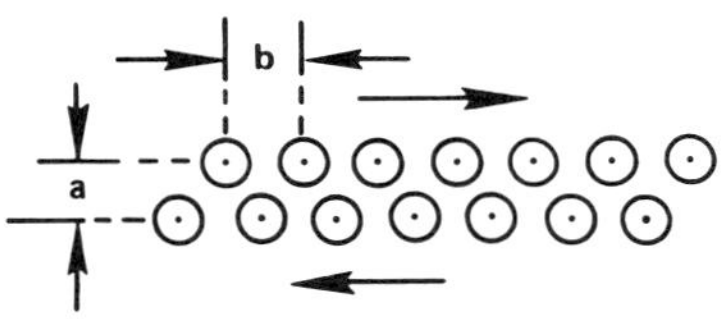

FIG. 2. Two rows of atoms subject to shear.

C. Dislocations and Low Shear Strength of Actual Crystals

The basic concept of dislocations is that plastic flow occurs in two stages: small regions of plastic shear first appear and then grow through the crystal, as shown in Fig. 3. A crystal subject to shearing stress undergoes elastic distortion until gliding starts. Part of the material slides with respect to the rest by unit atomic spacing. This occurs over the cross-hatched portion, but not over the entire slip plane. The line ABC which bounds the area over which slip has occurred, separating it from the unslid area, is called a dislocation line. The displacement across the slip plane changes discontinuously at this dislocation line from a unit amount to zero.

The slip over the hatched area can be described by a vector **b** which specifies the direction and distance by which atoms of the upper portion

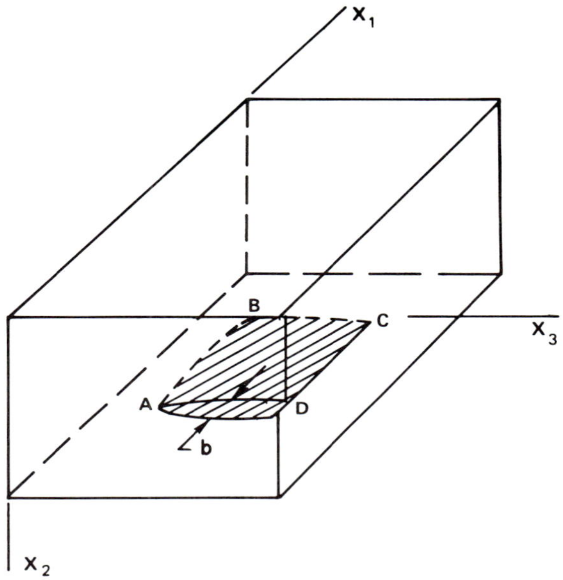

FIG. 3. Unit slip in area ABCD producing dislocation ABCDA.

have slid over the lower portion. The vector is known as Burger's vector (Burger, 1939). To define this vector more specifically (Hirth and Lothe, 1968), we consider a dislocation, as shown in Fig. 4, in which the positive direction of the dislocation points into the paper. We form a clockwise close circuit in the real crystal S123F, which lies entirely in good material and encloses a dislocation, and draw the same circuit

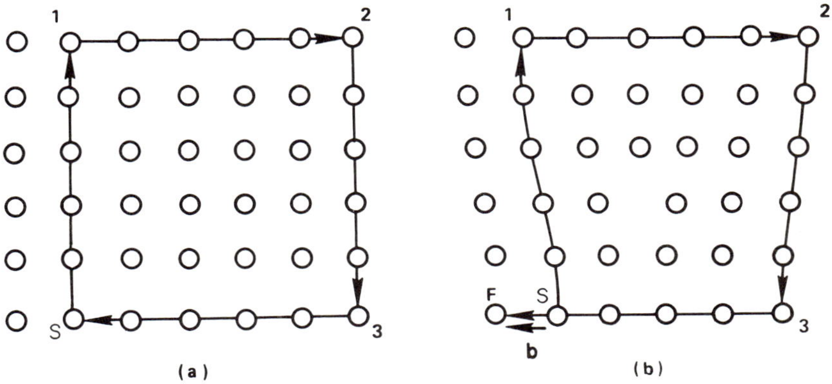

FIG. 4. Burger's circuits in a perfect reference crystal (a) and a real crystal (b).

in a perfect lattice as shown in Fig. 4b. The vector required to close the latter ciruit from the finishing point F to the starting point S defines the Burger vector **b**. The initiation of slip over a small local area and its gradual spreading to other areas requires a shear stress much less than that required for the simultaneous sliding of a complete layer over another. Figure 5a shows an edge dislocation in a simple cubic crystal.

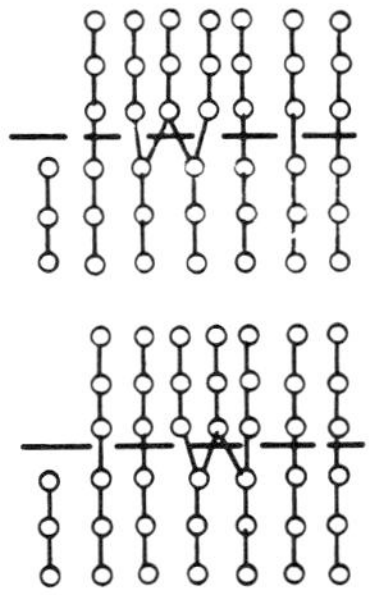

FIG. 5. Balanced forces at a dislocation.

This dislocation is in a symmetric position in a crystal so that the resistive forces on the left side of the dislocation are balanced by attractive forces on the right. When it moves half way to the next position, it takes on another symmetric configuration, Fig. 5b, and again all the forces are balanced. Thus, as indicated by Gilman (1960), a dislocation almost always has a system of balanced forces acting on it, and hence only a small biasing stress is needed to move the dislocation forward or backward. This explains the low shear strength of actual crystals.

D. Plastic Strain and Dislocation Movement

It has been pointed out by Dorn and Mote (1962) that, among the different mechanisms of plastic deformation in metals, the translation glide (slip) is the principal process of plastic deformation in fcc metals at low and intermediate temperatures. The present discussion is limited to plastic deformation caused by such a slip. Now consider several dislocation lines of various Burger's vectors $\mathbf{b}^{(1)}, \mathbf{b}^{(2)}, \ldots, \mathbf{b}^{(n)}$, with directions $\mathbf{t}^{(1)}, \mathbf{t}^{(2)}, \ldots, \mathbf{t}^{(n)}$ and displacements $\mathbf{V}^{(1)}, \mathbf{V}^{(2)}, \ldots, \mathbf{V}^{(n)}$ passing through a small surface ΔS in the deformed material. The total Burger's vector of n dislocations threading through surface ΔS is

$$\alpha_{hi}\nu_h \,\Delta S = \sum_{p=1}^{n} t_h^{(p)} b_i^{(p)} \nu_h \,, \tag{2.4}$$

where ν is the normal vector to ΔS. This α_{hi} is called the dislocation density tensor giving the i component of the Burger's vector per unit area normal to the x_h axis. The flow of the dislocation is denoted by

$$U_{lhi}\nu_h \, \Delta S = \sum_{m=1}^{n} V_l^{(m)} t_h^{(m)} b_i^{(m)} \nu_h \, . \tag{2.5}$$

The tensor U_{lhi} is called the dislocation flow tensor. Strain and rotation due to plastic deformation are closely related to dislocation density and flow tensors. Replacing a finite number of slip surfaces by an infinite number of them, each with infinitesimal gliding, as did Kroner (1958) and Eshelby (1956), we obtain a continuous displacement field V_i. The distortion of a plastically deformed body $u_{i,j}$ consists of the elastic distortion and plastic slip due to the infinitesimal plastic gliding

$$u_{i,j} = u'_{i,j} + u''_{i,j} \, , \tag{2.6}$$

where the single prime denotes the elastic part and the double prime the plastic part and the subscript j after a comma denotes differentiation with respect to the x_j axis. The plastic and elastic linear strains are

$$e''_{ij} = \tfrac{1}{2}(u''_{i,j} + u''_{j,i}) \tag{2.7}$$

and

$$e'_{ij} = \tfrac{1}{2}(u'_{i,j} + u'_{j,i}). \tag{2.8}$$

In a unit cube shown in Fig. 3, the Burger's vector is along the x_1 axis and the slip on the shaded area causes an average $u''_{1,2}$ in the body. This average $u''_{1,2}$ increases with the size of the shaded area. The dislocation line AB taken to be parallel to the Burger's vector is called a screw dislocation, and BC taken to be perpendicular to the Burger's vector is called an edge dislocation. AB corresponds to α_{11} and BC corresponds to α_{31}. The size of the slid area increases with the displacement of AB along the negative x_3 axis and with the displacement of BC along the x_1 axis. The plastic distortion $u''_{2,1}$ is hence caused by the displacement of α_{31} in the x_1 direction and/or by the displacement of α_{11} in the direction of decreasing x_3 :

$$u''_{2,1} = U_{131} - U_{311} \, . \tag{2.9}$$

In a general form, with the repetition of the subscript denoting summation from 1 to 3,

$$u''_{i,j} = -\epsilon_{imn} U_{mnj} \, , \tag{2.10}$$

where ϵ_{imn} is the skew-symmetric tensor defined by

$$\epsilon_{ijk} = \begin{cases} 0 \text{ when any two of the indices are equal,} \\ +1 \text{ when } i, j, k \text{ is a cyclic permutation of the numbers 1, 2, 3,} \\ -1 \text{ when } i, j, k \text{ is a cyclic permutation of the numbers 1, 3, 2.} \end{cases}$$

The plastic strain component is

$$e''_{ij} = \tfrac{1}{2}(u''_{i,j} + u''_{j,i}) = -\tfrac{1}{2}(\epsilon_{imn}U_{mnj} + \epsilon_{jmn}U_{mni}). \tag{2.11}$$

This gives the relation between plastic strain and the movement of the dislocations. This derivation follows that given by Mura (1967). Graphical representation of this relation for some simple dislocation movement has been given by Read (1953) and Gilman (1960).

Shear stress along the gliding direction on the gliding plane of the dislocation, known as the resolved shear stress, supplies the force that tends to cause dislocations to glide. This gliding is independent of other components of the applied stresses. The dependency on the resolved shear stress has been discussed theoretically by Nabarro (1952), and is demonstrated by the single crystals test, as shown by Schmid (1931), Jillon (1950), Taylor and Elam (1923), Schmid and Boas (1950), and Taylor (1938). It has also been experimentally observed that there exists a critical magnitude of this shear stress needed to move isolated dislocations in a lithium fluoride crystal (Johnston and Gilman, 1959; Lüche and Lange, 1952).

E. Macroscopic Slip in Single Crystals

Since the 1920's single crystals of sizes adequate for stress–strain measurements have been successfully produced, and much experimental work has been done to determine the macroscopic relationship between stress, strain, and the crystal axis of single crystals. These experiments show that crystals, in general, deform plastically by means of translation slip, with one part of the crystal sliding over another. This slip is anisotropic. It occurs along certain crystallographic directions on certain crystal planes. Each slip direction on a slip plane is called a slip system. These slip directions and planes are almost always those of maximum atomic density and correspond to those slip systems in which dislocations are most likely to move.

Many single crystal tensile specimens of different orientation have been tested (Schmid, 1931; Jillon, 1950; Taylor and Elam, 1923; Schmid and Boas, 1950; Barrett, 1952). The shear stress along the slip direction on the slip plane shown in Fig. 6 is

$$\tau = (P/A) \cos \phi \cos \lambda. \tag{2.11}$$

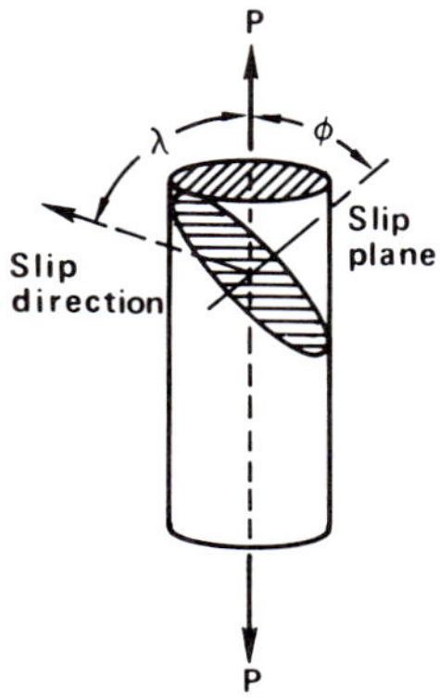

FIG. 6. Single crystal tensile test.

Tests show that slip begins when this resolved shear stress reaches a certain value called the critical shear stress. The initiation and continuation of one slip system, called single slip, has been found to depend on the resolved shear stress and to be independent of the normal pressure on the sliding plane, just as in the case of dislocation movement.

In a fcc crystal, as shown in Fig. 1, the plane cross-hatched is one slip plane. It has a normal with direction cosines $(1/\sqrt{3}, 1/\sqrt{3}, 1/\sqrt{3})$. The three slip directions on this plane are $(1/\sqrt{2}, -1/\sqrt{2}, 0)$. $(0, 1/\sqrt{2}, -1/\sqrt{2})$, and $(-1/\sqrt{2}, 0, 1/\sqrt{2})$. These correspond to the three edges of the shaded triangle. Due to the symmetry of the crystal lattice, there are three other slip planes. These give a total of four slip planes each with three slip directions, giving twelve slip systems. If the crystal is oriented such that the resolved shear stresses in two slip systems are equal, both systems slide when these shear stresses reach a certain critical value. This slip is called duplex slip. As shown by Taylor (1938), this critical shear stress for duplex slipping depends on the sum of the amounts of slip in these two systems in about the same way as the resolved shear stress is related to the amount of slip in single slipping. In the case of slip in more than two slip systems, the resolved shear stresses in all active slip systems have been assumed by Taylor (1938) to be the same and to depend on the sum of the amounts of slip occurring in all slip systems. This assumption is used in the following analysis.

The increase of the critical shear stress in the slid system with the amount of slip is called strain hardening and that in the unslid systems is called latent hardening. The rate of hardening per unit slip is higher for multiple slip than for single slip (Kocks, 1958, 1970). When a single crystal tensile specimen is distorted by sliding in a direction parallel to a crystal plane along a crystal direction, the orientation of the slip planes relative to the axis of extension varies as the straining proceeds. When the specimen axis reaches the position making equal angles with two slip planes, the resolved shear stresses in two slip systems—one on each plane along the same slip direction—are equal. For pure aluminum, the tendency for the specimen axis to move beyond the position of giving equal stresses on the two planes at the start of duplex slip is small (Taylor, 1938). This shows that for this metal, the strain hardening and latent hardening are approximately equal. Hence, equal hardening in active and latent systems is here assumed for multiple slip (Taylor, 1938).

F. Yield Surfaces of Single Crystals

When a slip system in a crystal with a set of parallel planes with normal α and a sliding direction β slides, the plastic strain produced by this slip is $e''_{\alpha\beta}$. Let $l_{\alpha i}$ and $l_{\beta j}$ denote the cosines of the angles between the α and i axes and the β and j axes, respectively. Noting $de''_{\alpha\beta} = de''_{\beta\alpha}$, we have

$$de''_{ji} = [l_{i(\alpha)}l_{j(\beta)} + l_{i(\beta)}l_{j(\alpha)}]\, de''_{(\alpha)(\beta)}\,, \tag{2.12}$$

where the parentheses on the subscript denote "no summation." Expressing the resolved shear stress $\tau_{\alpha\beta}$ in terms of the stress components along the x_i axes, we have

$$\tau_{\alpha\beta} = l_{\alpha i}l_{\beta j}\tau_{ij}\,.$$

During sliding, this resolved shear stress $\tau_{\alpha\beta}$ must be equal to the critical shear stress $C_{\alpha\beta}$. Hence, the condition for slip to initiate or to continue in this slip system is given by

$$F(\tau_{ij}) = l_{\alpha i}l_{\beta j}\tau_{ij} - (\pm C_{\alpha\beta}) = 0, \tag{2.13}$$

or equivalently,

$$F(\tau_{ij}) = \tfrac{1}{2}(l_{\alpha i}l_{\beta j} + l_{\alpha j}l_{\beta i})\tau_{ij} - (\pm C_{\alpha\beta}) = 0.$$

Considering τ_{ij} and τ_{ji} as distinct components, there are nine stress components. Taking each stress component as one coordinate, we have a nine-dimensional stress space. The yield surface, or loading surface, is defined as the surface in stress space with stress components as coordinates within which the stress may vary without any plastic strain increment and exterior to which an incremental plastic strain is produced. Here slip is taken to be the only source of plastic strain. Hence, (2.13) defines a yield surface. Since this equation is linear with respect to τ_{ij}, the surface consists of several pairs of planes called yield planes. Twelve slip systems in a fcc crystal give twelve pairs of yield planes. The reader should not confuse these yield planes with slip planes in a crystal. The component of the gradient along τ_{ij} axis is

$$\partial F/\partial \tau_{ij} = \tfrac{1}{2}(l_{\alpha i} l_{\beta j} + l_{\alpha j} l_{\beta i}). \tag{2.14}$$

Similarly, the plastic strain components e''_{ij} form a nine-dimensional space, which may be superposed on the corresponding stress space. It is seen from (2.12) and (2.14) that the incremental plastic strain vector $\mathbf{dE}''$ with de''_{ij} as components is *normal* to the yield planes given by (2.14). The yield planes of a fcc crystal form a 24-faced polyhedron called a yield polyhedron. When the stress vector with τ_{ij} as components is on a plane of the polyhedron, the incremental plastic strain vector $\mathbf{dE}''$ is normal to this plane. When the stress vector is on an edge of the polyhedron, the incremental plastic strain vector lies between the normals to the two adjacent yield planes. When the stress vector is at a vertex the incremental plastic strain vector lies within the cone bounded by the normals to the yield planes intersecting at the vertex (Bishop and Hill, 1951a, b; Kocks, 1970; Lin, 1968). The yield surface of single crystals is said to coincide with a surface of constant platic potential function, since its gradient gives the direction of the plastic strain increment just like the gradient of a Newtonian potential function gives the direction of the attractive force (MacMillan, 1930).

III. Homogeneous Strain Analysis of Polycrystals

A. Grain Boundary Effect

After the relation between the resolved shear stress and the amount of slip was experimentally obtained for single crystals, many attempts were made to deduce the laws of plasticity of a polycrystalline mass from the stress–strain relationship of single crystals by different investigators (Sachs, 1928; Cox and Sopwith, 1937; Taylor, 1938; Kochendorfer, 1941; Calnan and Clews, 1950; Kocks, 1958). Some of them (Sachs, 1928) simply imagined each crystal grain as being subject to sufficient uniaxial stress in the direction of the macroscopic stress to make it yield. Then the grains were assumed to be orientated at random and each grain considered to be unaffected by its neighbors. With this model, the sum of the forces due to each grain over any section of the polycrystal was assumed to equal the load on it. This model is far from being realistic, since neither the equilibrium nor the continuity condition is satisfied at the grain boundaries.

The main difference between a single crystal and a polycrystal is the presence of grain boundaries; these boundaries may have two effects. The first is the contribution of the grain boundaries themselves to the strength of the polycrystal, and the second is the constraints imposed on the plastic deformation of one grain by the differently oriented neighboring grains. The first effect has been experimentally studied by bicrystal tests (Kawada, 1951; Gilman, 1953; Livingston and Chalmers, 1957; Davis *et al.*, 1957; Fleisher and Backofen, 1960; Elbaum, 1960a,b). Let the component crystals of the bicrystal be designated as A and B, the tensile axis of the bicrystal be denoted by z, and the grain boundary lie in the xz plane, x, y, and z being cartesian coordinates. The condition of compatibility of the bicrystal is given as

$$e_{zz}^{\mathrm{A}} = e_{zz}^{\mathrm{B}}, \qquad e_{xx}^{\mathrm{A}} = e_{xx}^{\mathrm{B}}, \qquad e_{xz}^{\mathrm{A}} = e_{xz}^{\mathrm{B}}.$$

The bicrystal is said to be compatible if the above conditions are satisfied by homogeneous single slip in the most favorable slip system in each component crystal. The tensile stress–strain diagram of compatible bicrystals has been shown to be about the same as that of the component single crystals tested separately (Elbaum, 1960a; Gilman, 1953; Davis *et al.*, 1957). Hence, the contribution of grain boundaries of polycrystals is significant only in giving constraints to the plastic deformation of neighboring grains. This grain boundary has been estimated to be only about four atoms thick (Barrett, 1952; Dorn and Motes, 1962), and represents a transition region between two different orientations of two neighboring grains.

B. Taylor's Analysis of Rigid-Plastic Polycrystals

The first realistic model used to calculate the polycrystal uniaxial stress–strain relation from that of single crystals was that proposed by Taylor (1938). He regarded these grain boundaries merely as surfaces of zero thickness, across which crystal orientation changes from one to another.

It is known that metals undergo considerable deformation without forming cracks, so that the crystals originally in contact remain so during deformation. An aggregate of crystals of random orientations can deform without a gap occurring at the grain boundaries in an infinite number of ways. This actually takes place in a manner which corresponds to minimum work consistent with the given small incremental strain of the aggregate. Any other deformation satisfying the condition of no cavity at the grain boundaries must involve a greater amount of work.

An aggregate of randomly oriented fcc crystals under tension was considered by Taylor (1938). He assumed all grains to suffer the same homogeneous strain as that imposed on the aggregate, and indicated that his result gives an upper limit to the yield stress of the aggregate. Taylor assumed the crystals to be rigid plastic; hence the strain is caused purely by slip. Plastic strain caused by slip has no change in volume. Since a given strain can be defined by six components, with the condition of no change in volume, the plastic strain is reduced to five independent components which require slip in five independent slip systems. There are twelve slip systems in a fcc crystal. For determining which five out of the twelve are operative, Taylor (1938) applied the principle of virtual work. The imposed strain on each crystal is produced by those five slip systems which give the minimum sum of the amounts of slip in the slip systems. Let $d\gamma_{nk}$ be the incremental plastic shear strain caused by slip in the kth slip system of the nth crystal. The resolved shear stresses in all the active slip systems are the same, and are denoted by τ_n. The work done in the nth crystal of volume v_n is then

$$\left(\tau_n \sum_{k=1}^{m} d\gamma_{nk}\right) v_n ,$$

where m is the number of active slip systems. For N crystals in the aggregate, the work done on all the crystals is

$$d\overline{W} = \sum_{n=1}^{N} \left(\tau_n \sum_{k=1}^{m} d\gamma_{nk}\right) v_n . \tag{3.1}$$

The external work on the aggregate subject to a tensile stress P with incremental plastic tensile strain de'' is equated to the above, giving

$$P\, de'' \sum_{n=1}^{N} v_n = \sum_{n=1}^{N} \left(\tau_n \sum_{k=1}^{m} d\gamma_{nk}\right) v_n . \tag{3.2}$$

Taylor assumed the orientations of crystals to be randomly distributed.

All crystals of all orientations are assumed to be of the same size, then

$$P\,de'' = \frac{1}{N}\sum_{n=1}^{N}\left(\tau_n \sum_{k=1}^{n} d\gamma_{nk}\right). \tag{3.3}$$

For a given de'', the minimum sum of the amounts of slip for a given crystal orientation is calculated. From the relation between the critical shear stress and the sum of slips of single crystals, the critical shear stress τ_n and $(\tau_n \sum_{k=1}^{n} d\gamma_{nk})$ are determined for each crystal. Substituting these values in (3.3), the tensile stress–strain curve of the aggregate is readily determined. This general procedure was used by Taylor in his noteworthy paper, "Plastic Strain in Metals." The calculated aggregate tensile stress–strain curve has been shown to be close to the empirical curve (Taylor, 1938). Hence, the error due to the assumption of the same strain in all crystals seems to be small for such calculations.

C. Bishop and Hill's Principles of Maximum Work

The number of combinations of 5 out of 12, C_5^{12}, in 792. Even though the twelve slip systems are not all independent, the choice of the five to give minimum slip is still a very lengthy process. Bishop and Hill (1951a,b) have ingeniously presented the principle of maximum work by which the computation of these five slip systems is much simplified. Let $\mathbf{T}$ denote a stress vector in the stress space with τ_{ij} as coordinates, causing a given incremental plastic strain $\mathbf{dE}''$ with components de''_{ij}, and let $\mathbf{T}^*$ be any other stress vector with τ_{ij}^* as coordinates within the yield polyhedron of the crystal. Let τ_k and $\tau_k{}^*$ be the resolved shear stresses of the kth active slip system corresponding to τ_{ij} and τ_{ij}^*, respectively. The τ_k in the active slip systems is equal to the critical shear stress. Since τ_{ij}^* is within the yield surface, $\tau_k{}^*$ is less than the critical shear stress. Hence

$$(\tau_k - \tau_k{}^*)\,d\gamma_k \geqslant 0. \tag{3.4}$$

The incremental work due to the plastic strain increment dE'' is

$$d\overline{W} - d\overline{W}^* = (\mathbf{T} - \mathbf{T}^*)\cdot\mathbf{dE}'' = (d\tau_{ij} - d\tau_{ij}^*)\,de''_{ij}$$

$$= \sum_{k=1}^{m} (\tau_k - \tau_k{}^*)\,d\gamma_k \geqslant 0. \tag{3.5}$$

Hence

$$(\mathbf{T} - \mathbf{T}^*)\cdot\mathbf{dE}'' \geqslant 0. \tag{3.6}$$

This shows that among all stress states lying within the yield surface, the actual stress state giving **dE**″ is the one which lies on the yield surface and gives maximum work. The yield surface of a fcc crystal is a 24-face polyhedron. The stress state which produces a given **d**E″ is readily found. The yield planes intersecting this stress state give the active slip systems.

D. Aggregates of Elastic-Plastic Crystals

In both the calculations of Taylor (1938) and Bishop and Hill (1951a,b), the strain was assumed to be exclusively plastic. Elastic strain was neglected. This assumption is reasonably valid for large strains, but not valid in the small strain range where the plastic strain is of the same order of magnitude as the elastic strain. The effect of elastic strain on the Taylor and Bishop and Hill theories was considered by Lin (1957). The individual grains were assumed to suffer the same total strain (elastic plus plastic) as that of the aggregate. As the imposed strain increases the resolved shear strains in all slip systems in the crystal increase. The presence of elastic strain permits slip to take place in the slip system with the highest resolved shear stress as soon as the critical shear stress is reached in that slip system.

Let γ_n denote the slip in the nth slip system. Let this system have a sliding plane with normal α and sliding direction β. The plastic strain e''_{ij} caused by this γ_n is then

$$e''_{ij} = (l_{i\alpha}l_{j\beta} + l_{j\alpha}l_{i\beta})\gamma_n = \alpha_{ij(n)}\gamma_{(n)} , \tag{3.7}$$

where the parenthesis on the subscript n denote no summation and $\alpha_{ij(n)}$ denotes the terms in the parathensis. For more than one system slide,

$$e''_{ij} = \sum_n \alpha_{ij_n}\gamma_n . \tag{3.8}$$

The critical shear stress is a function of the sum of slips in all systems:

$$\tau_c = F\left(\sum_n \int | d\gamma_n |\right), \tag{3.9}$$

where $\int | d\gamma_n |$ is written in place of γ_n , since any reverse slip is assumed to have the same rate of hardening. From (2.6), the total strain is the sum of the elastic and plastic ones:

$$e_{ij} = e'_{ij} + e''_{ij} , \tag{3.10}$$

$$e'_{ij} = e_{ij} - \sum_n \alpha_{ij_n}\gamma_n . \tag{3.11}$$

The elastic resolved shear strain in the *m*th slip system is

$$\gamma'_m = \alpha_{ij_m}\left[e_{ij} - \sum_n (\alpha_{ij})_n \gamma_n\right], \tag{3.12}$$

and its resolved shear stress is

$$\tau_m = 2\mu\alpha_{ij_m}\left(e_{ij} - \sum \alpha_{ij_n}\gamma_n\right). \tag{3.13}$$

When e_{ij} is increased from zero, no slip occurs until the resolved shear stress in one slip system, say, τ_1, reaches the critical. Then this slip system slides, with the imposed strain e_{ij} subject to the condition that the resolved shear stress be equal to the critical, i.e., $\tau_1 = \tau_c$, or

$$2\mu\alpha_{ij_1}(e_{ij} - \alpha_{ij_1}\gamma_1) = F[\gamma_1]. \tag{3.14}$$

This single slip continues until the resolved shear stress in another slip system, say τ_2, reaches τ_c. Then this second system joins the first to slide and causes a duplex slip. The amounts of slip in these two systems are then controlled by the condition

$$\tau_1 = \tau_2 = \tau_c ,$$

or

$$2\mu\alpha_{ij_1}(e_{ij} - \alpha_{ij_1}\gamma_1 - \alpha_{ij_2}\gamma_2) = 2\mu\alpha_{ij_2}(e_{ij} - \alpha_{ij_1}\gamma_1 - \alpha_{ij_2}\gamma_2) = F[|\gamma_1| + |\gamma_2|]. \tag{3.15}$$

This duplex slip continues until the third, then the fourth, and then the fifth slip systems become active and join in sliding. For this multislip, the amounts of slip in the different active slip systems are governed by the condition

$$\tau_1 = \tau_2 = \cdots = \tau_p = \tau_c .$$

For any of the active slip systems

$$\tau_m = 2\mu\alpha_{ij_m}\left(e_{ij} - \sum \alpha_{ij_m}\gamma_n\right) = F\left[\sum_n \int |d\gamma_n|\right]. \tag{3.16}$$

Written in incremental form, with dots denoting the time rate, this becomes

$$2\mu\alpha_{ij_m}\left(\dot{e}_{ij} - \sum_{n=1}^{N} \alpha_{ij_n}\dot{\gamma}_n\right) = F'\left[\sum_{n=1}^{N} \int |d\gamma_n|\right] \sum_{p=1}^{N} \dot{\gamma}_p , \tag{3.17}$$

where N is the total number of slip systems having slid and F' is the derivative of F with respect to its argument.

Sliding causes no change in volume. Volumetric strain is purely elastic. Plastic strain has no change in volume and has five independent components. Hence, no more than five slip systems can be independent. There can be more than five slip systems having the resolved shear stresses equal to the critical simultaneously; hence the set of γ's which yields the same e' is not unique but $\sum \gamma$'s, e''_{ij}, and τ_i are unique (Budiansky and Wu, 1962; Hutchinson, 1964a; Hill, 1966). This corresponds to a stress state at a vertex of the 24-sided yield plyhedron, at which more than five yield planes intersect.

By the above described method, the amounts of slip in slip systems at any stage of loading can be determined. The stress tensor for a given imposed strain tensor can be obtained. Following this procedure, Czyzak *et al.* (1961) have numerically calculated the tensile stress–strain curve and the Bauschinger effect of a fcc polycrystal. This method is applicable to the cases where the ratios of the strain components are constant as well as those where these ratios vary (Lin, 1957; Payne, 1959).

This procedure is lengthier than that proposed by Bishop and Hill for determining the five active slip systems for uniaxial loading, but has the advantage of including the elastic strain and determining the sequence of active slip systems. The former is important in the study of the initial stage of plastic deformation where the elastic strain is not negligible, and the latter has been used by Lin and Lieb (1962) in selecting the active slip systems from those with the resolved shear stress equal to the critical by assuming those systems having slid previously to have preference over others in sliding.

However, the assumption of the same total strain (elastic and plastic) on all crystals does not satisfy the equilibrium condition across the grain boundaries, and hence this method is still not rigorous. A more rigorous treatment of the polycrystal plastic deformation has been developed by Lin and his associates and is presented in a later section.

IV. Simplified Slip Theories for Non-radial Loadings

The tensile stress–strain curve of a polycrystal may be readily obtained from tests. Hence, the theoretical calculations of this curve from single crystal slip characteristics is mainly of theoretical interest. However, for non-radial loadings where the ratios of the stress components vary, the five independent incremental plastic strain components (plastic dilatation is zero) are functions of the history of the six stress components. Tests to cover all possible loading conditions are not feasible. This incremental stress–strain relation for non-radial loadings is needed for stress analysis and, hence, is of both theoretical and practical interest Batdorf and Budiansky (1949, 1954) first applied the concept of slip to formulate the polycrystal plastic stress–strain relation under non-radial loading. They considered an aggregate of crystals, each of which has *one* slip system, and proposed a simplified slip theory of plasticity. They assumed that each crystal is subject to the same stress as that of the aggregate and the plastic shear in this slip system depends on the maximum magnitude of the resolved shear stress.

Lin (1954, 1958) considered an aggregate of crystals, each with three slip directions on a slip plane, and reformulated the simplified slip theory of Batdorf and Budiansky by considering the interactions of those crystals with the same slip plane. These two theories have yielded results in better agreement than other previously existing plasticity theories for a set of experimental data of a specimen under varying ratios of stress components. Similar to the above two theories, a theory called local strain theory was proposed by Malmeister (1956, 1965); Kliushinkov (1958); Knets and Kregers (1965). He assumed that plastic deformation is determined by shears in all possible planes, and in every such plane, the plastic shear strain is assumed to depend on the shear stress in this plane only. All these theories better represent the physical process of deformation than the existing mathematical theories of plasticity, but do not satisfy the compatibility condition between crystals with different slip planes.

V. Analysis Satisfying Both Equilibrium and Compatibility Conditions

Metals undergo considerable plastic deformation without the formation of cracks, so that crystals originally in contact with their neighbors remain so during deformation. This means that the equilibrium and compatibility conditions are satisfied throughout the aggregate. In the preceding two sections, two classes of simplified plasticity theories were presented. The first, given by Taylor, etc., satisfies the condition of compatibility but not the condition of equilibrium at the grain boundaries. The second, given by Batdorf, Budiansky, etc., satisfies the condition of equilibrium but not that of compatibility. Hence, both of these are simplified theories. In the present section, a rigorous method is shown to satisfy both the equilibrium and compatibility conditions.

A. Analogy between Body Force and Plastic Strain Gradient

Taylor (1934) has shown that sliding in a crystal by the movement of dislocations is such that perfect crystal structure is reformed after each atomic jump. The lattice structure of the bulk of the material remains essentially the same after the occurrence of slip. Hence, the elastic modulus of the crystal reamins the same. Strain e_{ij} is composed of two parts: the elastic part e'_{ij} caused by the elastic deformation of the lattice structure, and the plastic part e''_{ij} caused by slip:

$$e_{ij} = e'_{ij} + e''_{ij} .$$

The anisotropy of the elastic constants of single crystals varies from one metal to another. For pure aluminum this anisotropy is small and the present development is neglected. Then the elastic stress–strain relation is given by

$$\tau_{ij} = \delta_{ij}\lambda\theta' + 2\mu e'_{ij} , \tag{5.1}$$

where τ_{ij} is the stress component referring to a set of rectangular coordinates, δ_{ij} is the Kronecker delta, θ' is the elastic dilatation, and λ and μ are Lame's constants. With the elastic strain written as the difference between the total strain and the plastic strain, (5.1) has the form

$$\tau_{ij} = \delta_{ij}\lambda(\theta - \theta'') + 2\mu(e_{ij} - e''_{ij}), \tag{5.2}$$

where θ is the dilatation and θ'' is the plastic dilatation. The condition of equilibrium within a body of volume v is

$$\tau_{ij,j} + F_i = 0 \qquad \text{in} \quad v, \tag{5.3}$$

where the subscript after the comma denotes differentiation, the repetition of the subscript denotes summation from one to three, and F_i denotes the body force per unit volume along the x_i axis. At any point on the boundary Γ with normal $\boldsymbol{\nu}$, the i component of the surface traction per unit area $S_i^{(\nu)}$, can be written from the condition of equilibrium as

$$S_i^{(\nu)} = \tau_{ij}\nu_j \qquad \text{on} \quad \Gamma, \tag{5.4}$$

where ν_j is the cosine of the angle between the normal $\boldsymbol{\nu}$ and the x_j axis. Substituting (5.2) into (5.3) and (5.4), we obtain

$$\delta_{ij}\lambda\theta_{,j} + 2\mu e_{ij,j} - (\delta_{ij}\lambda\theta''_{,j} + 2\mu e''_{ij,j}) + F_i = 0, \tag{5.5}$$

$$S_i^{(\nu)} = \nu_j[\delta_{ij}\lambda\theta + 2\mu e_{ij} - (\delta_{ij}\lambda\theta'' + 2\mu e''_{ij})]. \tag{5.6}$$

It is seen that $-(\delta_{ij}\lambda\theta''_{,j} + 2\mu e''_{ij,j})$ and $(\delta_{ij}\lambda\theta'' + 2\mu\theta''_{ij})\,\nu_j$ are equivalent to F_i and $S_i^{(\nu)}$ in causing the strain field e_{ij}, and are here denoted by $\bar{F}_i$ and $\bar{S}_i^{(\nu)}$, respectively, giving

$$\delta_{ij}\lambda\theta_{,j} + 2\mu e_{ij,j} + F_i + \bar{F}_i = 0, \tag{5.7}$$

$$S_i^{(\nu)} + \bar{S}_i^{(\nu)} = \nu_j(\delta_{ij}\lambda\theta + 2\mu e_{ij}). \tag{5.8}$$

Hence, the strain distribution in a body with plastic strain under external load is the same as that in an elastic body (no plastic strain) with the additional equivalent body and surface forces $\bar{F}_i$ and $\bar{S}_i^{(\nu)}$. This reduces the solution of stress field of a body with known plastic strain distribution to the solution of an identical elastic body with an additional set of equivalent body and surface forces. Reisner (1931), Timoshenko (1934), and Eshelby (1957) have obtained the same results by different approaches.

If there is thermal strain instead of plastic strain, with thermal coefficient of expansion α and temperature T, we can write

$$e''_{ij} = \delta_{ij}\alpha T, \qquad \theta'' = e''_{ii} = 3\alpha T.$$

Then the equivalent body and surface forces become

$$\bar{F}_i = -(3\lambda + 2\mu)\alpha T_{,i}\,, \tag{5.9}$$

$$\overline{S_i^{(\nu)}} = \nu_i(3\lambda + 2\mu)\alpha T. \tag{5.10}$$

This is the well-known Duhamel's analogy (Duhamel, 1837, 1838; Neumann, 1841; Morgan, 1958; Rydewski, 1962) between temperature gradient and the body force in an elastic medium. Plastic strain due to slip has no dilatation, i.e., $\theta'' = 0$. The equivalent forces for it reduce to

$$\bar{F}_i = -2\mu e''_{ij,j}\,, \tag{5.11}$$

$$\overline{S_i^{(\nu)}} = 2\mu e''_{ij}\nu_j\,. \tag{5.12}$$

B. Displacement Field Caused by Slip Field (Lin, 1967, 1968)

Let u_i and $u_i{}^*$ be two sets of displacements which are single-valued and continuous throughout the body. Let F_i, $F_i{}^*$, S_i, and $S_i{}^*$ denote the corresponding body and surface forces. The conditions of equilibrium are given by Eqs. (5.3) and (5.4). Multiplying these two equations through by u_i^*, integrating over the volume v and the boundary surface Γ, respectively, and adding, we obtain

$$\int_v F_i u_i^* \, dv + \int_\Gamma S_i u_i^* \, d\Gamma + \int_v \tau_{ij,j} u_i^* \, dv = \int_\Gamma \tau_{ij} \nu_j u_i^* \, d\Gamma. \tag{5.13}$$

From the divergence theorem, we have

$$\begin{aligned} \int_\Gamma \tau_{ij} u_i^* \nu_j \, d\Gamma &= \int_v \tau_{ij,j} u_i^* \, dv + \int_v \tau_{ij} u_{i,j}^* \, dv \\ &= \int_v \tau_{ij,j} u_i^* \, dv + \int_v \tau_{ij} e_{ij}^* \, dv, \end{aligned} \tag{5.14}$$

since

$$\tau_{ij} u_{ij}^* = \tfrac{1}{2}(\tau_{ij} u_{i,j}^* + \tau_{ji} u_{j,i}^*) = \tau_{ij} \tfrac{1}{2}(u_{i,j}^* + u_{j,i}^*) = \tau_{ij} e_{ij}^* \, .$$

Substituting (5.13) into (5.14) yields

$$\int_v F_i u_i^* \, dv + \int_\Gamma S_i u_i^* \, d\Gamma = \int_v \tau_{ij} e_{ij}^* \, dv. \tag{5.15}$$

For isotropic elastic bodies,

$$\tau_{ij} e_{ij}^* = (\delta_{ij} \lambda e_{kk} + 2\mu e_{ij}) e_{ij}^* = \lambda e_{kk} e_{jj}^* + 2\mu e_{ij} e_{ij}^* = \tau_{ij}^* e_{ij} \, , \tag{5.16}$$

and from (5.15) and (5.16), we have

$$\int_v F_i u_i^* \, dv + \int_\Gamma S_i u_i^* \, d\Gamma = \int_v F_i^* u_i \, dv + \int S_i^* u_i \, d\Gamma. \tag{5.17}$$

This is the reciprocal theorem for elastic bodies. Now consider a body with inelastic strain e_{ij}''. This gives the equivalent body and surface forces

$$\bar{F}_i = -\delta_{ij} \lambda \theta''_{,j} - 2\mu e''_{ij,j}; \qquad \overline{S_i^{(\nu)}} = \delta_{ij} \lambda \theta'' \nu_j + 2\mu e''_{ij} \nu_j \, .$$

Replacing F_i and S_i in Eq. (5.17) by $F_i + \bar{F}_i$ and $S_i + \bar{S}_i$, respectively, yields

$$\begin{aligned} &\int_v (F_i - \delta_{ij} \lambda \theta''_{,j} - 2\mu e''_{ij,j}) u_i^* \, dv + \int_\Gamma (S_i + \delta_{ij} \lambda \theta'' \nu_j + 2\mu e''_{ij} \nu_j) u_i^* \, d\Gamma \\ &= \int_v F_i^* u_i \, dv + \int_\Gamma S_i^* u_i \, d\Gamma. \end{aligned} \tag{5.18}$$

Let the displacement due to inelastic strain be denoted by u_{i_s}. To determine u_{i_s}, we let F_i and S_i become zero and consider a continuous distribution of dislocation (Kroner, 1958; Eshelby, 1956) with inelastic strain e''_{ij}. Applying the divergence theorem to the second term on the left, we obtain

$$-2\mu \int_v e''_{ij,j} u_i{}^* \, dv + 2\mu \int_v (e''_{ij} u_i{}^*)_{,j} \, dv = 2\mu \int_v e''_{ij} u^*_{i,j} \, dv$$

$$\delta_{ij} \left[-\lambda \int_v \theta''_{,j} u_i{}^* \, dv + \lambda \int_v (\theta'' u_i{}^*)_{,j} \, dv \right] = \delta_{ij} \lambda \int \theta'' u^*_{i,j} \, dv = \lambda \int \theta'' u^*_{i,i} \, dv.$$

Substituting these in (5.18) and then writing θ'' as $\delta_{ij} e''_{ij}$, we obtain

$$\int_v \tau^*_{kl} e''_{kl} \, dv = \int_v F_i{}^* u_{i_s} \, dv + \int_\Gamma S_i{}^* u_{i_s} \, d\Gamma. \tag{5.19}$$

Now let $\tau^{(i)}_{kl}(x', x)$ denote the stress at x' caused by a unit concentrated force acting at x along the direction of the x_i axis. Then the right-hand side of (5.19) gives the displacement at x along the x_i direction caused by $e''_{kl}(x')$ throughout the body:

$$u_{i_s}(x) = \int_{v'} \tau^{(i)}_{kl}(x', x) e''_{kl}(x') \, dv'. \tag{5.20}$$

In the case of thermal strain only, the equivalent body and surface forces are given by Eqs. (5.9) and (5.10). The same procedure leads to

$$u_{i_s}(x) = \int_{v'} \tau^{(i)}_{kl}(x', x) \, \delta_{kl} \alpha T(x') \, dv' = \int_{v'} \tau^{(i)}_{kk}(x', x) \alpha T(x') \, dv'.$$

This was derived in a different way by Maysel (Nowacki, 1962) for displacement in a body caused by a steady temperature field in an isotropic body.

For plastic strain, $e''_{ii} = 0$. If we write $\tau^{(i)}_{kl}$ as the sum of the hydrostatic stress $\delta_{kl}(\frac{1}{3}\tau^{(i)}_{jj})$ and the deviated stress component $S^{(i)}_{kl}$ (5.20) becomes

$$\begin{aligned} u_{i_s}(x) &= \int_{v'} (S^{(i)}_{kl} + \delta_{kl}(\tfrac{1}{3}\tau^{(i)}_{jj})) e''_{kl}(x') \, dv' \\ &= \int_{v'} S^{(i)}_{kl} e''_{kl}(x') \, dv'. \end{aligned} \tag{5.21}$$

This gives the displacement field caused by a given distribution of plastic strain.

C. Macroscopic Aggregate Plastic Strain and Microscopic Slip Fields

Consider a unit cube of the aggregate of crystals with plastic strain distribution e''_{ij} in v'. Let the x_1, x_2, and x_3 axes be parallel to the sides of this cube. To find the plastic strain of the aggregate due to this e''_{ij}, we imagine a traction of $S_i^* = \tau_{ij}^* \nu_j$ applied to the boundary of the unit cube, where τ_{ij}^* is constant over the boundary surface Γ. The crystals are assumed to have isotropic and homogeneous elastic constants. The τ_{ij}^* corresponding to this S_i^* is constant throughout the body. From (5.19), with the displacement in the aggregate caused by the plastic strain denoted by U_i, we have

$$\tau_{ij}^* \int_{v'} e''_{ij}\, dv' = \int_{\Gamma} S_i^* U''_i \, d\Gamma = \tau_{ij}^* \int \nu_j U_i \, d\Gamma. \tag{5.22}$$

Let the macroscopic strain be defined as

$$E_{ij} = \tfrac{1}{2}\{[(\bar{U}_i)_{x_j=1} - (\bar{U}_i)_{x_j=0}] + [(\bar{U}_j)_{x_i=1} - (\bar{U}_j)_{x_i=0}]\}, \tag{5.23}$$

where $(\bar{U}_i)_{x_j=1}$ is the average displacement U_i over unit area of the face of the cube with $x_j = 1$. Then we have

$$\tau_{ij}^* \bar{E}_{ij} v = \int_{v'} \tau_{ii}^* e''_{jj}\, dv' = \tau_{ij}^* \int e''_{ij}\, dv'.$$

For the unit cube $v = 1$, and we have

$$\tau_{ij}^* E_{ij} = \tau_{ij}^* \int_{v'} e''_{ij}\, dv'. \tag{5.24}$$

This shows that for an aggregate of crystals with homogeneous isotropic elastic constants, the plastic strain of the aggregate is the average plastic strain of all crystals in the aggregate (Lin, 1967). The result is used in the calculation of the macroscopic aggregate plastic strains.

D. Stress Field Caused by a Given Plastic Strain Distribution

The crystal which has the maximum resolved shear stress slides first. This crystal is called the most favorably oriented crystal. It may be located near a free surface or at the interior of the aggregate. Now consider this sliding crystal located at the interior of a fine grained aggregate. The equivalent body force caused by slip in this crystal may then be considered to be applied in an infinite elastic medium. The displacement $u(x)$ in an infinite medium subject to a body force $F_i(x')$ acting through a finite volume v' has been given by Kelvin (Sokolinkoff, 1956; Love, 1927).

Let

$$A = (\lambda + \mu)/8\pi\lambda(\lambda + 2\mu), \qquad B = (\lambda + 3\mu)/(\lambda + \mu)$$

and

$$r = [(x_i - x_i')(x_i - x_i')]^{1/2}.$$

The displacement along direction k is given as

$$u_k(x) = A \int_{v'} \{B\, \delta_{ki}[F_i(x')/r] - (1/r)_{,k}(x_i - x_i')F_i(x')\}\, dv', \qquad (5.25)$$

where $F_i(x')$ is the i component of the body force in terms of the variables of integration x_i'. The derivative of u_k with respect to x_l can be shown to converge uniformly both inside and outside v', so the differentiation may be carried out under the integral sign giving

$$u_{k,l}(x) = A \int_{v'} [BF_i(x')\, \delta_{ik}(1/r)_{,l} - (1/r)_{,kl}(x_i - x_i')F_i(x') - (1/r)_{,k}\, \delta_{il}F_i(x')\, dv'], \qquad (5.26)$$

$$e_{kl}(x) = \tfrac{1}{2}(u_{k,l} + u_{l,k})$$

$$= A \int_{v'} \{\tfrac{1}{2}(B - 1)[F_i(x')\, \delta_{ik}(1/r)_{,l} + F_i(x')\, \delta_{il}(1/r)_{,k}]$$

$$- (1/r)_{,kl}(x_i - x_i')F_i(x')\}\, dv'. \qquad (5.27)$$

For unit force $F_i(x')\, dv' = 1$ applied at x',

$$e_{kl}^{(i)}(x) = A \cdot \tfrac{1}{2}(B - 1)[\delta_{ik}(1/r)_{,l} + \delta_{il}(1/r)_{,k}] - A(1/r)_{,kl}(x_i - x_i'), \qquad (5.28)$$

$$e_{mm}^{(i)}(x) = A(B - 1)(1/r)_{,i}\,, \qquad (5.29)$$

$$\tau_{kl}^{(i)}(x) = \delta_{kl}\lambda e_{mm}(x) + 2\mu e_{kl}(x)$$

$$= \delta_{kl}\lambda A(B - 1)(1/r)_{,i} + A\mu(B - 1)[\delta_{ik}(1/r)_{,l} + \delta_{il}(1/r)_{,k}]$$

$$- 2\mu A(1/r)_{,kl}(x_i - x_i').$$

Writing this out in the long form, we have

$$\tau_{kl}^{(i)}(x) = -\delta_{kl}\frac{\lambda}{4\pi(\lambda+2\mu)}\frac{x_i - x_i'}{r^3} - \frac{\mu}{4\pi(\lambda+2\mu)}$$
$$\times\left[\delta_{ik}\frac{(x_l - x_l')}{r^3} + \delta_{il}\frac{(x_k - x_k')}{r^3}\right]$$
$$- 2\mu\frac{\lambda+\mu}{8\pi\mu(\lambda+\mu)}\left[\frac{3(x_i - x_i')(x_k - x_k')(x_l - x_l')}{r^5} - \frac{(x_i - x_i')}{r^3}\delta_{kl}\right], \tag{5.30}$$

or

$$\tau_{kl}^{(i)}(x) = -\frac{3(\lambda+\mu)}{4\pi(\lambda+2\mu)}\frac{(x_i - x_i')(x_k - x_k')(x_l - x_l')}{r^5}$$
$$+ \frac{\mu}{4\pi(\lambda+\mu)}\left[\delta_{kl}\frac{(x_i - x_i')}{r^3} - \delta_{ik}\frac{(x_l - x_l')}{r^3} - \delta_{il}\frac{(x_k - x_k')}{r^3}\right]. \tag{5.31}$$

Substituting this for $\tau_{kl}^{(i)}$ in Eq. (5.20), we have

$$u_{i_s} = \int\left\{-\frac{3(\lambda+\mu)}{4\pi(\lambda+2\mu)}\frac{(x_i - x_i')(x_k - x_k')(x_l - x_l')}{r^5}\right.$$
$$\left.+ \frac{\mu}{4\pi(\lambda+\mu)}\left[\delta_{kl}\frac{x_i - x_i'}{r^3} - \delta_{ik}\frac{(x_l - x_l')}{r^3} - \delta_{il}\frac{(x_k - x_k')}{r^3}\right]\right\} e_{kl}''\,dv', \tag{5.32}$$

$$e_{ij_s} = \tfrac{1}{2}[(\partial u_{i_s}/\partial x_j) + (\partial u_{j_s}/\partial x_i)]$$
$$= \int_{v'} \psi_{ijkl}(x, x')e_{kl}''\,dv', \tag{5.33}$$

where

$$\psi_{ijkl}(x, x') = -\frac{3(\lambda+\mu)}{8\pi(\lambda+2\mu)}\left\{\frac{\partial}{\partial x_j}\frac{(x_i - x_i')(x_k - x_k')(x_l - x_l')}{r^5}\right.$$
$$\left.+ \frac{\partial}{\partial x_i}\frac{(x_j - x_j')(x_k - x_k')(x_l - x_l')}{r^5}\right\}$$
$$+ \frac{\mu}{8\pi(\lambda+\mu)}\left\{\left[\frac{\partial}{\partial x_j}\left(\frac{x_i - x_i'}{r^3}\right) + \frac{\partial}{\partial x_i}\left(\frac{x_j - x_j'}{r^3}\right)\right]\right.$$
$$- \delta_{ik}\left[\frac{\partial}{\partial x_j}\left(\frac{x_l - x_l'}{r^3}\right) + \frac{\partial}{\partial x_i}\left(\frac{x_l - x_l'}{r^3}\right)\right]$$
$$\left.- \delta_{il}\left[\frac{\partial}{\partial x_j}\left(\frac{x_k - x_k'}{r^3}\right) + \frac{\partial}{\partial x_i}\left(\frac{x_k - x_k'}{r^3}\right)\right]\right\}, \tag{5.34}$$

and

$$e_{mm_s} = \int \psi_{mmkl}(x, x')e''_{kl}\, dv', \tag{5.35}$$

with

$$\begin{aligned}\psi_{mmkl} = &-\frac{3(\lambda+\mu)}{4\pi(\lambda+2\mu)}\left\{\frac{\partial}{\partial x_m}\frac{(x_m - x_m')(x_k - x_k')(x_l - x_l')}{r^5}\right\}\\ &+\frac{\mu}{4\pi(\lambda+\mu)}\left\{\delta_{kl}\frac{\partial}{\partial x_m}\left(\frac{x_m - x_m'}{r^3}\right)\right.\\ &\left.-\delta_{mk}\frac{\partial}{\partial x_m}\left(\frac{x_l - x_l'}{r^3}\right) - \delta_{ml}\frac{\partial}{\partial x_m}\left(\frac{x_k - x_k'}{r^3}\right)\right\}.\end{aligned} \tag{5.36}$$

The repetition of subscript m again denotes summation from 1 to 3. Then

$$\tau_{ij_s}(x) = \int_{v'} [\lambda\, \delta_{ij}\psi_{mmkl}(x, x') + 2\mu\psi_{ijkl}(x, x')]e''_{kl}(x)\, dv'. \tag{5.37}$$

This gives the stress field caused by a given plastic strain distribution e''_{kl} in an infinite medium. With e_{ij_s} denoting the total strain due to slip, the elastic strain due to slip is given by

$$e'_{ij_s} = e_{ij_s} - e''_{ij}\,.$$

After the aggregate is unloaded, the plastic strain remains and the residual stress is then

$$\tau_{ij_r} = \tau_{ij_s} - 2\mu e''_{ij}\,. \tag{5.38}$$

VI. Self-Consistent Theories of Polycrystal Plastic Deformation

A. Eshelby's Solution of Ellipsoidal Inclusion (Eshelby, 1957)

Eshelby considered a region called an inclusion, in an infinite homogeneous isotropic elastic medium, to undergo a change of shape and size that would be an arbitrary homogeneous strain e^T_{ij} if the surrounding material were absent. From (5.2), the stress is then

$$\tau_{ij} = \lambda\, \delta_{ij}(e_{mm} - e^T_{mm}) + 2\mu(e_{ij} - e^T_{ij}).$$

From Section V, A the equivalent body and surface forces caused by this strain are

$$\overline{F}_i = -\lambda\, \delta_{ij} e^T_{mm,i} - 2\mu e^T_{ij,j}\,,$$

$$\overline{S_i^{(\nu)}} = (\lambda\, \delta_{ij} e^T_{mm} + 2\mu e_j{}^T)\nu_j\,. \tag{6.1}$$

Since e^T_{ij} is homogeneous in the inclusion, the strain gradient is zero. Hence, $\overline{F}_i = 0$. Let

$$p_{ij} = \lambda\, \delta_{ij} e^T_{mm} + 2\mu e^T_{ij}\,, \tag{6.2}$$

$$\overline{S_j^{(\nu)}} = p_{jk}\nu_k\,. \tag{6.3}$$

Expressing λ in terms of μ and Poisson's ratio ν, we can write (5.25) in an alternate form

$$u_i(x) = \frac{1}{16\pi\mu(1-\nu)} \int_{v'} \frac{F_j(x')}{r} \left[(3-4\nu)\,\delta_{ij} + \frac{(x_i - x_i')(x_j - x_j')}{r^2}\right] dv'. \tag{6.4}$$

Substituting (6.3) into (6.4) yields

$$u_i(x) = \frac{1}{16\pi\mu(1-\nu)} \int_\Gamma \frac{p_{ik}\nu_k}{r} \left[(3-4\nu)\,\delta_{ij} + \frac{(x_i - x_i')(x_j - x_j')}{r^2}\right] d\Gamma,$$

where Γ denotes the inclusion surface. By the divergence theorem, and noting that p_{jk} is homogeneous, we have

$$u_i(x) = \frac{p_{jk}}{16\pi\mu(1-\nu)} \int_{v'} \left[(3-4\nu)\frac{\delta_{ij}}{r} + \frac{(x_i - x_i')(x_j - x_j')}{r^3}\right]_{,k} dv'. \tag{6.5}$$

Let $\mathbf{l}$ be a unit vector along $\mathbf{r}$ and $l_i = (x_i - x_i')/r$. Equation (6.5) can be written as

$$u_i(x) = \frac{p_{jk}}{16\pi\mu(1-\nu)} \int_{v'} \frac{dv'}{r^2} f_{ijk}(\mathbf{l}) = \frac{e^T_{jk}}{8\pi(1-\nu)} \int \frac{dv'}{r^2} g_{ijk}(\mathbf{l}), \tag{6.6}$$

where

$$f_{ijk} = (1-2\nu)(\delta_{ij} l_k + \delta_{ik} l_j) - \delta_{jk} l_i + 3 l_i l_j l_k\,,$$

$$g_{ijk} = (1-2\nu)(\delta_{ij} l_k + \delta_{ik} l_j - \delta_{jk} l_i) + 3 l_i l_j l_k\,, \tag{6.7}$$

$$u_i(x) = e^T_{jk} g_{ijk} v'/8\pi(1-\nu)r^2. \tag{6.8}$$

Let us redefine the direction cosines l_i to be those of a line drawn from point x to x'. This involves a change of signs of the integrals.

Integrating over an elementary cone $d\omega(\mathbf{1})$, centered on the direction $\mathbf{1}$ with its vertex at x, gives

$$d\omega(\mathbf{1}) = dv'/r^2. \tag{6.9}$$

Then (77) becomes

$$8\pi\mu(1-\nu)u_i(x) = -e_{jk}^T \int_{4\pi} r(\mathbf{1})\, d\omega(\mathbf{1}) g_{ijk}(\mathbf{1}). \tag{6.10}$$

If this inclusion is ellipsoidal in shape, a point on the surface must satisfy the following (Fig. 7) condition:

$$[(x + rl_1)/a]^2 + [(y + rl_2)/b]^2 + [(z + rl_3)/c]^2 = 1, \tag{6.11}$$

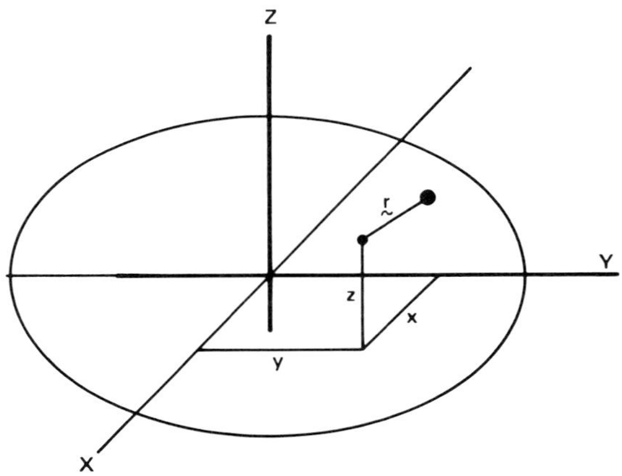

FIG. 7. An ellipsoidal inclusion.

where a, b, and c are the semiaxes of the ellipsoid. This gives

$$gr^2 + 2fr = e, \tag{6.12}$$

where

$$g = l_1{}^2/a^2 + l_2{}^2/b^2 + l_3{}^2/c^2,$$
$$f = l_1 x/a^2 + l_2 y/b^2 + l_3 z/c^2, \tag{6.13}$$
$$e = 1 - x^2/a^2 - y^2/b^2 - z^2/c^2,$$
$$r = (f/g) + (f^2/g^2 + e/g)^{1/2}. \tag{6.14}$$

Substituting this into (6.10) yields

$$8\pi\mu(1-\nu)u_i(x) = -e_{jk}^T \int_{4\pi} [-(f/g) + (f^2/g^2 + e/g)^{1/2}]\, d\omega(\mathbf{1}) g_{ijk}(\mathbf{1}).$$

Since the square root term is even in **1** while g_{ijk} is odd in **1**, this term does not contribute to the integral and hence may be omitted. Then we have

$$u_i(x) = \frac{e_{jk}^T}{8\pi(1-\nu)} \int_{4\pi} \frac{f}{g} g_{ijk}(\mathbf{1})\, d\omega(\mathbf{1}). \tag{6.15}$$

Let

$$\lambda_1 = l_1/a^2, \qquad \lambda_2 = l_2/b^2, \qquad \text{and} \quad \lambda_3 = l_3/c^2. \tag{6.16}$$

We have

$$u_i(x) = \frac{x_m e_{jk}^T}{8\pi(1-\nu)} \int_{4\pi} \frac{\lambda_m g_{ijk}}{g}\, d\omega, \tag{6.17}$$

$$e_{il}(x) = \frac{e_{jk}}{16\pi(1-\nu)} \int_{4\pi} \frac{\lambda_i g_{ljk} + \lambda_l g_{ijk}}{g}\, d\omega. \tag{6.18}$$

This shows that the strain *inside* the ellipsoidal inclusion is *uniform* and depends only on the shape of the ellipsoid. This result is clearly of great physical interest. Equation (6.18) can be written as

$$e_{il}(x) = S_{ilmn} e_{mn}^T \,. \tag{6.19}$$

The expressions for S_{ilmn} have been given by Routh (1922). One of these is given here:

$$S_{1212} = \frac{3}{16\pi(1-\nu)} (a^2 + b^2) I_{ab} + \frac{1-2\nu}{16\pi(1-\nu)} (I_a + I_b),$$

where

$$I_a = 2\pi abc \int_0^\infty \frac{du}{(a^2+u)\,\Delta}, \qquad I_b = 2\pi abc \int_0^\infty \frac{du}{(b^2+u)\,\Delta},$$

$$\Delta = (a^2+u)^{1/2}(b^2+u)^{1/2}(c^2+u)^{1/2},$$

$$I_{ab} = \frac{2}{3}\pi abc \int_0^\infty \frac{du}{(a^2+u)(b^2+u)\,\Delta}\,.$$

For spherical inclusions, the above reduces to

$$\Delta\ (a^2+u)^{3/2},$$

$$I_a = I_b = 4\pi a^3/3,$$

$$I_{ab} = \tfrac{4}{15}\pi a^3,$$

and

$$S_{1212} = (4-5\nu)/15(1-\nu). \tag{6.20}$$

B. Spherical Slid Crystals

When a polycrystalline aggregate of randomly oriented crystals of homogeneous isotropic elastic constants is uniformly loaded, the stress is uniform throughout the aggregate before slip occurs. Since different crystals have different orientations, the resolved shear stresses in different crystals are different. All crystals are assumed to have the same initial critical shear stress. The crystal with maximum resolved shear stress will first reach this critical shear stress and slide. At the initial stage of plastic deformation of the aggregate, very few crystals slide. The distance between two nearest slid crystals is large, so the effect of slip of one slid crystal on another is small. The stress and strain fields in and around each slid crystal are essentially the same as those of only one slid crystal in an infinite aggregate. This reduces the problem to an inclusion problem. From the preceding section, when an ellipsoidal inclusion undergoes a spontaneous change in form which, if the surrounding material were absent, would be some homogeneous shear deformation, the stress in the inclusion is also homogeneous. If the first slid crystal were of ellipsoidal shape, this crystal would undergo uniform slip, since the stress caused by this uniform slip is uniform within the slid crystal, by Eshelby's results, and the resultant of this stress and the applied uniform aggregate stress would give uniform resolved shear stress in the slid crystal. This agrees with the single crystal slip characteristics of uniform slip for uniform resolved shear stress. Hence, this uniform slip in the slid crystal satisfies the equilibrium continuity conditions of the aggregate as well as the single crystal slip characteristics. Budiansky *et al.* (1960; Budiansky and Wu, 1962) assumed the slid crystals to be of spherical shape. The shear strain e_{ij}^c in the slid crystal caused by uniform slip e_{ij}'' in this crystal, from Eshelby's solution, is then

$$e_{ij}^c = 2[(4 - 5\nu)/15(1 - \nu)]e_{ij}'' = be_{ij}'' , \tag{6.21}$$

since $e_{ij}'' = e_{ji}''$ and $S_{ijkl} = S_{ijlk}$.

Let there be a stress τ_{ij_0} at infinity. Then the strain in the inclusion becomes

$$e_{ij}^c = L[\tau_{ij_0}] + be_{ij}'' ,$$

where $L[]$ is the isotropic Hookian operator of the isotropic polycrystal. The elastic strain in this slid crystal is

$$e_{ij}' = L[\tau_{ij_0}] - (1 - b)e_{ij}'' .$$

This gives a uniform stress τ_{ij} in the slid crystal given by

$$\tau_{ij} = \tau_{ij_0} - 2\mu(1 - b)e''_{ij} . \tag{6.22}$$

Writing this in incremental form

$$\dot{\tau}_{ij} = \dot{\tau}_{ij_0} - 2\mu(1 - b)\dot{e}''_{ij} , \tag{6.23}$$

where dots denote the time rate. Let ij denote the active slip system for single slip. The resolved shear stress τ_{ij} is a function of the amount of slip e''_{ij} . With a given $\dot{\tau}_{ij_0}$ and the relation between the critical shear stress and the amount of slip, $\dot{\tau}_{ij}$ and $\dot{e}''_{ij}$ can be obtained from (6.23). This inclusion solution neglects the interaction effect of neighboring slid crystals. The validity of these results was limited to the very initial stage of plastic deformation. Error increases with the density of slid crystals in the aggregate.

C. Average Interaction Effect of Slid Crystals

We now consider a polycrystalline aggregate under tension. The crystal orientations are assumed to be randomly distributed in all sections of the aggregate. The sum of the loads carried by all the individual crystals cut by a section must balance the applied tensile load on the aggregate. The stress relieved by the slid crystals must be carried by other crystals. Hence, the sliding of one group of crystals increases the average load taken by other groups of crystals. Kroner (1961) took this average interaction effect between groups of crystals into consideration in developing an analytical procedure to calculate the polycrystal stress–strain relationship from single crystal characteristics. Budiansky and Wu (1962) rederived Kroner's scheme of incorporating this average interaction effect by a different physical reasoning. They imagined the slid crystal to be embedded not in an elastic matrix, but in a matrix satisfying the following stress–strain law:

$$e_{ij} = L[\tau_{ij}] + E''_{ij} , \tag{6.24}$$

where E''_{ij} is the average value of e''_{ij} throughout the polycrystal and is its macroscopic plastic strain. Now a uniform slip e''_{ij} in the slid crystal causes a uniform strain inside given by

$$e^c_{ij} = b(e''_{ij} - E''_{ij}). \tag{6.25}$$

If this medium is subject to an applied stress of τ_{ij_0}, the strain within this crystal is

$$e_{ij} = L[\tau_{ij_0}] + E''_{ij} + b(e''_{ij} - E''_{ij}). \tag{6.26}$$

Writing this in terms of stress and noting (6.22) and (6.24) we have

$$\tau_{ij} = \tau_{ij_0} - 2\mu(1 - b)(e''_{ij} - E''_{ij}) \tag{6.27}$$

and

$$\dot{\tau}_{ij} = \dot{\tau}_{ij_0} - 2\mu(1 - b)(\dot{e}''_{ij} - \dot{E}''_{ij}). \tag{6.28}$$

Equation (6.27) satisfies the condition that the average τ_{ij} in all crystals be equal to τ_{ij_0} applied to the aggregate. This equation was first given by Kroner (1961). Equation (6.28) links the macroscopic stress rate $\dot{\tau}_{ij_0}$ to the macroscopic plastic strain rate $\dot{E}''_{ij}$ through the relation between the local stres rate $\dot{\tau}_{ij}$ and local plastic strain rate $\dot{e}''_{ij}$ of the particular crystals. This interaction effect is essential in satisfying the equilibrium condition in using inclusion solutions to calculate the polycrystal plastic deformation beyond its very early stage. The stress–strain history of the spherical slid crystal is dependent on the orientation of the crystal axes relative to the specimen axes. The orientation distribution is assumed to be random; hence, the aggregate plastic strain is the average of the over-all orientations of the plastic strains in the spherical slid crystals. Thus

$$E''_{ij} = (e''_{ij})_{\text{av.}} = \frac{1}{4\pi^2} \iint_H \sin \eta \, d\eta \, d\beta \int_{-\pi}^{\pi} e''_{ij} \, d\phi, \tag{6.29}$$

where η, β, and ϕ are the Euler angles illustrated in Fig. 8, which define the crystal axes with respect to the specimen axes, and where H denotes the hemisphere shown in the figure. In general, for a given $\dot{\tau}_{ij_0}$, $\dot{E}''_{ij}$ is not known. It is first assumed, then $\dot{e}_{ij}$ is calculated for all crystals. If the average $\dot{e}''_{ij}$ does not equal the assumed $\dot{E}''_{ij}$, a different $\dot{E}''_{ij}$ is to be

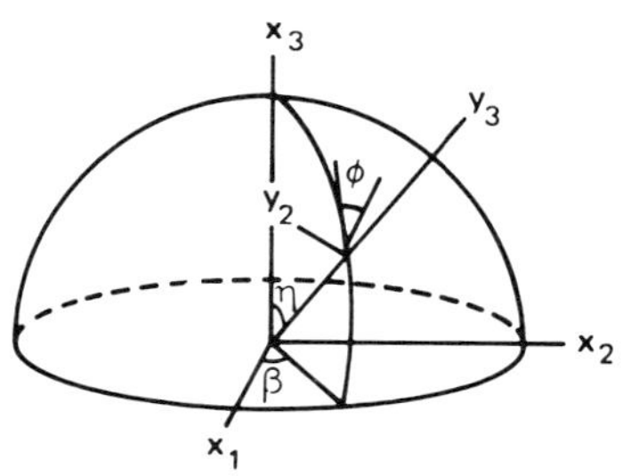

FIG. 8. Euler angles ϕ, β, η relating specimen axes x_1, x_2, x_3 and crystal axes y_1, y_2, y_3.

assumed again. This involves an iterative scheme. However, this procedure can be simplified for some simple loadings. Using this method, Budiansky and Wu (1962) have obtained numerical results of stress–strain curves in simple shear and simple tension for strain-hardening fcc polycrystals. In their calculation each individual crystal is assumed to have kinematic hardening. The initial yield surface of each crystal is allowed to translate (without rotation) as a rigid body in the stress space. This yield surface, which is a 24-sided polyhedron for a fcc crystal, is imagined to be pushed but not pulled by a frictionless peg situated at the loading point. This type of hardening was proposed by Ishlinskii (1954) and Prager (1955, 1956).

In 1964, Hutchinson (1964a) considered fcc polycrystals in which each crystal follows the hardening rule of Taylor, i.e., the initial yield surface expands uniformly. This is known as isotropic hardening. Based on the same model used by Budiansky and Wu (1962) and Kroner (1961), Hutchinson (1964a) calculated stress–strain curves for pure shear and simple tension. He also analyzed polycrystals composed of perfectly plastic single crystals for reversed loadings. These results clearly show the Bauschinger effect. Similar calculations of the stress–strain curves were made by Hutchinson (1964b) for bcc (body centered cubic) polycrystals.

In the above self-consistent analyses, the slid crystal was assumed to be embedded in an isotropic elastic matrix. In a slid crystal of spherical shape, the strain e_{ij}^c is related to its plastic slip e_{ij}'' by the factor b in Eq. (6.21). This factor was obtained from Eshelby's solution for an inclusion in an infinite isotropic elastic matrix. Actually slip has occured in a significant portion of the matrix. This causes the matrix to have pronounced directional weaknesses due to the constraint of the matrix of an already yielded aggregate. A theory taking into consideration this directional weakness has been proposed by Hershey (1954) and Hill (1965). Numerical calculations of uniaxial stress–strain curves based on this theory have been recently developed and performed by Hutchinson (1970) for fcc polycrystals.

D. Correspondence of Self-Consistent Theory with Homogeneous Strain Model

Let $R_{ij_0} = \tau_{ij_0} + 2\mu(1 - b)\, E_{ij}''$. Substituting this and (3.8) into (6.27), we have

$$R_{ij_0} = \tau_{ij} + 2\mu(1 - b) \sum_{n=1}^{N} \alpha_{ij_n}\gamma_n \,, \tag{6.30}$$

where n denotes an active slip system. The resolved shear stress in this system must be equal to the critical shear stress

$$\tau_{ij_m} = \alpha_{ij_m}\tau_{ij} = \alpha_{ij_m}\left[R_{ij_0} - 2\mu(1-b)\sum_{n=1}^{N}\alpha_{ij_n}\gamma_n\right] = \tau_c . \qquad (6.31)$$

Substituting (3.9) into the above and then writing it in incremental form, and assuming linear strain hardening, we have

$$\dot{\tau}_{ij_m} = \alpha_{ij_m}\dot{R}_{ij_0} - 2\mu(1-b)\alpha_{ij_m}\sum_{n=1}^{N}\alpha_{ij_n}\dot{\gamma}_n = c_1\sum_{n=1}^{N}\dot{\gamma}_n , \qquad (6.32)$$

where N is the number of systems slid and c_1 is the rate of hardening.

With the circumflex denoting the variables in Lin's model of uniform total strain, (3.17) becomes

$$\alpha_{ij_m}\left[2\mu\hat{\dot{e}}_{ij} - 2\mu\sum_{n}\alpha_{ij_n}\hat{\dot{\gamma}}_n\right] = c_1\sum_{n=1}^{N}\dot{\gamma}_n . \qquad (6.33)$$

Comparing these two equations, it is seen that they are equivalent if we let

$$\dot{R}_{ij_0} = 2\mu\hat{\dot{e}}_{ij} \qquad \text{and} \qquad \hat{\dot{\gamma}}_n = (1-b)\dot{\gamma}_n . \qquad (6.34)$$

This gives $\hat{\dot{e}}''_{ij} = (1-b)\,\dot{e}''_{ij}$ and hence $\hat{E}''_{ij} = (1-b)\,E_{ij}$. The polycrystalline stress of Lin's model is taken to be the average of the stresses in crystals of all orientations, i.e., $(\hat{\dot{\tau}}_{ij})_{\text{av.}} = \tau_{ij_0}$. This enables one to obtain the stress–strain curve for one model from that calculated for the other. This interesting result was first pointed out by Hutchinson (1964a).

VII. Calculation of Heterogeneous Stress and Slip Fields

At the beginning of plastic deformation, Eshelby's results show the homogeneity of stress and slip in an ellipsoidal slid crystal in an elastic medium, and the heterogeneity of stress outside the slid crystal. Since a three-dimensional continuum cannot be occupied by ellipsoids alone, nonellipsoidal slid crystals exist. It is clear that beyond the very early stage of plastic deformation, the slip and stress fields are heterogeneous. To consider this heterogeneity, analyses more rigorous than the self-consistent theories are needed. In this section, this heterogeneity of stress and slip is considered and the slid crystals are not confined to ellipsoidal shapes.

A. Determination of Slip Distribution

Let (y_1, y_2, y_3) be a set of rectangular coordinates with y_1 measured in a direction normal to the slip plane and y_2 measured along the slip direction of a slip system at a source point where slip has taken place. The plastic shear strain caused by this slip in the pth slip system is denoted by $[\gamma_{y_1 y_2}(x')]_p$, where x' denotes the location of the point in the slid crystal. This plastic strain referring to the x coordinate is then

$$e''_{kl}(x') = \left[\gamma_{y_1 y_2}(x') \left(\frac{\partial x_k'}{\partial y_1} \frac{\partial x_l'}{\partial y_2} + \frac{\partial x_k'}{\partial y_1} \frac{\partial x_l'}{\partial y_2}\right)\right]_p$$

$$= [\gamma(x')\alpha_{kl}]_p , \tag{7.1}$$

where α_{kl} denotes the second parathensis and the subscripts $y_1 y_2$ are dropped in the last bracket, since γ always refers to the slip system. If there are N slip systems in the crystal, we have

$$e''_{kl}(x') = \sum_{p=1}^{N} [\alpha_{kl}(x')\gamma(x')]_p . \tag{7.2}$$

Let$(z_1, z_2, z_3)_n$ be a set of rectangular axes with z_{1_n} normal to the slip plane and z_{2_n} along the slip direction of the nth slip system at point x, at which the resolved shear stress is to be found. This resolved shear stress in the nth system caused by slip throughout the aggregate is obtained by transforming the stress $\tau_{ij_s}(x)$ given by (5.37) in x coordinates into the corresponding components in z coordinates:

$$[\tau_{z_1 z_{2_s}}(x)]_n = \left(\frac{\partial x_i}{\partial z_1} \frac{\partial x_j}{\partial z_2}\right)_n \tau_{ij_s}(x), \tag{7.3}$$

where the subscript n outside the bracket refers to the z axis of the nth slip system. If there are N slip systems slid at a field point x, the resolved shear plastic strain in the nth system due to slip in a different slip system is

$$[e''_{z_1 z_2}(x)]_n = \sum_{m=1}^{N} \left[\frac{\partial y_1}{\partial z_1} \frac{\partial y_2}{\partial z_2} + \frac{\partial y_1}{\partial z_2} \frac{\partial y_2}{\partial z_1}\right]_{nm} [\gamma(x)]_m , \tag{7.4}$$

where the subscripts n and m outside the first bracket on the right-hand side refer to the z axes of the nth slip system and to the y axes of the mth slid system, respectively, at point x.

Let the resolved shear stress caused by the uniform applied aggregate stress τ_{ij_0} in the nth slip system at point x be denoted by $[\tau_{z_1 z_{2_0}}(x)]_n$.

The resolved shear stress in the nth slip system at point x then becomes

$$[\tau_{z_1z_2}(x)]_n = [\tau_{z_1z_{2_0}}(x)]_n + [\tau_{z_1z_{2_s}}(x)]_n - 2\mu[e''_{z_1z_2}(x)]_n\,, \tag{7.5}$$

where the last two terms on the right-hand side are given by (7.3) and (7.4). The critical shear stress is given by (3.9). For the nth slip system to slide, we have

$$[\tau_{z_1z_{2_0}}(x)]_n + [\tau_{z_1z_{2_s}}(x)] - 2\mu[e''_{z_1z_2}(x)]_n = F\left[\sum_{p=1}^{N}\int |\, d\gamma_p\,|\right]. \tag{7.6}$$

At the initial stage of plastic deformation, only the highest stressed slip system of the most favorably oriented crystal slides. This is single slip, and (7.6) reduces to

$$F[\gamma_{z_1z_2}(x)] = \tau_{z_1z_{2_0}}(x) + \tau_{z_1z_{2_s}}(x) - 2\mu\gamma_{z_1z_2}(x). \tag{7.7}$$

During this initial stage of sliding $F[\gamma(x)]$ may be approximated by a linear relation:

$$F(\gamma_{y_1y_2}) = C_0 + C_1\gamma_{y_1y_2}\,. \tag{7.8}$$

$\tau_{z_1z_{2_s}}(x)$ is a linear function of the amounts of slip at different points in the slid crystal. Hence, (7.7) gives a set of linear equations in terms of $\gamma_{z_1z_2}$ at different points in the crystal. There are as many linear equations as there are unknowns and these $\gamma_{z_1z_2}$'s are readily solved. As discussed earlier, if this first slid crystal in this infinite medium is ellipsoidal in shape, Eshelby's solution shows that the initial slip in this crystal is uniform. However, actual crystals exist in all shapes and they cannot be all ellipsoidal. Lin *et al.* (1961) have applied (7.7) and (7.8) to calculate the slip distribution in the first slid crystal of cubic shape in a fine grained aggregate. The calculated $\gamma_{z_1z_2}$ is not uniform, being considerably larger in the interior than at the boundary, and the stress caused by sliding fades rapidly with the distance from the slid crystal. At a distance of about three times the linear dimension of the crystal, this stress practically vanishes. Hence, when the average distance between two adjacent slid crystals is three times their linear dimension, the interaction effect between the two slid crystals may be neglected. This corresponds to the very initial stage of plastic deformation of polycrystals.

As the loading is increased, the number of slid crystals and the number of the slid systems in each of these crystals increase. We let P denote the number of currently active slip systems at point x during an increment of applied aggregate stress $\Delta\tau_{ij_0}$. Writing (5.37), (7.3), (7.4) and (7.6) in incremental form, we have

$$\Delta\tau_{ij_s}(x) = \int_{v'} [\lambda\, \delta_{ij}\psi_{mmkl}(x, x') + 2\mu\psi_{ijkl}(x, x')]\, \Delta e''_{kl}\, dv' \tag{7.9}$$

$$[\Delta\tau_{z_1z_2}(x)]_n = \left[\frac{\partial x_i}{\partial z_1}\frac{\partial y_j}{\partial z_2}\right]_n \Delta\tau_{ij_s}(x), \tag{7.10}$$

$$[\Delta e''_{z_1z_2}(x)]_n = \sum_{p=1}^{P}\left[\frac{\partial y_1}{\partial z_1}\frac{\partial y_2}{\partial z_2} + \frac{\partial y_1}{\partial z_2}\frac{\partial y_2}{\partial z_1}\right]_{nm} [\Delta\gamma(x)]_m\,, \tag{7.11}$$

$$[\Delta\tau_{z_1z_{2_0}}(x)]_n + [\Delta\tau_{z_1z_{2_s}}(x)]_n - 2\mu[\Delta e''_{z_1z_2}(x)]_n$$

$$= F'\left[\sum_{n=1}^{N}\int |\, d\gamma(x)|\right]_n \left[\sum_{p=1}^{P} |\,\Delta\gamma(x)|\right]_p . \tag{7.12}$$

At every point where slip takes place, there are P (unknown) $\Delta\gamma$'s and P equations given by (7.12). These $\Delta\gamma$'s of all the active slip systems at all sliding points in the aggregate are then determined by (7.12). The incremental slip distribution $[\Delta\gamma(x)]_n$ obtained in this way *satisfies the equilibrium and compatibility conditions throughout the aggregate.* From this incremental slip field, the incremental plastic strain field is obtained by

$$\Delta e''_{ij}(x) = \sum_{n=1}^{P} [\alpha_{ij}(x)]_n[\Delta\gamma(x)]_n\,.$$

The incremental stress field may then be obtained by

$$\Delta\tau_{ij}(x) = \Delta\tau_{ij_0} + \Delta\tau_{ij_s}(x) - 2\mu\, \Delta e''_{ij}(x).$$

B. Latent Elastic Energy

When a polycrystalline metal is plastically deformed, part of the work done reappears in the form of heat and the remaining part remains latent and is called latent energy. Quinney and Taylor (1937) and others (Farren and Taylor, 1925; Titchener and Bever, 1958) have studied this phenomenon experimentally for large strains and have found this latent energy ranges between 5 and 15% of the work done. After the polycrystal is loaded beyond the elastic range and then the load is removed, a plastic strain field remains. The work done by this plastic strain is equal to that done by slip in the different slip systems. This work is assumed to have transformed into heat. Let H be the heat per unit volume

$$H = \int_{e''_{ij}} \tau_{ij}\, de''_{ij} = \int \tau \sum_n |\, d\gamma_n \,|, \tag{7.13}$$

where τ is the resolved shear stress. From (5.38) the residual stress is

$$\tau_{ij_r} = \tau_{ij_s} - 2\mu e''_{ij} \,. \tag{7.14}$$

For isotropic elastic materials (Lin, 1968)

$$e'_{ij_r} = (\tau_{ij_r}/2\mu) - \delta_{ij}[\lambda/2\mu(3\lambda + 2\mu)]\tau_{kk_r} \,. \tag{7.15}$$

The elastic energy of the residual stress per unit volume is

$$U = \tfrac{1}{2}\tau_{ij_r} e'_{ij_r} = (1/4\mu)\tau_{ij_r}\tau_{ij_r} - [\lambda/4\mu(3\lambda + 2\mu)](\tau_{kk_r})^2. \tag{7.16}$$

This elastic energy is stored in the metal and hence is the latent energy which has been calculated by Lin and Ito (1967) and Ito (1968).

C. Theoretical Initial Yield Surfaces of Polycrystals

Let us consider an aggregate of crystals which have the same isotropic elastic constants. When the aggregate is loaded uniformly whithin the initial yield surface, no plastic strain occurs. Both the stress and the strain in all crystals are homogeneous and are the same as the aggregate. As the load on the aggregate is increased, the resolved shear stress of one slip system in the most favorably oriented crystal reaches the critical shear stress and this crystal starts to slide. By definition, the yield surface of the aggregate is the surface in stress space, with stress components applied to the aggregate as coordinates, within which the stress vector may change without causing incremental plastic strain and beyond which incremental plastic strain is produced. Slip in any crystal causes plastic strain of the aggregate. Hence, the stress initiating slip in some crystal must lie on the initial yield surface. Since the stress in each crystal is the same as that of the aggregate, the yield polyhedrons of all crystals in the aggregate are the same as if these crystals were loaded independently. Hence, the initial yield surface of the aggregate is bounded by the yield planes of the individual crystals.

The crystal orientations in the aggregate are assumed to be randomly distributed and the initial critical shear stress is assumed to be the same in all crystals. The slip systems of the individual crystals of a fine-grained aggregate are assumed to cover all possible directions on all planes. When the maximum shear stress of the aggregate along a direction β on a plane with normal α reaches this critical shear stress, the crystal with this $\alpha\beta$-slip system also reaches this critical shear stress, and starts to slide. Hence, this maximum shear stress lies on the initial yield surface of the aggregate. Therefore, the theoretical initial yield surface of an aggregate of crystals with the same isotropic elastic constants and the same initial critical shear stress coincides with Trescas's yield surface of maximum shear stress.

D. Subsequent Loading Surfaces of Polycrystals

When a polycrystal is loaded under a combined loading beyond the initial yield surface, the new yield surface, also known as the loading surface, changes both in size and shape. Consider again an aggregate of crystals with homogeneous and isotropic elastic constants. After slip occurs in some crystals the stress in the aggregate is no longer homogeneous. As the loading is further increased, more crystals slide and the slip and stress distributions are heterogenous. Consider that the aggregate is then unloaded. The slip remains and causes heterogeneous redisual stress τ_{ij_r}. Now imagine that the aggregate is reloaded. No additional plastic strain occurred during the imaginary process of unloading and reloading. Denoting this aggregate reloading stress by τ_{ij_A}, the stress in the aggregate is heterogeneous and equal to

$$\tau_{ij} = \tau_{ij_r} + \tau_{ij_A}. \tag{7.17}$$

Now consider in the aggregate one crystal in the $\alpha\beta$-slip system. The resolved shear stress in this slip system is

$$\tau_{\alpha\beta} = l_{\alpha i} l_{\beta j}(\tau_{ij_A} + \tau_{ij_r}). \tag{7.18}$$

The initial critical shear stress in this slip system is denoted by C_0 and the increase of the critical shear stress by $\Delta C_{\alpha\beta}$. The condition for this slip system to initiate or to continue sliding is

$$F(\tau_{ij_A}) = | l_{\alpha i} l_{\beta j}(\tau_{ij} + \tau_{ij_r})| - (C_0 + \Delta C_{\alpha\beta}) = 0. \tag{7.19}$$

The initial yield planes of each slip system in each crystal are represented by

$$| l_{\alpha i} l_{\beta j} \tau_{ij} | = C_0. \tag{7.20}$$

After slip occurs, the yield planes of the crystal are given by (7.19). In this equation, $\Delta C_{\alpha\beta}$ has the effect of increasing the distance between the pair of parallel yield planes, while τ_{ij_r} causes rigid translation of the pair of yield planes in the stress space. A fcc crystal has twelve slip systems giving twelve pairs of parallel yield planes. As indicated previously, the initial yield surface of a fcc polycrystal is bounded by the initial yield polyhedrons of all the individual crystals. After slip occurs, the yield polyhedrons of the slid crystals are expanded and translated, while those of the unslid crystals are translated. The new yield surface (also called the loading surface) of the aggregate is bounded by the faces of these new yield polyhedrons and, hence, must be convex. As the loading increases beyond the initial yield surface this surface does not undergo a uniform expansion, commonly known as isotropic hardening, such as the uniform expansion of the Huber-Mises circle of initial yielding into another concentric circle on the π-plane (Huber, 1904; von Mises, 1913; Hencky, 1924; Hill, 1950; Drucker, 1950; Naghdi, 1960), nor does it translate as a rigid body, commonly known as kinematic hardening (Ishlinski, 1954; Prager, 1955, 1956; Ziegler, 1959). Different yield planes bounding the loading surface move differently, similarly to the loading function composed of a finite number of linear functions of stress components proposed by Sanders (1954).

At a stage of loading beyond the initial yielding, some crystal has slid in the $\alpha\beta$-slip system. We then have

$$|\tau_{\alpha\beta}| = |\tau_{\alpha\beta_0} + \tau_{\alpha\beta_r}| = C_0 + \Delta C_{\alpha\beta}\,. \tag{7.21}$$

The incremental aggregate stress, giving no further slip in the aggregate, causes no change in the residual stress $\tau_{\alpha\beta_r}$ and produces no increase in the resolved shear stress in the active slip systems. This requires

$$d|\tau_{\alpha\beta_A}| \leqslant 0. \tag{7.22}$$

This condition for each of these sliding systems in each crystal gives one pair of parallel bounding planes to the yield surface. One out of each of these pairs of planes passes through the loading point in stress space. There are a number of these yield planes of different orientations passing through the same loading point. They clearly form a vertex at the loading point on the loading surface, as shown by Lin and Ito (1965, 1966).

Consider one polycrystal to be loaded under combined axial compression τ_{11} and shear τ_{12} beyond initial yielding. Among the slid crystals, one slides in the $\alpha\beta$-slip system. For this crystal to have no further slip,

$$d\tau_{\alpha\beta_A} = l_{\alpha 1}l_{\beta 1}\, d\tau_{11_A} + (l_{\alpha 1}l_{\beta 2} + l_{\alpha 2}l_{\beta 1})\, d\tau_{12_A} \leqslant 0. \tag{7.23}$$

Since there are only two nonvanishing stress components applied to the aggregate, the stress space reduces to a stress plane with τ_{11} and τ_{12} as two perpendicular coordinates. The yield locus is bounded by straight lines given by (7.19) passing through the loading point for all sliding crystals. There are many sliding crystals in the aggregate, so there are many of these lines passing through the loading point. The two closest to the τ_{11} axis give the vertex of the yield locus. When this uniaxial compression is applied beyond the initial yielding and the compression is further increased, more crystals in the aggregate slide and hence more of these bounding straight lines pass through the loading point and the angle between the two nearest to the τ_{11} axis becomes less and less. This shows theoretically that as compression proceeds, the corner in the yield locus becomes sharper. This was shown by Lin (1960) and agrees well with test data obtained by different investigators (Naghdi *et al.*, 1958; Phillips 1960; Phillips and Gray, 1958; Phillips and Kaechele, 1958; Hu and Bratt, 1958; Hsu, 1966; Bertsch and Findlay, 1962; Gill and Parker, 1958).

E. Normality of Incremental Plastic Strain Vector to Loading Surfaces

Now consider an aggregate of crystals with either isotropic or anisotropic elastic constants loaded beyond the initial yield surface. A new loading surface is formed. Let $\mathbf{T}$ be a stress vector in stress space that causes a given incremental plastic strain vector $d\mathbf{E}''$ and let $\mathbf{T}^*$ be any other stress vector within this loading surface. Let τ_{ij} and τ_{ij}^* denote the stress field corresponding to $\mathbf{T}$ and $\mathbf{T}^*$, respectively. Further, let τ_k and $\tau_k{}^*$ be the resolved shear stresses in the kth active slip system of a crystal corresponding to τ_{ij} and τ_{ij}^*, respectively. From (3.5), the incremental difference in work due to the plastic strain increment is given by

$$d\overline{W} - d\overline{W}^* = (\mathbf{T} - \mathbf{T}^*) \cdot d\mathbf{E}''v = \int (\tau_{ij} - \tau_{ij}^*)\, de_{ij}''\, dv$$

$$= \int \sum_{k=1}^{m} (\tau_k - \tau_k{}^*)\, d\gamma_k dv \geqslant 0.$$

Hence, we obtain the same expression

$$(\mathbf{T} - \mathbf{T}^*) \cdot d\mathbf{E}'' \geqslant 0 \qquad (7.24)$$

for polycrystals. This gives the same result as Drucker's postulate (1951). Consider $(\mathbf{T} - \mathbf{T}^*)$ as a stress vector in stress space with $\mathbf{T}$ on the yield surface and $\mathbf{T}^*$ within the yield surface. $d\mathbf{E}''$ is the incremental plastic strain vector of the aggregate. $(\mathbf{T} - \mathbf{T}^*) \cdot d\mathbf{E}''$ is the scalar product of these two vectors, and θ is the angle between them. Equation (7.24) may be written as $|\mathbf{T} - \mathbf{T}^*| \times |d\mathbf{E}''| \cos\theta \geqslant 0$. The intersection of the plane containing these two vectors and the loading surface gives a yield locus as shown in Fig. 9. Consider two stress vectors $\mathbf{T}_1{}^*$ and $\mathbf{T}_2{}^*$

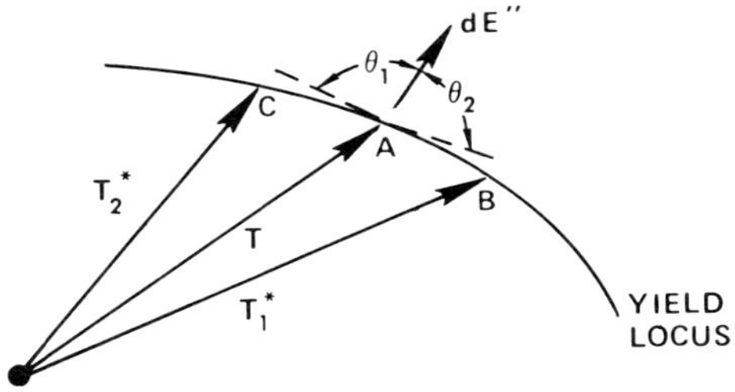

FIG. 9. Normality of incremental plastic strain vector to a smooth loading surface.

in this plane and very close to $\mathbf{T}$: one on the right giving $\mathbf{T} - \mathbf{T}_1{}^* = \mathbf{BA}$, and one on the left of T giving $\mathbf{T} - \mathbf{T}_2{}^* = \mathbf{CA}$, as shown in the figure. θ_1 and θ_2 corresponding to $\mathbf{T}_1{}^*$ and $\mathbf{T}_2{}^*$ are also indicated in the figure. From (7.25)

$$\theta_1 \leqslant 90° \qquad \text{and} \qquad \theta_2 \leqslant 90°. \tag{7.25}$$

If this portion of the yield surface is smooth, BAC is a smooth curve and approaches the tangent line as C and B approach A, giving

$$\theta_1 + \theta_2 = 180°. \tag{7.26}$$

In order to satisfy conditions (7.25) and (7.26),

$$\theta_1 = \theta_2 = 90°. \tag{7.27}$$

Therefore, the plastic strain increment vector $d\mathbf{E}''$ must be normal to the yield surface at a smooth point. If the stress vector lies at a vertex of the loading surface as shown in Fig. 10, the incremental plastic strain vector must lie within the cone bounded by normals to the yield planes intersecting at the vertex. The direction of the incremental plastic strain vector is uniquely defined on a smooth portion of the loading surface but not at the vertex.

Incremental plastic strain with the condition of zero dilatation has five independent components. A vector is completely defined is its direction and magnitude are known. A smooth loading surface gives the direction of this vector and hence gives the ratios of the five incremental plastic strain components. Hence, the loading surface in stress space not only gives the range of stress vectors in which elasticity holds but also gives four of the five unknowns in the incremental plastic strain components when the stress vector is at a smooth point of the surface. However, when the stress vector is at the vertex, the direction of the incremental plastic strain vector is no longer uniquely determined.

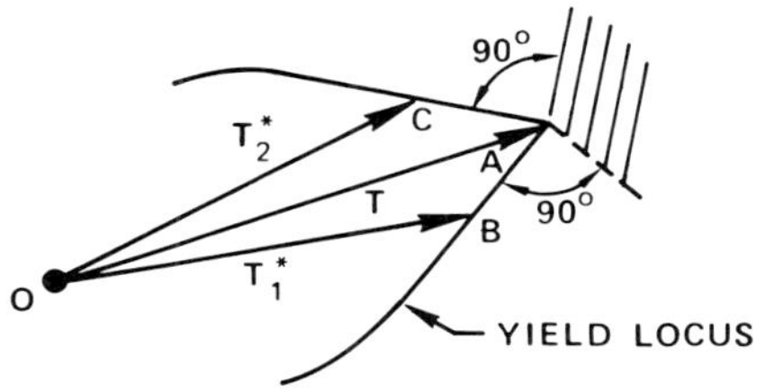

FIG. 10. A vertex on the loading surface.

F. Numerical Calculation of Incremental Stress–Strain Relations of Two Polycrystals

Lin and Ito in 1966 considered a fine grained aggregate composed of identical basic cubic blocks of 64 cube-shaped crystals having different orientations, as shown in Fig. 11. Each crystal is assumed to have only one slip plane with its normal denoted by a unit vector **n**. On this plane there are three equally spaced slip directions. This slip plane corresponds to the basal plane of a close-packed hexagonal crystal, for which it is known the elastic constants are anisotropic. However, this anisotropy is here neglected to simplify the calculation. These crystals are assumed to have a Young's modulus $E = 1.46 \times 10^7$ psi, a Poisson's ratio $\nu = 0.3$, an initial critical shear stress $\tau_c = 35$ psi, and an initial rate of strain hardening $c_1 = 1400$ psi per unti slip strain. The 64 slip plane orientations—one for each of the 64 crystals in the basic block—are chosen to approximate a uniform distribution of planes tangent to a unit hemisphere. This infinite medium is now subject to a uniform loading. The method developed in Section V, D and VII, A is applied to determine the incremental slip distribution and then the incremental stress–strain relation of the medium. This infinite medium is filled with the basic cubic blocks of crystals. The values of $\Delta\gamma_{y_1y_2}$ and its gradient at (x_1', x_2', x_3') are exactly the same at all points defined by $(x_1' - n_1a, x_2' - n_2a, x_3' - n_3a)$, where n_1, n_2, n_3 are integers and a is the linear dimension of the cubic block. The integral in (7.9) over a three dimensional infinite space reduces to the sum of a triply infinite number of

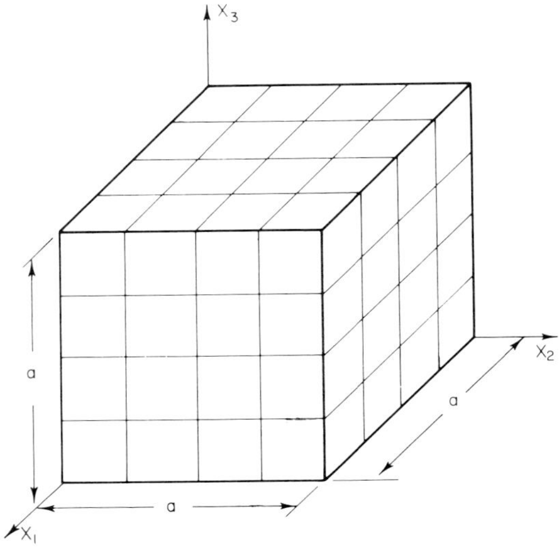

FIG. 11. Basic block of 64 cube-shaped crystals.

integrals over only one basic block with $\psi_{mmkl}(x, x')$ and $\psi_{ijkl}(x, x')$ as the corresponding functions of $(x_1, x_2, x_3; x_1' - na, x_2' - n_2a, x_3' - n_3a)$, where n_1, n_2, and n_3 cover all integers:

$$\begin{aligned}\Delta\tau_{ij_s}(x) = & \sum_{n_1=-\infty}^{\infty} \sum_{n_2=-\infty}^{\infty} \sum_{n_3=-\infty}^{\infty} \int_0^a \int_0^a \int_0^a \\ & \times [\lambda\, \delta_{ij}\psi_{mmkl}(x_1, x_2, x_3; x_1' - n_1a, x_2' - n_2a, x_3' - n_3a) \\ & + 2\mu\psi_{ijkl}(x_1, x_2, x_3; x_1' - n_1a, x_2' - n_2a, x_3' - n_3a)] \\ & \times \Delta e''_{kl}(x')\, dv'. \end{aligned} \tag{7.28}$$

Two out of the three slip directions on a slip plane are independent, so that only two slip directions can be active at one time. Hence N in (7.12) may be 3 but P in (7.11) and (7.12) cannot exceed 2. At every sliding point there can be two unknowns $[\Delta\gamma_{y_1y_2}(x)]_p$ with two equations as given by (7.12). These unknowns can be solved for any number of chosen grid points. However, the computation increases rapidly with the number of sliding points considered. Hence, in the computation, only one point located at the center of each of the 64 crystals was considered and it was assumed that the plastic strain calculated at this point represents the average plastic strain in the crystal. Equation (7.12) was applied to each sliding point ant the unknowns $\Delta\gamma_{y_1y_2}(x)$ solved numerically. The incremental slip field of the aggregate was obtained. Then the incremental plastic strain field was obtained from (7.11) and the incremental stress field from (7.12).

The aggregate was loaded to $\tau_{x_1x_1} = 120$ psi. The theoretical yield locus at this load was calculated and is shown in Fig. 12. It is clearly

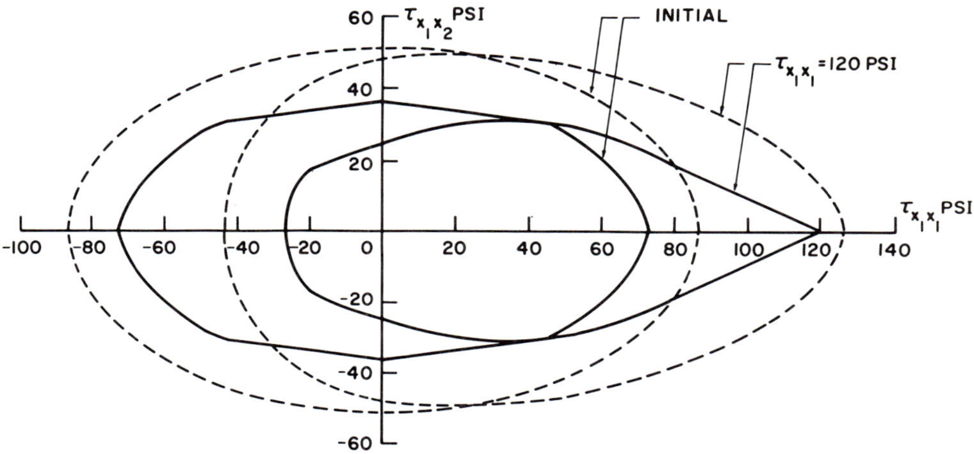

FIG. 12. Yield surfaces: initial and tensile loading at $\tau_{11} = 120$ psi. ——, $\Delta e = 0^+$ (infinitesimal); – – – $\Delta e = 0.1$ microinch/inch, where $\Delta e = [(\Delta e''_{x_1x_1})^2 + (\Delta e''_{x_1x_2})^2]^{1/2}$. From Lin and Ito (1966).

seen that there is a vertex at the loading point as predicted. There is no unique normal at this vertex; the condition of normality of the incremental plastic strain vector to the loading surface does not give a unique direction to this vector. This is quite different from the von Mises' assumption of the coincidence of the loading surface with the surface of constant second deviated stress invariant ($\bar{J}_2 =$ constant), which gives a unique incremental plastic strain vector.

G. Causes of the Discrepancy between the Calculated Theoretical and Experimental Results

The theoretical yield surfaces denote infinitesimal incremental plastic strain. Experimentally, however, the incremental plastic strain is not observed until it grows to a measurable value. Hence, an experimental loading surface can only approach the theoretical when the measuring instrument can detect infinitesimal plastic strain. This instrument, of course, has never been achieved. Hence, Lin and Ito (1966) calculated the loading surface giving an incremental plastic strain of the magnitude of 0.1 microinch/inch for the aggregate considered. It is seen from Fig. 12 that the loading surface giving a finite incremental plastic strain has no vertex, but the curvature at the loading point is increased. This agrees well with experimental observations (Naghdi *et al.*, 1958; Naghdi and Rowley, 1954; Phillips, 1960; Phillips and Gray, 1958; Phillips and Kaechele, 1958; Hu and Bratt, 1958; Hsu, 1966; Bertsch and Findley, 1962; Gill and Parker, 1958).

When the stress vector is on a smooth portion of the loading surface, the direction of the incremental plastic strain vector is uniquely determined regardless of the direction of the incremental stress vector. Hence, if the direction of the incremental plastic strain vector varies with the incremental stress vector, the stress vector is taken to be at a vertex of the loading. This relation was recently used by Hecker (1970) to check the existence of a vertex. His experiments indicate a definite correlation between the change in direction of the incremental stress vector and the incremental plastic strain. He concluded the existence of a vertex at the loading point. Then he measured the yield surface giving a finite incremental plastic strain of 5 to 10 microinch/inch and found this yield surface to be smooth. This clearly agrees well with the previous theoretical predictions.

The theoretical plastic buckling strength of a perfect plate under edge compression (Handleman and Prager, 1949) depends on the initial shear modulus of the material immediately after being compressed beyond the elastic limit for different ratios of the incremental strain components $\Delta e_{12}/\Delta e_{11}$, i.e., for different incremental stress components $\Delta\tau_{x_1x_2}/\Delta\tau_{x_1x_2}$. A smooth yield surface symmetrical to $\tau_{x_1x_1}$, such as that given by the von Mises theory, predicts this shear modulus to be elastic and gives an excessive overestimate of the plastic buckling strength of the plate (Handleman and Prager, 1949). The existence of the vertex reduces this initial shear modulus and, hence, may account for the observed low plastic buckling strength of the plates (Ilyushin, 1944, Bijlaard, 1949; Pride and Heimerl, 1952; Stowell, 1948; Gerard and Becker, 1957).

Many experiments have been conducted to check the normality of the incremental plastic strain vector to the loading surface. The loading surfaces measured are clearly not those for infinitesimal plastic strain increment, but those for finite strain increments. The tangent to these loading surfaces and the direction of the finite incremental plastic strain vector have been theoretically calculated for this idealized aggregate of hexagonal crystals for different ratios of incremental stress components $\Delta\tau_{x_1x_2}/\Delta\tau_{x_1x_2}$, as shown in Figs. 13–15. The direction of the incremental

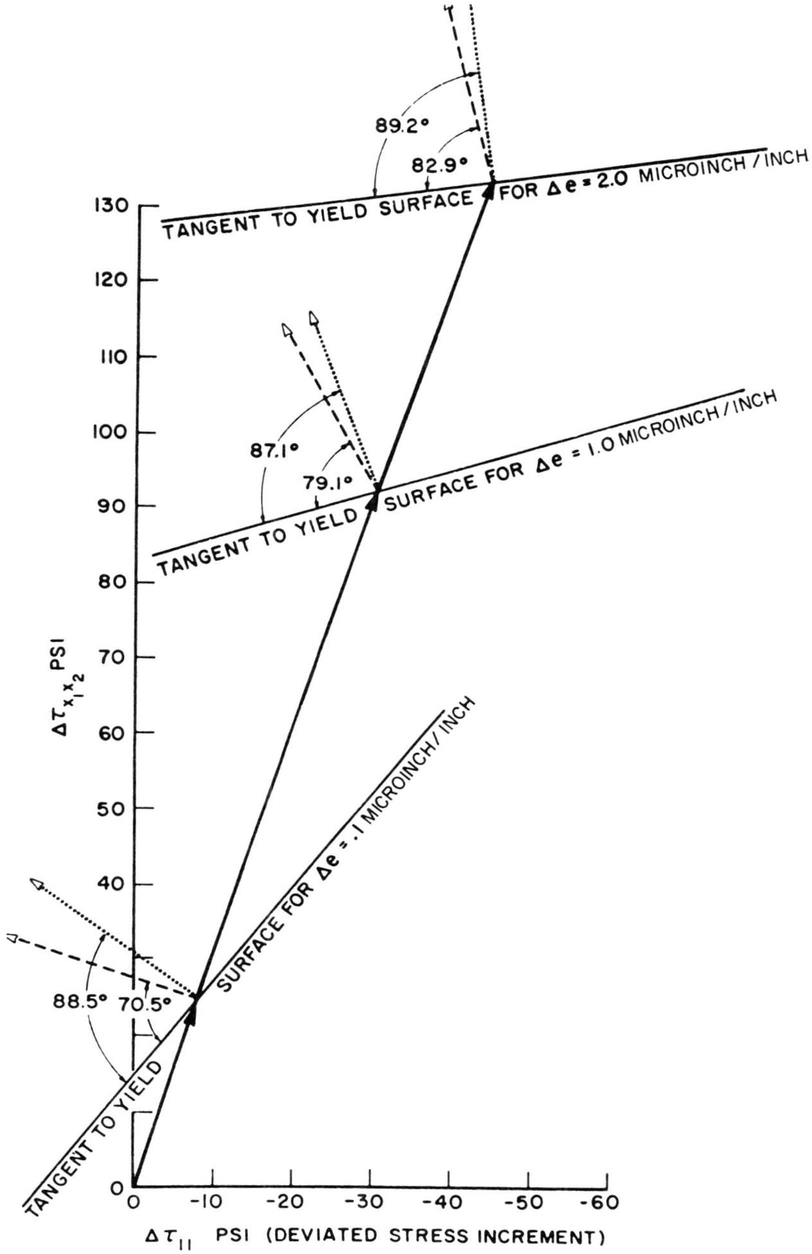

FIG. 13. The direction of plastic strain vector: $\Delta\tau_{11}/\Delta\tau_{12} = 1/2$. Initial stress $\tau''_{x_1x_1} = 120$ psi. Key: $\Delta e = [(\Delta e''_{11})^2 + (\Delta e''_{12})^2]^{1/2}$. Solid arrow, incremental deviated stress vector; dotted arrow, theoretical direction of the slope $\Delta e''_{12}/\Delta e''_{11}$; dashed arrow, von Mises direction of the slope $\Delta e''_{12}/\Delta e''_{11}$. From Lin and Ito (1966).

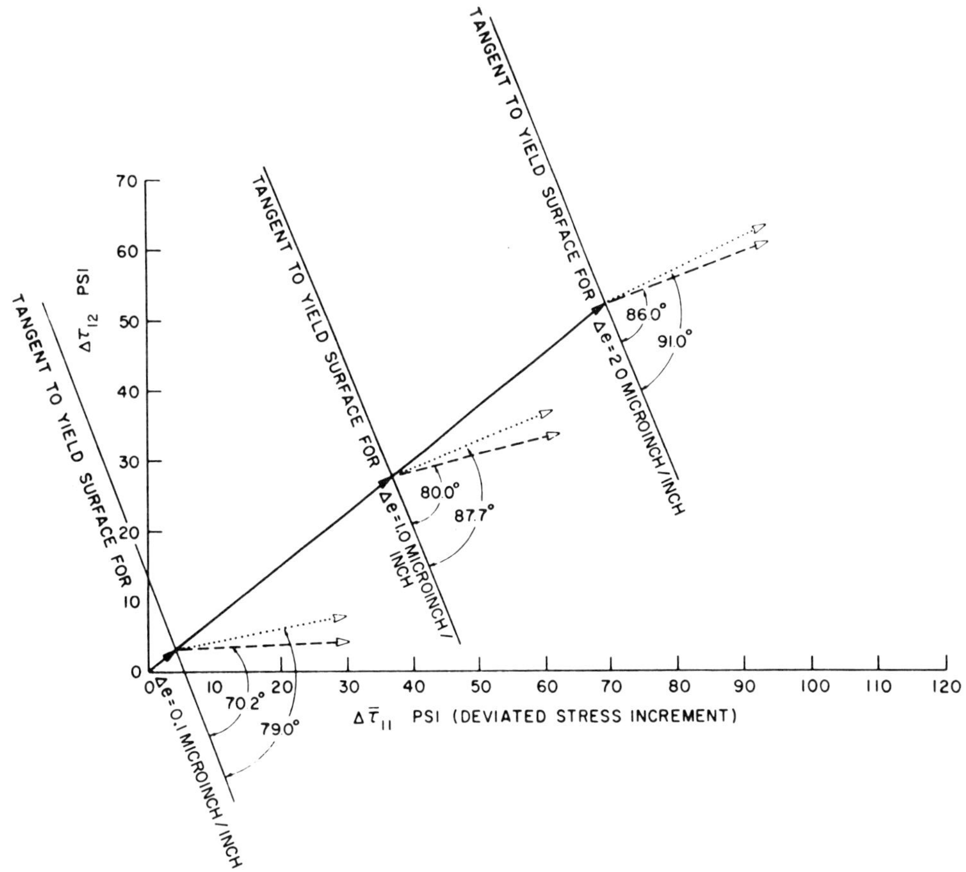

FIG. 14. The direction of plastic strain vector: $\Delta\tau_{11}/\Delta\tau_{12} = 2/1$. Initial stress $\tau_{11} = 120$ psi. For key see caption for Fig. 13. From Lin and Ito (1966).

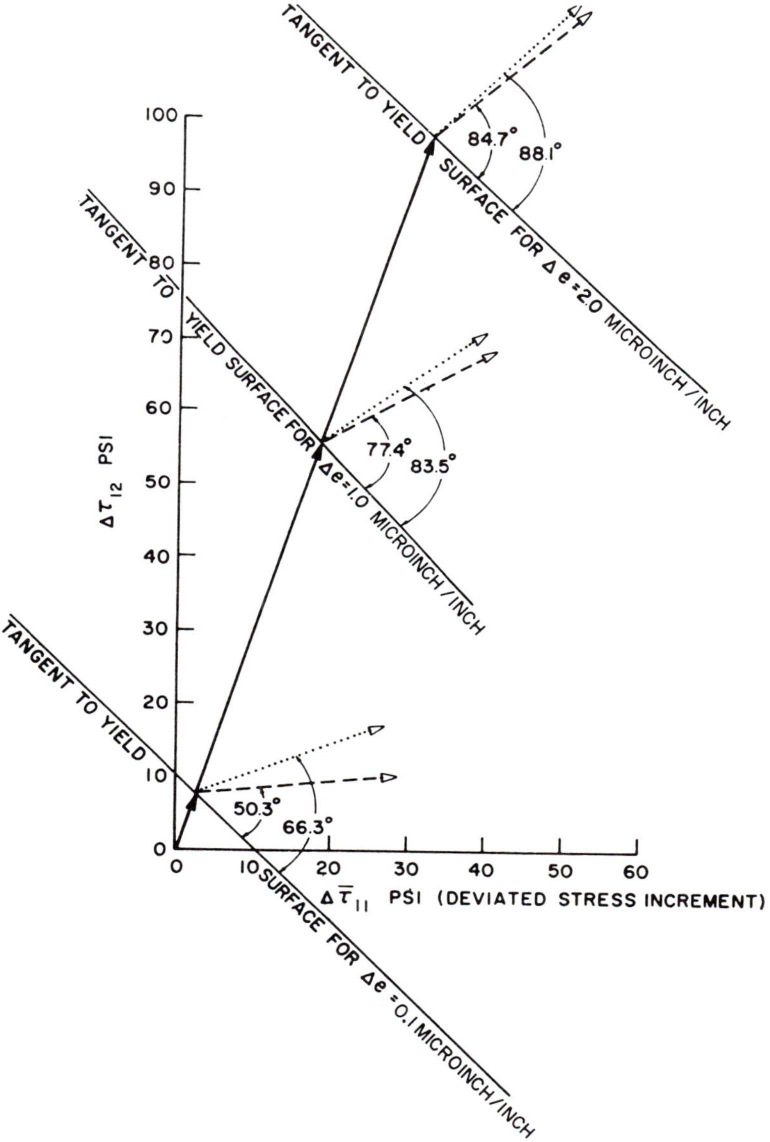

FIG. 15. The direction of plastic strain vector: $\Delta\tau_{11}/\Delta\tau_{12} = -1/2$. Initial stress $\tau_{11} = 120$ psi. For key see caption for Fig. 13. From Lin and Ito (1966).

plastic strain vector based on von Mises' theory is also shown in these figures for comparison. It is seen that the error in the direction of the incremental plastic strain vector by the von Mises theory is largest when the magnitude of the incremental plastic strain is infinitesimal. This error decreases as the magnitude of this incremental strain increases.

Recently a fine-grained aggregate of randomly oriented fcc crystals was considered by Ito and Lin (1968). In their theory a basic cubic block is taken to be composed of 64 differently oriented fcc crystals. The three dimensional space is filled with these identical basic cubic blocks. The nature of the calculation is the same as for the aggregate of hexagonal crystals, except that now twelve slip systems are considered instead of three in each crystal. All the crystals are assumed to be elastically isotropic and homogeneous. The material properties of these crystals are assumed to correspond to those of commercially pure aluminum with Young's modulus $E = 10^7$ psi, Poisson's ratio $\nu = 0.3$, the initial critical shear stress $c_0 = 53.5$ psi, and the initial rate of hardening $c_1 = 1.3 \times 10^6$ psi per unit slip. The orientations of these 64 crystals were chosen to represent random distribution. This polycrystal was again considered to be subjected to the axial $\tau_{x_1x_1}$ and shear $\tau_{x_1x_2}$ stresses. The theoretical initial yield locus causing infinitesimal plastic strain as well as that for finite incremental plastic strain was calculated for radial loadings or combined axial and shear loadings. As mentioned before, the observed incremental plastic strain is always finite, and the experimental yield locus always corresponds to finite incremental plastic strain. These calculated initial yield loci of the aggregate are shown in Fig. 16 and compared with those given by the Tresca and von Mises theories. It is seen that the locus giving finite plastic strain agrees well with that given by von Mises theory, while the locus giving infinitesimal plastic strain is close to that given by the Tresca theory. Experimental results of Taylor and Quinney (1931) on copper, steel, and aluminum are shown in Fig. 17, together with Tresca's von Mises' curves. These data, along with most other test results are in much better agreement with von Mises' than with Tresca's criterion. The experimental locus corresponds to a theoretical one with finite incremental plastic strain and, hence, is close to von Mises' prediction. The loading surfaces of the aggregate after being uniaxially loaded to $\tau_{x_1x_1} = 120$ psi for an infinitesimal and a finite incremental plastic strain of 0.1 microinch/inch have been calculated. It is found again that a vertex exists at the loading point of the yield locus corresponding to an infinitesimal incremental plastic strain but no vertex exists on the locus corresponding to a finite incremental plastic strain. This explains why a vetex is not explicitly observed in experiments. From the present analysis the yield planes

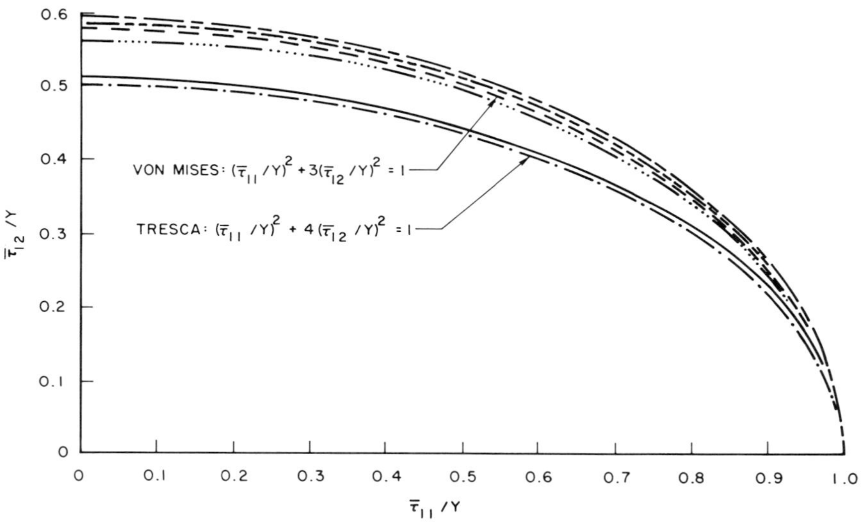

FIG. 16. Initial yield locus of a fcc polycrystal. ——, Initial yield locus; — ··· —, initial offset yield locus, $\epsilon = 0.1$ microinch per inch; — — —, initial offset yield locus, $\epsilon = 0.2$ microinch per inch; — - —, initial offset yield locus, $\epsilon = 0.3$ microinch per inch; Y = tensile yield stress; $\epsilon = [(\epsilon''_{11})^2 + (\epsilon''_{12})^2]^{1/2}$. From Ito (1968).

intersecting at the vertex interact with each other. Displacement of one yield plane causes displacements of others. The displacements of these yield planes are not independent (Lin and Ito, 1966). Hence, Koiter's derivation of the uniqueness and variational theorems (1953) based on independent yield planes is not applicable to the present theory of plasticity.

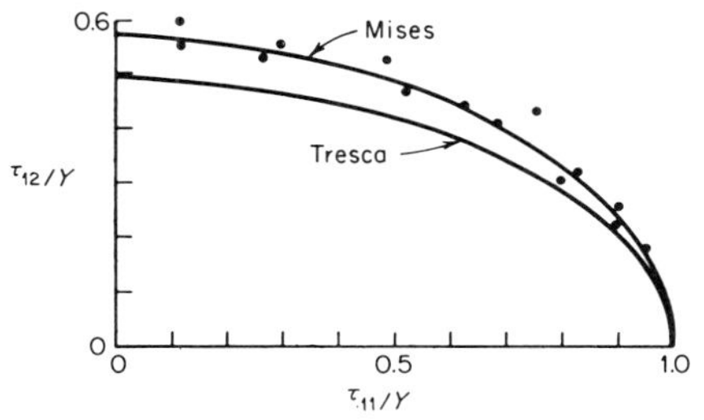

FIG. 17. Yield curves under combined stresses τ_{12} and τ_{11}; Y = yield stress under pure tension. Data from Taylor and Quinney (1931).

H. Correlations between Physical and Mathematical Theories of Plasticity

It is readily seen that convexity of loading surfaces and the normality of the incremental plastic strain vector to the loading surface, commonly assumed in mathematical theories of plasticity, can be derived from the slip characteristics of crystals. These slip characteristics predict the existence of a vertex at the loading point of the loading surface. That seems to account for the large disparity between the calculated plastic buckling strength based on von Mises' criterion and the experimental buckling strength of a plate under compression. The calculated loading surfaces based on infinitesimal incremental plastic strain have a vertex, but this vertex disappears in those surfaces calculated on the basis of finite measurable incremental plastic strain. This explains why experimental observations have not been able to directly prove nor disapprove the existence of this vertex on the loading surface. The theoretical initial yield surface for an infinitesimal plastic strain is the same as the Tresca's yield surface of maximum shear stress. However, the theoretical initial yield surface corresponding to a finite measurable plastic strain is close to von Mises' initial yield surface. The loading surface does not expand uniformly, as predicted by isotropic hardening, nor translate like a rigid body, as predicted by kinematic hardening.

I. General Remarks

The derivations given in Section V are completely general and rigorous for an aggregate of elastically isotropic homogeneous crystals. The method of calculation of the slip field satisfies the conditions of equilibrium, compatibility, and the given single crystal characteristics and hence is rigorous. It is seen that there are many correlations between the physical theory and the mathematical theories of plasticity. It seems likely that in the not too distant future, the physical theory may be used to yield a more accurate stress–strain relationship in the analyses of structure.

The above calculations are based on the assumption that there is no initial stress field in the single crystal or in the aggregate. It is known that there are dislocations in the actual crystals. These dislocations cause an initial stress field, which causes the heterogeneous plastic deformations as indicated by slip lines observed on metal surfaces (Barrett, 1952). This has to be considered in explaining the Bauschinger effect in single crystals and the fatigue band in metals (Lin and Ito, 1968, 1969a,b).

ACKNOWLEDGMENTS

This work was mainly supported by the National Science Foundation under Grant GK 15348.

REFERENCES

BARRETT, C. S. (1952). "Structure of Metals," Chapter 1, p. 345. McGraw-Hill, New York.

BATDORF, S. B., and BUDIANSKY, B. (1949). A mathematical theory of plasticity based on slip. *Nat. Adv. Comm. Aeronaut*, Tech. Notes 1871.

BATDORF, S. B., and BUDIANSKY, B. (1954). Polyaxial stress-strain relations of strain-hardening metal. *J. Appl. Mech.* **21**, 323–326.

BERTSCH, P. K., and FINDLEY, W. N. (1968). An experimental study of subsequent yield surfaces-corners, normality, Bauschinger and allied effects. *Proc. U.S. Nat. Congr. Appl. Mech., 4th*, 1962 pp. 893–907.

BIJLAARD, P. P. (1949). Theory and tests on plastic stability of plastics and shells. *J. Aerosp. Sci.* **16**, 529–541.

BISHOP, J. F. W., and HILL, R. (1951a). A theory of the plastic distortion of a polycrystalline aggregate under combined stress. *Phil. Mag.* [7] **42**, 414–427.

BISHOP, J. F. W., and HILL, R. (1951b). A theoretical derivation of the plastic properties of a polycrystalline face-centered metal. *Phil. Mag.* [7] **42**, 1298–1307.

BUDIANSKY, B., and WU, T. Y. (1962). Theoretical prediction of plastic strains of polycrystals. *Proc. U.S. Nat. Congr. Appl. Mech., 4th*, 1962 p. 1175.

BUDIANSKY, B., HASHIN, Z., and SANDERS, J. L., JR. (1960). The stress field of a slipped crystal and the early plastic behavior of polycrystalline materials. *Proc. Symp. Nav. Struct. Mech., 2nd*, 1960 p. 239.

BURGER, J. M. (1939). Some considerations on the fields of stress connected with dislocations in a regular crystal lattice. *Proc. Kon. Ned. Akad. Wetensch.* **42**, 293 and 378.

CALNAN, E. A., and CLEWS, C. J. B. (1950). Deformation textures in free-centered cubic metals. *Phil. Mag.* [7] **41**, 1085–1100.

COX, H. L., and SOPWITH, D. E. (1937). Effect of orientation on stresses in single crystals and of random orientation on strength of polycrystalline aggregates. *Proc. Phys. Soc., London* **49**, 134.

CZYZAK, S. J., BOW, N., and PAYNE, H. (1961). On the tensile stress-strain relation and the Bauschinger effect for polycrystalline materials from Taylor's model. *J. Mech. Phys. Solids* **9**, 63.

DAVIS, R. S., FLEISCHER, R. L., LIVINGSTON, J. D., and CHALMERS, B. (1957). Effect of orientation on the plastic deformation of aluminum single crystals and bicrystals. *Trans. AIME* **209**, 136.

DORN, J. F., and MOTE, J. D. (1962). On the plastic behavior of polycrystalline aggregates. *Mater. Sci. Res.* 1, 11–56.

DRUCKER, D. C. (1950). A survey of theory and experiment. Rep. A11-S1. Grad. Div. Appl. Math., Brown University.

DRUCKER, D. C. (1951). A more fundamental approach to plastic stress-strain relations. *Proc. U.S. Nat. Congr. Appl. Mech., 1st*, 1951 p. 487.

DRUCKER, D. C. (1962). Basic concept, plasticity and viscoelasticity. *In* "Handbook of Engineering Mechanics" (W. Flügge, ed.), pp. 46-3–46-16. McGraw-Hill, New York.

DUHAMEL, J. M. C. (1837). Seconde mémoire sur les phénomènes thermoméchaniques. *J. École Polytech.* **15**, 1–57.

DUHAMEL, J. M. C. (1838). Mémoire sur le calcul des actions moléculaires dévéloppées par les changements de température dans le corps. *Mem. Inst. Fr.* **5**, 440–498.

ELBAUM, C. (1960a). The relation between the plastic deformation of single crystals and polycrystals. *Proc. Symp. Nav. Struct. Mech., 2nd,* 1960 pp. 107–120.

ELBAUM, C. (1960b). Plastic deformation of aluminum multicrystals. *Trans. AIME* **218**, 444–448.

ESHELBY, J. D. (1956). The continuum theory of lattice defects. *Solid State Phys.* **3**, 79–127.

ESHELBY, J. D. (1957). The determination of the elastic field of an ellipsoidal inclusion and related problems. *Proc. Roy. Soc., Ser. A* **241**, 398.

FARREN, W. S., and TAYLOR, G. I. (1925). The heat developed during plastic extension of metals. *Proc. Roy. Soc., Ser. A* **107**, 422–451.

FLEISHER, R. L., and BACKOFEN, W. A. (1960). Effect of grain boundaries in tensile deformation at low temperature. *Trans. AIME* **218**, 243–251.

FRANKEL, J. (1926). Zur Theorie der Elastizitätsgrenze und der Festigkeit Kristallinischer Körper. *Z. Phys.* **37**, 572.

GERARD, G., and BECKER, H. (1957). Buckling of flat plates. *Nat. Adv. Comm. Aeronaut., Tech. Notes* **3781**.

GILL, S. S., and PARKER, J. (1958). Plastic stress-strain relationships—some experiments on the effect of loading paths and loading history. *J. Appl. Mech.* **25**, 77–87.

GILMAN, J. J. (1953). Deformation of symmetric zinc bicrystals. *Acta Met.* **1**, 426–427.

GILMAN, J. J. (1960). Physical nature of plastic flow and fracture plasticity. *Proc. Symp. Nav. Struct. Mech., 2nd,* 1960 pp. 43–99.

HANDLEMAN, E. H., and PRAGER, W. (1949). Plastic buckling of a rectangular plate under edge thrusts. *Nat. Adv. Comm. Aeronaut., Tech. Rep.* **946**.

HECKER, S. S. (1970). Private communication.

HENCKY, H. (1924). Zur Theorie Plastischer Deformationen und der hierdurch im Material hervorgerufenen Nachspannungen. *Z. Angew. Math. Mech.* **4**, 323.

HERSHEY, A. V. (1954). The plasticity of an isotropic aggregate of anisotropic face-centered cubic crystals. *J. Appl. Mech.* **21**, 241–249.

HILL, R. (1950). "The Mathematical Theory of Plasticity," p. 14. Oxford Univ. Press (Clarendon), London and New York.

HILL, R. (1965). Continuum micro-mechanics of elastoplastic polycrystals. *J. Mech. Phys. Solids* **13**, 89–101.

HILL, R. (1966). Generalized constitutive relations for incremental deformation of metal crystals by multislip. *J. Mech. Phys. Solids* **14**, 99–102.

HIRTH, J. P., and LOTHE, J. (1968). "Theory of Dislocations," p. 21. McGraw-Hill, New York.

HSU, T. C. (1966). Definition of the yield point in plasticity and its effect on the shape of the yield locus. *J. Strain Anal.* **1**, 331–338.

HU, L. W., and BRATT, J. F. (1958). Effect of tensile plastic deformation on yield condition. *J. Appl. Mech.* **23**, 411.

HUBER, M. T. (1904). *Czas. Tech., Lemberg* **22**, 81.

HUTCHINSON, J. W. (1964a). Plastic stress-strain relations of f.c.c. polycrystalline metals hardening according to Taylor rule. *J. Mech. Phys. Solids* **12**, 11–24.

HUTCHINSON, J. W. (1964b). Plastic deformation of f.c.c. polycrystals. *J. Mech. Phys. Solids* **12**, 24–33.

HUTCHINSON, J. W. (1970). Elastic-plastic behavior of polycrystalline metals and composites. *Proc. Roy. Soc.* (*Ser. A*) **319**, 247–272.

ILYUSHIN, A. A. (1944). Stability of plates and shells stressed beyond the proportional limit. *Prikl. Mat. Mekh.* **8**, 337. *Nat. Adv. Comm. Aeronaut., Tech. Memo.* 1196.

ILYUSHIN, A. A. (1955). Contemporary problems of the theory of plasticity (in Russian). *Vestn. Mosk. Gos. Univ.* No. 4/5, pp. 101–113.

ISHLINSKI, A. (1954). The general theory of plasticity with linear hardening. *Ukr. Mat. Zh.* **6**, 314.

ITO, Y. M. (1968). Theoretical plastic behavior of f.c.c. polycrystalline aggregate. Ph.D. Dissertation, University of California, Los Angeles, California.

ITO, Y. M., and LIN, T. H. (1968). Theoretical plastic strain of a facecentered cubic polycrystal. *Progr. Eng. Sci.* **6**, 169–188.

JILLON, D. C. (1950). Quantitative stress-strain studies on zinc single crystals in tension. *Trans. AIME* **188**, 1129.

JOHNSTON, W. G., and GILMAN, J. J. (1959). Dislocation velocities, dislocation densities and plastic flow in lithium fluoride crystals. *J. Appl. Phys.* **30**, 129.

KAWADA, T. (1951). The plastic deformation of zinc bicrystals. *Proc. World Met. Congr., 1st*, 1951, p. 591.

KLIUSHINKOV, V. D. (1958). On plasticity laws for work-hardening materials. *Prikl. Mat. Mekh.* **22**, 97–118.

KNETS, I. V., and KREGERS, A. F. (1965). Relation between the strain and stress tensors in successive biaxial tension. *Mekh. Polim.* **1**, 43–51.

KOCHENDORFER, A. (1941). "Plastische Eigenschaften von Kristallen." Springer, Berlin.

KOCKS, U. F. (1958). Polyship in polycrystals. *Acta Met.* **6**, 85.

KOCKS, U. F. (1970). "The Relation Between Polycrystal Deformation and Single Crystal Deformation," Metallurgical Transactions, Vol. 1, pp. 1121–1143.

KOITER, W. T. (1953). Stress-strain relations, uniqueness and variational theorems for elastic-plastic materials with a singular yield surface. *Quart. Appl. Math.* **11**, 350–353.

KRÖNER, E. (1958). "Kontinuumstheorie der Versetzungen und Eigenspannungen" (Continuum Theory of Dislocations and Initial Stress). Springer, Berlin.

KRÖNER, E. (1961). Zur plastischen verformung des vielkristalls. *Acta Met.* **9**, 155.

LIN, T. H. (1954). A proposed theory of plasticity based on slip. *Proc. U.S. Nat. Congr. Appl. Mech., 2nd*, 1954 pp. 461–468.

LIN, T. H. (1957). Analysis of elastic and platic strains of a free-centered cubic crystal. *J. Mech. Phys. Solids* **5**, 143–159.

LIN, T. H. (1958). On stress-strain relations based on slips. *Proc. U.S. Nat. Congr. Appl. Mech., 3rd*, 1958, pp. 581–588.

LIN, T. H. (1960). On the associated flow rule of plasticity based on crystal slip. *J. Franklin Inst.* **270**, 291–300.

LIN, T. H. (1968). "Theory of Inelastic Structures," pp. 32 and 51–54. Wiley, New York.

LIN, T. H. (1967). Reciprocal theorem of displacement of inelastic bodies. *J. Compos. Mater.* **1**, 144–151.

LIN, T. H., and ITO, Y. M. (1965). Theoretical plastic distortion of a polycrystalline aggregate under combined and reversed stresses. *J. Mech. Phys. Solids* **13**, 103.

LIN, T. H., and ITO, Y. M. (1966). Theoretical plastic stress-strain relationship of a polycrystal and comparisons with von Mises' and Tresca's plasticity theories. *Int. J. Eng. Sci.* **4**, 543–561.

LIN, T. H., and ITO, Y. M. (1967). Latent elastic strain energy due to the residual stresses in a plastically deformed polycrystal. *J. Appl. Mech.* **34**, 606–611.

LIN, T. H., and ITO, Y. M. (1968). Mechanism of fatigue crack nucleation based on microstresses caused by slip. Rep. No. 68-19. University of California, Los Angeles, California.

LIN, T. H., and ITO, Y. M. (1969a). Fatigue crack nucleation in metals. *Proc. Nat. Acad. Sci. U.S.* **62**, 631–635.
LIN, T. H., and ITO, Y. M. (1969b). Mechanics of a fatigue crack nucleation mechanism. *J. Mech. Phys. Solids* **17**, 511–523.
LIN, T. H., and LIEB, B. (1962). Rotation of crystals under axiel strain. *J. Mech. Phys. Solids* **10**, 65–70.
LIN, T. H., UCHIYAMA, S., and MARTIN, D. (1961). Stress field in metals at initial stage of plastic deformation. *J. Mech. Phys. Solids* **9**, 200–209.
LIVINGSTON, J. D., and CHALMERS, B. (1957). Multiple slip in bicrystal deformation. *Acta Met.* **5**, 322.
LOVE, A. E. H. (1927). "A Treatise on the Mathematical Theory of Elasticity," pp. 183–185. Dover, New York.
LÜCHE, K., and LANGE, H. (1952). Über die Form der Verfestigungskurve von Reinstaluminiumkristallen und die Bildung von Deformationsbändern. *Z. Metallk.* **43**, 55.
MACMILLAN, W. D. (1930). "The Theory of the Potential," pp. 24–25. Dover, New York.
MALMEISTER, A. (1956). Plastichnost' kvazilineinago tela (Plasticity of a quasilinear body). Sbovnik Voprosy dihamiki i dinamicheskoy prochosti (Collection Problems of Dynamics and Dynamic Strength). *Latv. PSR* **4**, 37–48.
MALMEISTER, A. (1965). Principles of theory of local strains. *Polym. Mech. (USSR)* **1**, No. 4, 12.
MORGAN, A. J. A. (1958). A proof of Duhamel's analogy for thermal stresses. *J. Aerosp. Sci.* **25**, 466–467.
MURA, T. (1967). "Continuum Theory of Dislocations and Plasticity." Springer, Berlin.
NABARRO, F. R. N. (1952). The mathematical theory of stationary dislocations. *Advan. Phys.* **1**, 299.
NAGHDI, P. M. (1960). Stress-strain relations in plasticity and thermoplasticity plasticity. *Proc. Symp. Nav. Struct. Mech., 2nd*, 1960 p. 121.
NAGHDI, P. M., and ROWLEY, J. C. (1954). An experimental study of biaxial stress-strain relations in plasticity. *J. Mech. Phys. Solids* **3**, 63–80.
NAGHDI, P. M., ESSENBURG, F., and KOFF, W. (1958). An experimental study of initial and subsequent yield surfaces in plasticity. *J. Appl. Mech.* **25**, 201.
NEUMANN, F. (1841). Die Gesetze der Doppelbrechung des Lichts in Comprimirten oder ungleichförmig erwärmten unkrystallinischen kröpern. *Abh. Akad. Wiss. Berlin*, Part 2, p. 1.
NOWACKI, W. (1962). "Thermoelasticity," pp. 32–38. Pergamon Press, Oxford.
OLSZAK, W., MROZ, M., and PERGGNA, P. (1963). "Recent Trends in the Development of the Theory of Plasticity," p. 6. Pergamon Press, Oxford.
OROWAN, E. (1934). Zur kristallplastizität I. Tieftemperaturplastizität und Beckersche Formerl. *Z. Phys.* **89**, 605.
PAYNE, H. (1959). The slip theory of plasticity for crystalline aggregates. *J. Mech. Phys. Solids* **7**, 126–134.
PHILLIPS, A. (1960). Pointed vertices in plasticity. *Proc. Symp. Nav. Struct. Mech.*, 1960 pp. 202–214.
PHILLIPS, A., and GRAY, G. (1958). Experimental investigation of corners in the yield surface. ONR Tech. Rep. No. 5. Yale University.
PHILLIPS, A., and KAECHELE, L. (1958). Combined stress tests in plasticity. *J. Appl. Mech.* **23**, 43.
POLANYI, M. (1934). Über eine Art Gitterstörung, die einen Kristall plastisch machen konnte. *Z. Phys.* **89**, 660.

PRAGER, W. (1955). The theory of plasticity: A survey of recent achievements. *Proc. Inst. Mech. Eng. (London)* 1955, **169**, 41–51.

PRAGER, W. (1956). Method of analyzing stresses and strains in workhardening plastic solids. *J. Appl. Mech.* **23**, 493.

PRIDE, R. A., and HEIMERL, E. J. (1952). Plastic buckling of simply supported plates in compression. *J. Aerosp. Sci.* **19**, 69–70.

QUINNEY, H., and TAYLOR, G. I. (1937). The emission of the latent energy due to previous cold working when a metal is heated. *Proc. Roy. Soc., Ser. A* **163**, 157–181.

READ, T. A. (1953). "Dislocations in Crystals," p. 36. McGraw-Hill, New York.

REISNER, H. (1931). Eigenspannungen und Eigenspannung squellen. *Z. Angew. Math. Mech.* **11**, 1–8.

ROUTH, E. J. (1922). "Analytical Statics," Vol. II, pp. 107–108. Cambridge Univ. Press, London and New York.

RYDEWSKI, J. R. (1962). "Introduction to Structural Problems in Nuclear Engineering," pp. 211–257. Macmillan, New York.

SACHS, G. (1928). Zur ableilung einer fleissbedingung. *VDI (Ver. Deut. Ing.) Z.* **72**, 734.

SANDERS, J. L., JR. (1954). Plastic stress-strain relations based on linear loading functions. *Proc. U.S. Nat. Congr. Appl. Mech., 2nd,* 1954, p. 455.

SCHMID, E. (1931). Beiträge Zur Physik und Metallographie Des Magnesiums. *Z. Elektrochem.* **37**, 447.

SCHMID, E., and BOAS, W. (1950). "Plasticity of Crystals," pp. 125 and 152. Hughes, London.

SOKOLNIKOFF, I. S. (1956). "Mathematical Theory of Elasticity," p. 336. McGraw-Hill, New York.

STOWELL, E. Z. (1948). A unified theory of plastic buckling of columns and plates. *Nat. Adv. Comm. Aeronaut., Tech. Rep.* **898**.

TAYLOR, G. I. (1934). The mechanism of plastic deformation of crystals. *Proc. Roy. Soc., Ser. A* **165**, 362–404.

TAYLOR, G. I. (1938). Plastic strain in metals. *J. Inst. Metals* **62**, No. 1, 307–324.

TAYLOR, G. I. (1956). Strains of crystalline aggregates. *Proc. Colloq. Deformation Flow Solids,* 1956, pp. 3–12.

TAYLOR, G. I., and ELAM, C. F. (1923). The distortion of an aluminum crystal during a tensile test. *Proc. Roy. Soc., Ser. A* **102**, 647–667.

TAYLOR, G. I., and QUINNEY, H. (1931). The plastic distortion of metals. *Phil. Trans. Roy. Soc. London* **230**, 323.

TIMOSHENKO, S. (1934). "Theory of Elasticity," pp. 213–215. McGraw-Hill, New York.

TITCHENER, A. L., and BEVER, M. B. (1958). The stored energy of cold work. *Progr. Metal Phys.* **7**, 247–338.

VON MISES, T. (1913). *Gottingen Nachr., Math.-Phys. Kl.,* p. 582.

ZIEGLER, H. (1959). A modification of Prager's hardening rule. *Quart. Appl. Math.* **17**, 55.

Thermodynamic Theory of Viscoplasticity

PIOTR PERZYNA

Institute of Fundamental Technical Research, Polish Academy of Sciences, Warsaw, Poland

I. Introduction

Four different approaches to a thermodynamic theory of a continuum can be distinguished, which differ from each other in the fundamental postulates on which the theory is based. None of these approaches is concerned with the atomic structure of the material; therefore, they represent a purely phenomenological approximation. All these theories are characterized by the same fundamental requirement that the results should be obtained without having recourse to statistical or kinetic methods.

Within each of these four approaches there are several methods of describing dissipative effects. The oldest and simplest way is to introduce a viscous stress which depends on the rate of strain. In this case the constitutive equations are differential relations. Another description of dissipation assumes constitutive equations of the rate type. This method is generally used in the theory of plasticity. A more general method introduces the assumption that the entire past history of the strain influences the stress in a manner compatible with the principle of fading memory. The constitutive equations here are of the functional type. A fourth method postulates the existence of internal state variables which influence of free energy and whose rate of change is governed by the differential equations in which the strain appears.

The principal postulates of the first approach, initiated by Onsager's work and usually called the classical thermodynamics of irreversible processes, are as follows (cf. Onsager, 1931a,b; Casimir, 1945; Eckart, 1940, 1948; Meixner, 1941a,b, 1942, 1943; see also De Groot, 1966; Meixner and Reik, 1959; De Groot and Mazur, 1962): (1) The principle of local state is assumed to be valid. (2) The Gibbs relation is satisfied. (3) The equation of entropy balance is assumed to involve a term expressing the entropy production which can be represented as a sum of products of fluxes and forces. This term is zero for a state of equilibrium and positive for an irreversible process. (4) The fluxes are functions of the forces, not necessarily linear. However, the Onsager–Casimir reciprocity relations concern only coefficients of the linear terms of the series expansions. Using this approach, a thermodynamic description of elastic, rheologic, and plastic materials was obtained (see papers by Bridgman, 1950; Biot, 1954, 1956, 1958; Drucker, 1964; Wehrli and Ziegler, 1962; Ziegler, 1957, 1958, 1961, 1962a,b, 1963a,b, 1964, 1966; Vakulenko, 1958, 1959; Dillon, 1963; Kluitenberg, 1962a,b,c, 1963, 1964, 1966, 1971; Eringen, 1960; Koh and Eringen, 1966; Kestin, 1966, 1968; Besseling, 1968; Onat, 1968). The description of the dissipative effects within the framework of the first approach, postulating the existence of internal state variables, was carried out by Valanis (1966, 1967, 1968). By applying the technique of the linearized thermodynamic theory of irreversible processes Valanis obtained a system of differential equations describing the behavior of a viscoelastic material.*

The second approach, called the rational thermodynamics or the thermodynamic theory of materials with memory, was initiated by the work of Coleman and Noll (1963). The fundamental postulates of this approach are as follows: (1) The temperature and entropy functions are assumed to exist for nonequilibrium states. (2) The principal restriction imposed on the constitutive equations is the Clausius–Duhem inequality. (3) The notion of the thermodynamic state is modified by assuming that the state of a given particle at time t is characterized, in general, by the time history of the local configuration of that particle. It should be

* A similar system of differential constitutive equations but only for small deformations has been given by Schapery (1964).

emphasized, however, that in particular cases the history of the local configuration of a particle can be determined by giving the actual values of this configuration and its time derivatives (cf., for instance, Coleman and Mizel, 1964). (4) No limitations are introduced for the processes considered. The constitutive equations are in general nonlinear. Within the framework of this approach, thermodynamic foundations for rheologic materials were established (cf. Coleman, 1964a,b; Coleman and Mizel, 1964; Truesdell, 1966a; Truesdell and Noll, 1965; Green and Adkins, 1960; Green and Zerna, 1968; Christensen and Naghdi, 1967; Wang and Bowen, 1966. The same was done for plastic materials also (cf. Green and Naghdi, 1965a,b, 1967, 1968; Green *et al.*, 1968; Dillon, 1967). The general thermodynamic theory with internal state variables within the framework of the second approach has been presented by Coleman and Gurtin 1967.

The third approach was developed by Meixner 1965, 1966, 1968a and is called the thermodynamic theory of passive systems. It is based on the following postulates: (1) The introduction of the notion of entropy is avoided for nonequilibrium states and the principle of local state is not assumed. (2) The Clausius–Duhem inequality is replaced by an inequality expressing the fundamental property of passivity. This inequality follows from the second law of thermodynamics and the condition of thermodynamic stability, and is known to apply only to states of equilibrium. (3) The temperature concept is assumed to exist for nonequilibrium states. (4) As a consequence of the fundamental inequality, the class of processes under consideration is limited to small deviations from the equilibrium conditions. This permits full linearization of the constitutive equations. An important feature of this approach is the clear physical meaning of all the quantities introduced.

Each of the three approaches above has its weaknesses and none is commonly accepted.* The first is subject to excessive limitations in the form of the assumptions of the Onsager–Casimir relations. Its present development does not appear promising for the difficulties that are encountered in nonlinear mechanics.† The second approach is criticized principally from the point of view of physical foundations (cf. Meixner, 1968a). Indeed, we must agree with the opinion that the problem of physical interpretation of quantities such as temperature or entropy has not found a detailed treatment within the framework of this approach.

* This fact was pointed out by a detailed discussion at the IUTAM Symposium on Irreversible Aspects of Continuum Mechanics in Vienna, June 1966 (cf. Meixner, 1968a; Truesdell, 1966b,c).

† A detailed critical analysis of this approach can be found in the monograph of Truesdell and Toupin (1960).

The advantage of the second approach is its well-developed mathematical foundation which offers the possibility of analysis of many interesting processes. It also works for nonlinear materials, and it is worth mentioning that the theories of elastic and viscoelastic materials can be derived as particular cases of the theory of materials with memory (cf. Coleman, 1964a,b). This theory permits the description of many important mechanical phenomena, such as elastic instability and phenomena accompanying wave propagation (cf. Truesdell, 1966b,c). The applicability of the third approach is, on the other hand, limited to linear problems. It does not seem likely that further generalization to nonlinear problems is possible within its framework. The results obtained concern problems of linear viscoelasticity only (cf. Meixner, 1964, 1965, 1966, 1968a).

Recent work concerned with axiomatic foundations of continuum thermodynamics has shown clearly the correctness of the conception of the second approach.* Although it did not remove the objections against the physical foundations of the theory, it has formulated in a mathematically accurate manner the conditions for the application of the methods of rational thermodynamics.

The fourth approach has recently been proposed by Meixner (1968b,c, 1969a,b,c). By considering electrical networks as a special example of thermodynamic systems, Meixner (1968c) proved that the entropy of an electrical network is not unique. From this result he concluded that the entropy concept in nonequilibrium becomes doubtful for all thermodynamic systems. Both classical thermodynamics of irreversible processes and rational thermodynamics used the concept of entropy in nonequilibrium. In connection with this Meixner (1969c) asks two questions: "How is it possible that classical thermodynamics of irreversible processes... has led to many reasonable and experimentally verified results if the concepts of entropy and entropy production, basic concepts of this theory, are not unique or not even applicable? If it is in fact true that these concepts are questionable, is it then possible to derive the laws of continuum physics without having recourse to these concepts?" By considering processes in simple thermodynamic materials Meixner (1969c) showed that the laws of continuum physics can be obtained by using the first and second laws of thermodynamics without speaking of a nonequilibrium entropy. This approach is based on a fundamental inequality which has the following features: (1) It does not

* The axiomatic foundations of continuum thermodynamics have been presented in the papers by Gurtin and Williams (1966, 1967a,b). They are a generalization to thermodynamic problems of the earlier conceptions of Noll (1958, 1959) concerning the purely mechanical theory of a continuum (cf. also Giles, 1964).

involve entropy in nonequilibrium. (2) In its derivation no thermodynamic stability conditions are used. (3) The state variables of the natural state at $t = -\infty$ do not explicitly enter the integrand. (4) In the integrand an auxiliary function is introduced which represents the entropy of a fictitious equilibrium state computed for the actual specific internal energy and for the actual deformation gradient. (5) It contains the temperature function defined in nonequilibrium.

Meixner (1969c) has shown that the classical thermodynamics of irreversible processes may be treated as a well defined special case of this *theory of processes in simple thermodynamic materials*. Note that this general theory developed by Meixner is valid for solid materials under arbitrary deformations.

The aim of the present paper is to review the thermodynamic foundations of the theory of viscoplasticity. The essential feature of viscoplasticity is the simultaneous description of rheologic and plastic effects of a material. The necessity for simultaneous consideration of viscoelastic and plastic properties of a material is indicated by the experimental investigations of dynamic loads. They clearly show that during dynamic loading of a test piece the plastic and viscoplastic effects are coupled and have equal importance. The viscous properties of the material introduce a time dependence of the states of stress and strain. The plastic properties, on the other hand, make these states depend on the deformation path. Different result will be obtained for different deformation paths and for different durations of the process.

The thermodynamic theory of elastic–viscoplastic materials presents for finite strains two basic difficulties.* The first of these is connected with the kinematic description of plastic deformation. The second difficulty concerns the problem of choice of thermodynamic variables of state. The important question arising here is whether the plastic deformation tensor may be treated as a thermodynamic state variable.

It appears that by investigating thermodynamic processes for finite deformation in viscoplastic materials, characterized by a nonlinearity resulting from dependence on time and path, a description within the framework of the second approach is possible.

In a series of papers, Green and Naghdi (1965a,b, 1967, 1968; Green *et al.*, 1968) developed the thermodynamic theory of plasticity for finite strains using the rate-type theory, and generalized the inviscid theory of plasticity to nonisothermal finite deformations. They used the principle of material frame indifference which they called "invariance

* For the notion of elastic-viscoplastic material, see Naghdi and Murch (1963) and Perzyna (1966a).

requirements under superposed rigid body motions," and fully explored the thermodynamical restrictions. In the first two papers Green and Naghdi (1965a,b) assumed the postulate of additivity of the elastic and inelastic parts of the deformation tensor. Green, McInnis, and Naghdi (1968) then pointed out that it is possible to construct an equivalent theory without the explicit introduction of the additivity postulate (cf. also Beckman, 1964; Lee and Liu, 1967).

Pipkin and Rivlin (1965) have presented the constitutive equations for rate-independent materials with memory, i.e., materials for which the stress at any instant of time depends on the deformation history, but not on the rate at which the deformation has been executed. They have shown that a general theory of elastic–plastic materials arises as a special case of the theory of rate-independent materials. This is a generalization of an idea first announced by Ilyushin (1954, 1961). Further development of the concept of rate independence was undertaken by Owen and Williams (1968) and by Owen (1968).

Perzyna (1968a) described an elastic–viscoplastic material as a material with memory, for which the history of the local configuration depends on time as well as the path.

Viscoplastic materials of the rate type, for which the path dependency is characterized by different sets of constitutive equations for loading and unloading processes, were discussed by Perzyna (1966b,c), Perzyna and Wojno (1966), Wojno (1967), and Green and Naghdi (1967).

The formulation of the thermodynamic theory of a rate-sensitive plastic material within the framework of thermodynamics of a material with internal state variables was given by Perzyna and Wojno (1968). In this thermodynamic theory of an inelastic material both difficulties mentioned before were taken into account. The deformation tensor and temperature were considered as thermodynamic state variables, while the components of the inelastic deformation tensor appear as internal state parameters (hidden parameters). No connection between the deformation tensor and the inelastic deformation tensor was postulated. The deformation tensor arises from the kinematics of the given body motion; the inelastic deformation tensor is determined by the solution of an initial-value problem for an ordinary first order differential equation.

A similar theory of rate-dependent plasticity was formulated independently by Kratochvil and Dillon (1968, 1969). They assumed that quantities which are related to motion and arrangement of dislocations in the material play the role of internal state variables.

II. Thermodynamics of a Material with Internal State Variables

Let us consider a body $\mathscr{B}$ with particles X and assume that this body may deform and conduct heat. We shall assume that couple stresses and body couples are absent.

The thermodynamic process of a body $\mathscr{B}$ is described by a set of functions $\{\boldsymbol{\chi}, \mathbf{T}, \mathbf{b}, \psi, \mathbf{q}, r, \eta, \vartheta, \alpha^{(i)}, \mathbf{A}^{(j)}\}$ of the particle X and time t. These functions have the following meaning. The function of the motion $\boldsymbol{\chi}(X, t)$ determines the spatial position occupied by the material point X at time t, which in the reference configuration $\mathscr{R}$ occupied the position $\mathbf{X}$, i.e.,

$$\mathbf{x} = \boldsymbol{\chi}(X, t) \tag{2.1}$$

The components of the function $\boldsymbol{\chi}$ are assumed to be continuously differentiable. The function $\mathbf{T}(X, t)$ is a symmetric Cauchy stress tensor, $\mathbf{b}(X, t)$ is the body force per unit mass, $\psi(X, t)$ denotes specific free energy per unit mass, $\mathbf{q}(X, t)$ is the heat flux vector, $r(X, t)$ is the heat supply per unit mass and unit time, $\eta(X, t)$ is the specific entropy, $\vartheta(X, t)$ is the local absolute temperature, and $\{\alpha^{(i)}, \mathbf{A}^{(j)}\}$ $(i = 1, 2,..., m;$ $j = 1, 2,..., n)$ are the internal state variables; $\alpha^{(i)}$ denotes the scalar internal parameters and $\mathbf{A}^{(j)}$ the internal state tensors. We assume that all $\mathbf{A}^{(j)}$ are symmetric second-order tensors.

Since we identify the material point X with its position $\mathbf{X}$ in the reference configuration $\mathscr{R}$, the deformation gradient $\mathbf{F}$ is determined by

$$\mathbf{F} = \text{grad}\, \boldsymbol{\chi}(\mathbf{X}, t) \tag{2.2}$$

where grad refers to the material coordinates $\mathbf{X}$. We shall introduce the second Piola–Kirchhoff stress tensor

$$\mathbf{T}_{\mathscr{R}} \equiv J\mathbf{F}^{-1}\mathbf{T}(\mathbf{F}^{-1})^T \tag{2.3}$$

where $J = \det \mathbf{F} > 0$. Similarly, the heat flux vector per unit surface in the reference configuration $\mathscr{R}$ will be defined as follows

$$\mathbf{q}_{\mathscr{R}} \equiv J\mathbf{F}^{-1}\mathbf{q} \tag{2.4}$$

The set of functions

$$\boldsymbol{\chi}, \mathbf{T}_{\mathscr{R}}, \mathbf{b}, \psi, \mathbf{q}_{\mathscr{R}}, r, \eta, \vartheta, \alpha^{(i)}, \mathbf{A}^{(j)},$$

defined for every particle X in $\mathscr{B}$ and for any time t, is called the thermodynamic process in $\mathscr{B}$ if, and only if, it is compatible with the condition for the balance of linear momentum (Cauchy's first law of motion)

$$\text{div}(\mathbf{F}\mathbf{T}_{\mathscr{R}}) + \rho_{\mathscr{R}}\mathbf{b} = \rho_{\mathscr{R}}\ddot{\mathbf{x}}, \tag{2.5}$$

and with the balance of energy (the first law of thermodynamics)

$$\tfrac{1}{2}\,\mathrm{tr}(\mathbf{T}_{\mathscr{R}}\dot{\mathbf{C}}) - \mathrm{div}\,\mathbf{q}_{\mathscr{R}} - \rho_{\mathscr{R}}(\dot{\psi} + \vartheta\dot{\eta} + \dot{\vartheta}\eta) + \rho_{\mathscr{R}} r = 0, \tag{2.6}$$

where the operator div refers to the material coordinates $\mathbf{X}$, $\rho_{\mathscr{R}}$ is the mass density in the reference configuration $\mathscr{R}$, the dot denotes *material differentiation* with respect to time, and $\mathbf{C}$ is the right Cauchy–Green deformation tensor

$$\mathbf{C} = \mathbf{F}^T\mathbf{F}. \tag{2.7}$$

In order to define a thermodynamic process in $\mathscr{B}$, it suffices to prescribe the set of functions omitting $\mathbf{b}$ and r. These two functions are then uniquely determined by the Eqs. (2.5) and (2.6).

We shall require that for a thermodynamic process in $\mathscr{B}$ the thermodynamic postulate be satisfied at any time t. This is equivalent to the following inequality*

$$-\dot{\psi} - \dot{\vartheta}\eta + (1/2\rho_{\mathscr{R}})\,\mathrm{tr}(\mathbf{T}_{\mathscr{R}}\dot{\mathbf{C}}) - (1/\rho_{\mathscr{R}}\vartheta)\,\mathbf{q}_{\mathscr{R}}\cdot\mathbf{g} \geqslant 0 \tag{2.8}$$

which must be satisfied for every particle X in $\mathscr{B}$. (In inequality (2.8) $\mathbf{g}$ stands for grad ϑ.)

Similarly, we shall assume that all constitutive equations describing the physical properties of a material satisfy the principle of material frame indifference as formulated by Noll (cf. Noll, 1958; Truesdell and Noll, 1965).

For definiteness, we assume here that the internal state parameters $\{\alpha^{(i)}, \mathbf{A}^{(j)}\}$ remain invariant upon change of frame.

A material with internal state variables is characterized at particle X by constitutive equations as follows†

$$\begin{aligned}
\psi &= \hat{\psi}(\mathbf{C}, \vartheta, \mathbf{g}, \alpha^{(i)}, \mathbf{A}^{(j)}),\\
\eta &= \hat{\eta}(\mathbf{C}, \vartheta, \mathbf{g}, \alpha^{(i)}, \mathbf{A}^{(j)}),\\
\mathbf{T}_{\mathscr{R}} &= \hat{\mathbf{T}}_{\mathscr{R}}(\mathbf{C}, \vartheta, \mathbf{g}, \alpha^{(i)}, \mathbf{A}^{(j)}),\\
\mathbf{q}_{\mathscr{R}} &= \hat{\mathbf{q}}_{\mathscr{R}}(\mathbf{C}, \vartheta, \mathbf{g}, \alpha^{(i)}, \mathbf{A}^{(j)}),\\
\dot{\alpha}^{(i)} &= \hat{\dot{\alpha}}^{(j)}(\mathbf{C}, \vartheta, \mathbf{g}, \alpha^{(i)}, \mathbf{A}^{(j)}),\\
\dot{\mathbf{A}}^{(j)} &= \hat{\dot{\mathbf{A}}}^{(j)}(\mathbf{C}, \vartheta, \mathbf{g}, \alpha^{(i)}, \mathbf{A}^{(j)}).
\end{aligned} \tag{2.9}$$

* This inequality is implied by the Clausius–Duhem inequality (cf. Coleman and Noll, 1963).

† See Coleman and Gurtin (1967) and Valanis (1968). In the basic system of constitutive equations describing material with internal state variables, Coleman and Gurtin (1967)

Obviously we use in (2.9) the principle of equipresence formulated by Truesdell (cf. Coleman and Mizel, 1964; Truesdell and Noll, 1965).

We shall say that the thermodynamic process in $\mathscr{B}$ described by the system of functions $\{\boldsymbol{\chi}, \mathbf{T}_{\mathscr{R}}, \psi, \mathbf{q}_{\mathscr{R}}, \eta, \vartheta, \alpha^{(i)}, \mathbf{A}^{(j)}\}$ is called an admissible process in $\mathscr{B}$ if it is compatible with the constitutive equations (2.9) at each point X of $\mathscr{B}$ and for all time t.

To investigate the restrictions imposed on the constitutive equations (2.9) by the thermodynamic postulate (2.8) we compute

$$\dot{\psi} = \operatorname{tr}(\partial_{\mathbf{C}}\hat{\psi}\dot{\mathbf{C}}) + \partial_{\vartheta}\hat{\psi}\dot{\vartheta} + \partial_{\mathbf{g}}\hat{\psi}\cdot\dot{\mathbf{g}} + \sum_{i=1}^{m} \partial_{\alpha^{(i)}}\hat{\psi}\dot{\alpha}^{(i)} + \sum_{j=1}^{n} \operatorname{tr}(\partial_{\mathbf{A}^{(j)}}\hat{\psi}\dot{\mathbf{A}}^{(j)}) \tag{2.10}$$

Substitution of (2.10) into (2.8) gives

$$(1/2\rho_{\mathscr{R}}) \operatorname{tr}[(\mathbf{T}_{\mathscr{R}} - 2\rho_{\mathscr{R}}\partial_{\mathbf{C}}\hat{\psi})\dot{\mathbf{C}}] - (\partial_{\vartheta}\hat{\psi} + \eta)\dot{\vartheta} - \partial_{\mathbf{g}}\hat{\psi}\cdot\dot{\mathbf{g}}$$

$$- \sum_{i=1}^{m} \partial_{\alpha^{(i)}}\hat{\psi}\dot{\alpha}^{(i)} - \sum_{j=1}^{n} \operatorname{tr}[\partial_{\mathbf{A}^{(j)}}\hat{\psi}\dot{\mathbf{A}}^{(j)}] - (1/\rho_{\mathscr{R}}\vartheta)\,\mathbf{q}_{\mathscr{R}}\cdot\mathbf{g} \geqslant 0. \tag{2.11}$$

By choosing arbitrary values $\dot{\mathbf{C}}$, $\dot{\vartheta}$, and $\dot{\mathbf{g}}$ it is possible to determine an admissible thermodynamic process in $\mathscr{B}$ (cf. Coleman and Gurtin, 1967; Valanis, 1968). Hence, to satisfy the inequality (2.11) we must assume

$$\partial_{\mathbf{g}}\hat{\psi} = \mathbf{0}, \tag{2.12}$$

$$\mathbf{T}_{\mathscr{R}} = 2\rho_{\mathscr{R}}\partial_{\mathbf{C}}\hat{\psi}(\mathbf{C}, \vartheta, \alpha^{(i)}, \mathbf{A}^{(j)}), \tag{2.13}$$

$$\eta = -\partial_{\vartheta}\hat{\psi}(\mathbf{C}, \vartheta, \alpha^{(i)}, \mathbf{A}^{(j)}), \tag{2.14}$$

$$\sum_{i=1}^{m} \partial_{\alpha^{(i)}}\hat{\psi}\dot{\alpha}^{(i)} + \sum_{j=1}^{n} \operatorname{tr}[\partial_{\mathbf{A}^{(j)}}\hat{\psi}\dot{\mathbf{A}}^{(j)}] + (1/\rho_{\mathscr{R}}\vartheta)\,\hat{\mathbf{q}}_{\mathscr{R}}(\mathbf{C}, \vartheta, \mathbf{g}, \alpha^{(i)}, \mathbf{A}^{(j)})\cdot\mathbf{g} \leqslant 0. \tag{2.15}$$

The relation (2.12) implies that the response function of free energy $\hat{\psi}$ is independent of $\mathbf{g} = \operatorname{grad}\vartheta$. The equations (2.13) and (2.14) determine the response functions for stress tensor $\hat{\mathbf{T}}_{\mathscr{R}}$ and for entropy $\hat{\eta}$ by means of the response function $\hat{\psi}(\mathbf{C}, \vartheta, \alpha^{(i)}, \mathbf{A}^{(j)})$. The inequality (2.15) is called the general dissipation inequality for a material with internal state variables.

---

and Valanis (1968) assumed the vector parameter $\boldsymbol{\alpha} = (\alpha_1, \ldots, \alpha_n)$ defining the internal state. The introduction of the tensorial internal parameters is due to Perzyna and Wojno (1968) and Kratochvil and Dillon (1968, 1969). For our purpose, it is convenient to introduce simultaneously the scalar and tensorial internal parameters $\{\alpha^{(i)}, \mathbf{A}^{(j)}\}$ (cf. Kratochvil and Dillon, 1968, 1969).

It will be useful to introduce the definition of the internal dissipation function as follows

$$\sigma = \hat{\sigma}(\mathbf{C}, \vartheta, \mathbf{g}, \alpha^{(i)}, \mathbf{A}^{(j)}) \tag{2.16}$$

$$= -\frac{1}{\vartheta}\left\{\sum_{i=1}^{m} \partial_{\alpha^{(i)}}\hat{\psi}\hat{\dot{\alpha}}(\mathbf{C}, \vartheta, \mathbf{g}, \alpha^{(i)}, \mathbf{A}^{(j)}) + \sum_{j=1}^{n} \operatorname{tr}[\partial_{\mathbf{A}^{(j)}}\hat{\psi}\hat{\dot{\mathbf{A}}}^{(j)}(\mathbf{C}, \vartheta, \mathbf{g}, \alpha^{(i)}, \mathbf{A}^{(j)})]\right\}.$$

The general dissipation inequality (2.15) implies that for $\mathbf{g} = 0$ the internal dissipation inequality is satisfied

$$\hat{\sigma}(\mathbf{C}, \vartheta, \mathbf{0}, \alpha^{(i)}, \mathbf{A}^{(j)}) \geqslant 0. \tag{2.17}$$

III. General Description of an Elastic–Viscoplastic Material

Experimental investigations of dynamical properties of materials have shown that many materials behave differently under dynamic loading and static loading. These results have also proved that the basic reason for these differences is the strain-rate sensitivity of the material investigated. The influenced of the strain-rate effect may be taken into account within the framework of viscoplasticity. Every material displays more or less definite viscous properties. For many materials, however, these properties are more pronounced after the plastic state has been reached. In these cases it may be assumed that the material displays viscous properties in the plastic range only. We assume here that before yielding, material has only elastic properties, and after yielding has viscoelastic and plastic properties. This is why we shall call the rate-sensitive plastic material an elastic–viscoplastic one.

General foundations for the study of problems connected with rate-sensitive plastic material were given by Hohenemser and Prager (1932). Further development of this idea is contained in the papers (Bodner, 1968; Olszak and Perzyna, 1966; Perzyna, 1963a,b, 1964, 1966a, 1968a,b,c; Perzyna and Wierzbicki, 1964; Perzyna and Wojno, 1966, 1968).*

The object of the present section is to discuss a general description of an elastic–viscoplastic material within the framework of thermodynamics with internal state variables.

From the relation for internal dissipation (2.16) it can be seen that the internal parameters $\{\alpha^{(i)}, \mathbf{A}^{(j)}\}$ are introduced to describe the internal

* Perzyna (1966a) includes an account of the previous developments with infinitesimal deformations, as well as related experimental aspects of the subject.

dissipation of the material. For an elastic–viscoplastic material, such parameters will be the work-hardening parameter $\varkappa$, the symmetric, inelastic deformation tensor $\mathbf{P}$, and the dislocation density tensors $\mathbf{\Gamma}^{(k)}$. Thus, we assume

$$\{\alpha^{(i)}, \mathbf{A}^{(j)}\} = \{\varkappa, \mathbf{P}, \mathbf{\Gamma}^{(k)}\},$$

$$i = 1; \qquad j = 1, 2,..., n; \qquad k = 1, 2,..., n - 1.$$

From previous research in viscoplasticity, we know that the rate of the inelastic deformation tensor $\dot{\mathbf{P}}$ is proportional to the function $\Phi(\mathscr{F})$, where $\mathscr{F}$ is the static yield function (cf. Perzyna, 1963a,b). Since an elastic–viscoplastic material before yielding has only elastic properties, the initial yield condition can be assumed to be similar to that used in the inviscid theory of plasticity. Thus, the static yield condition can be taken as follows:

$$\mathscr{F} = (1/\varkappa) f(\mathbf{T}_{\mathscr{R}}, \vartheta, \mathbf{P}, \mathbf{\Gamma}^{(k)}) - 1, \tag{3.1}$$

where the isotropic work-hardening parameter $\varkappa$ is determined by the differential equation

$$\dot{\varkappa} = \operatorname{tr}\{\mathbf{N}(\mathbf{T}_{\mathscr{R}}, \vartheta, \varkappa, \mathbf{P}, \mathbf{\Gamma}^{(k)})\dot{\mathbf{P}}\}. \tag{3.2}$$

This equation for $\varkappa$ satisfies the condition

$$\dot{\varkappa} = 0 \qquad \text{when} \quad \dot{\mathbf{P}} = \mathbf{0}, \tag{3.3}$$

which is generally assumed in the theory of plasticity. It means that the work-hardening effect disappears when there is no increment of plastic deformation. The functions f and $\mathbf{N}$ are given tensor functions. The function $\Phi(\mathscr{F})$ may be chosen to represent the results of tests of the dynamical behavior of materials. At the same time, the proper choice of $\Phi(\mathscr{F})$ provides a description of the influence of the rate of deformation and the temperature on the yield limit of the material. It is postulated that the following differential equation determines the internal state parameter, tensor $\mathbf{P}$, for an elastic–viscoplastic material:

$$\dot{\mathbf{P}} = \gamma(\vartheta)\langle\Phi(\mathscr{F})\rangle\, \mathbf{M}(\mathbf{T}_{\mathscr{R}}, \vartheta, \mathbf{g}, \varkappa, \mathbf{P}, \mathbf{\Gamma}^{(k)}), \tag{3.4}$$

where $\gamma(\vartheta)$ is a temperature-dependent viscosity coefficient of the material, the symbol $\langle\Phi(\mathscr{F})\rangle$ is defined as follows

$$\langle\Phi(\mathscr{F})\rangle = \begin{cases} 0 & \text{for} \quad \mathscr{F} \leqslant 0, \\ \Phi(\mathscr{F}) & \text{for} \quad \mathscr{F} > 0, \end{cases} \tag{3.5}$$

and $\mathbf{M}$ denotes a symmetric tensor function (i.e., $\mathbf{M} = \mathbf{M}^T$). To ensure that the internal state tensor $\mathbf{P}$ is invariant upon a change of frame, it is sufficient to assume that the tensor function $\mathbf{M}$ is invariant.

Equation (3.4) postulates that the rate of change of the inelastic deformation tensor $\dot{\mathbf{P}}$ is a function of the excess of stress over the static yield condition. In Section VIII we shall show that the Eq. (3.4) is a generalization of the known relation for shear strain rate in the physical theory of dislocation, for face centered cubic crystals, based on the thermally activated process.

The constitutive assumption (3.4) yields the following dynamical yield criterion

$$f(\mathbf{T}_{\mathscr{R}}, \vartheta, \mathbf{P}, \mathbf{\Gamma}^{(k)}) = \varkappa\{1 + \Phi^{-1}[(\operatorname{tr} \dot{\mathbf{P}}^2)^{1/2}/(\gamma(\vartheta))(\operatorname{tr} \mathbf{M}^2)^{-1/2}]\}. \tag{3.6}$$

This relation may be interpreted as a description of actual change of the yield surface during the thermodynamic process. This change is caused by isotropic and anisotropic work-hardening effects, and by the influence of the rate of change of the inelastic deformation tensor and temperature on the yield limit of a material. We note here that the relation (3.6) constitutes a basis for experimental investigations which seek to examine the theoretical assumptions.

During the yielding process the usual arrangement of dislocations is changed. The applied stress must in general be increased for additional yielding to occur. These changes in the dislocation arrangement can be expressed in terms of the changes of the quantities $\mathbf{\Gamma}^{(k)}$. Thus we assume

$$\dot{\mathbf{\Gamma}}^{(k)} = \mathbf{L}^{(k)}(\mathbf{C}, \vartheta, \mathbf{g}, \varkappa, \mathbf{P}, \mathbf{\Gamma}^{(k)})[\dot{\mathbf{P}}]. \tag{3.7}$$

The proportionality of $\dot{\mathbf{\Gamma}}^{(k)}$ and $\dot{\mathbf{P}}$ in (3.7) expresses the fact that changes in the dislocation arrangement are produced during plastic deformation only.

The dislocation arrangement tensors $\mathbf{\Gamma}^{(k)}$ were first introduced into the theory of plasticity for the description of the structural state by Kröner (1960, 1962, 1966); cf. also Kratochvil and Dillon (1968, 1969). These tensors are interpreted as the excess dislocation density $\mathbf{\Gamma}^{(1)}$, the excess dislocation loop density $\mathbf{\Gamma}^{(2)}$, the excess dislocation pair loop density $\mathbf{\Gamma}^{(3)}$, etc. One could continue in this manner and get the infinite and complete set of state variables which would describe any detail of the dislocation arrangement. Since it is not likely that all the particulars on a microscopic scale are important for the macroscopic viscoplastic behavior, but rather that certain averages only are macroscopically effective, one is inclined to introduce a finite set of the variables $\mathbf{\Gamma}^{(k)}$.

In accordance with the previous results for a material with internal state variables and the constitutive assumptions introduced, the full system of constitutive equations for an elastic–viscoplastic material may now be written as follows:

$$\psi = \hat{\psi}(\mathbf{C}, \vartheta, \varkappa, \mathbf{P}, \boldsymbol{\Gamma}^{(k)}), \tag{3.8}$$

$$\mathbf{T}_{\mathscr{R}} = 2\rho_{\mathscr{R}}\partial_{\mathbf{C}}\hat{\psi}, \tag{3.9}$$

$$\eta = -\partial_{\vartheta}\hat{\psi}, \tag{3.10}$$

$$\mathbf{q}_{\mathscr{R}} = \hat{\mathbf{q}}_{\mathscr{R}}(\mathbf{C}, \vartheta, \mathbf{g}, \varkappa, \mathbf{P}, \boldsymbol{\Gamma}^{(k)}), \tag{3.11}$$

$$\dot{\mathbf{P}} = \gamma(\vartheta)\langle\Phi(\mathscr{F})\rangle\, \mathbf{M}(\mathbf{T}_{\mathscr{R}}, \vartheta, \mathbf{g}, \varkappa, \mathbf{P}, \boldsymbol{\Gamma}^{(k)}), \tag{3.12}$$

$$\dot{\varkappa} = \operatorname{tr}\{\mathbf{N}(\mathbf{T}_{\mathscr{R}}, \vartheta, \varkappa, \mathbf{P}, \boldsymbol{\Gamma}^{(k)})\dot{\mathbf{P}}\}, \tag{3.13}$$

$$\dot{\boldsymbol{\Gamma}}^{(k)} = \mathbf{L}^{(k)}(\mathbf{C}, \vartheta, \mathbf{g}, \varkappa, \mathbf{P}, \boldsymbol{\Gamma}^{(k)})[\dot{\mathbf{P}}]. \tag{3.14}$$

Thus, an elastic–viscoplastic material is described in a thermodynamic process by the response functions $\hat{\psi}$, $\hat{\mathbf{q}}_{\mathscr{R}}$, $\Phi(\mathscr{F})$, $\mathbf{M}, \mathbf{N}$, $\mathbf{L}^{(k)}$, and by the viscosity coefficient $\gamma(\vartheta)$.

As a result of the relation (3.9), the differential equations (3.12)–(3.14) which determined the internal state parameters $\varkappa$, $\mathbf{P}$, and $\boldsymbol{\Gamma}^{(k)}$ for an elastic–viscoplastic material can be written in the following form:

$$\dot{\mathbf{P}} = \mathbf{G}(\mathbf{C}, \vartheta, \mathbf{g}, \varkappa, \mathbf{P}, \boldsymbol{\Gamma}^{(k)}), \tag{3.15}$$

$$\dot{\varkappa} = \beta(\mathbf{C}, \vartheta, \mathbf{g}, \varkappa, \mathbf{P}, \boldsymbol{\Gamma}^{(k)}), \tag{3.16}$$

$$\dot{\boldsymbol{\Gamma}}^{(k)} = \boldsymbol{\Omega}^{(k)}(\mathbf{C}, \vartheta, \mathbf{g}, \varkappa, \mathbf{P}, \boldsymbol{\Gamma}^{(k)}). \tag{3.17}$$

Now we can discuss the important features of the constitutive equations for an elastic–viscoplastic material, (3.8)–(3.14). The internal parameters $\varkappa$ and $\boldsymbol{\Gamma}^{(k)}$ describe the inviscid plastic properties of a material; hence the differential equations which define them, (3.13), (3.14), are independent of a change of the time scale. The inelastic deformation tensor $\mathbf{P}$ represents the tensorial internal parameter responsible for the viscous effects generated by thermal activations during plastic deformation. Thus, the differential equation (3.12) depends on a change of the time scale and involves viscoplastic effects. As a result of simultaneous descriptions of viscous and plastic properties of a material we have the differential equations for internal state parameters in the form (3.15)–(3.17). All these equations depend on a change of the time scale and are typical for a rate-sensitive plastic material.

The equations (3.15)–(3.17) may be considered as the definitions of the inelastic deformation tensor $\mathbf{P}$, the work-hardening parameter $\varkappa$, and the dislocation arrangement tensors $\boldsymbol{\Gamma}^{(k)}$, respectively. These equations show that the present theory of an elastic–viscoplastic material takes into account the history of the deformation tensor **C**, the temperature ϑ, and the temperature gradient **g**. This is implied by the following fact: In order to integrate the differential equations (3.15)–(3.17), and to determine the actual values of the inelastic deformation tensor $\mathbf{P}(t)$, the work-hardening parameter $\varkappa(t)$, and the dislocation arrangement tensors $\mathbf{\Gamma}^{(k)}$ at X in $\mathscr{B}$, we must know the initial values $\mathbf{P}^0$, $\varkappa^0$, and $\mathbf{\Gamma}^{(k)0}$ and the full histories of **C**, ϑ, and **g**.

Due to the fact that the response function for the heat flux vector $\hat{\mathbf{q}}_{\mathscr{R}}$ depends on the history of the deformation gradient **g**, the present theory can describe the heat conduction in an elastic–viscoplastic material with finite wave speeds for thermal disturbances.*

In the present theory, the general dissipation inequality has the form

$$\partial_\varkappa\hat{\psi}\dot{\varkappa} + \mathrm{tr}[\partial_{\mathbf{P}}\hat{\psi}\dot{\mathbf{P}}] + \sum_{k=1}^{n-1} \mathrm{tr}(\partial_{\mathbf{\Gamma}^{(k)}}\hat{\psi}\dot{\mathbf{\Gamma}}^{(k)}) + (1/\rho_{\mathscr{R}}\vartheta)\,\hat{\mathbf{q}}_{\mathscr{R}}\cdot\mathbf{g} \leqslant 0, \tag{3.18}$$

and ensures fulfillment of the thermodynamic postulate (2.8).

The internal dissipation function σ is determined by the relation

$$\begin{aligned}\sigma &= \hat{\sigma}(\mathbf{C}, \vartheta, \mathbf{g}, \varkappa, \mathbf{P}, \mathbf{\Gamma}^{(k)}),\\ &= -(1/\vartheta)\left\{\partial_\varkappa\hat{\psi}\dot{\varkappa} + \mathrm{tr}[\partial_{\mathbf{P}}\hat{\psi}\dot{\mathbf{P}}] + \sum_{k=1}^{n-1}\mathrm{tr}[\partial_{\mathbf{\Gamma}^{(k)}}\hat{\psi}\dot{\mathbf{\Gamma}}^{(k)}]\right\},\\ &= -(1/\vartheta)\left\{\partial_\varkappa\hat{\psi}\beta + \mathrm{tr}[\partial_{\mathbf{P}}\hat{\psi}\mathbf{G}] + \sum_{k=1}^{n-1}\mathrm{tr}[\partial_{\mathbf{\Gamma}^{(k)}}\hat{\psi}\mathbf{\Omega}^{(k)}]\right\}.\end{aligned}$$

IV. Elastic–Plastic Material

The dynamical yield condition (3.6) implies that an elastic–viscoplastic material loses its strain rate sensitivity if, and only if, the viscosity coefficient $\gamma(\vartheta) \to \infty$. In this case the static yield condition

$$\mathscr{F} = 0 \tag{4.1}$$

is satisfied, the material loses its viscosity, and behaves as an elastic–plastic material. From the definition of the symbol $\langle\Phi(\mathscr{F})\rangle$ [cf. the

* This can be compared with a general theory of heat conduction with finite wave speed for nonlinear materials with memory, presented by Gurtin and Pipkin (1968).

relation (3.5)], we see that the differential equation determining the plastic deformation tensor **P** takes the following form

$$\dot{\mathbf{P}} = \Lambda \mathbf{M}(\mathbf{T}_{\mathscr{R}}, \vartheta, \mathbf{g}, \varkappa, \mathbf{P}, \mathbf{\Gamma}^{(k)}), \tag{4.2}$$

where the parameter $\Lambda = \gamma(\vartheta)\langle\Phi(\mathscr{F})\rangle$ may be determined from the condition that the point in the temperature–stress space representing the actual state of temperature and stress lies on the yield surface

$$f(\mathbf{T}_{\mathscr{R}}, \vartheta, \mathbf{P}, \mathbf{\Gamma}^{(k)}) = \varkappa. \tag{4.3}$$

Here the work-hardening parameter $\varkappa$ is determined by the relation [cf. the definition (3.2)]

$$\dot{\varkappa} = \mathrm{tr}\{\mathbf{N}(\mathbf{T}_{\mathscr{R}}, \vartheta, \varkappa, \mathbf{P}, \mathbf{\Gamma}^{(k)})\dot{\mathbf{P}}\}. \tag{4.4}$$

We assume that the dislocation arrangement tensors $\mathbf{\Gamma}^{(k)}$ are determined by the differential equations (3.7).

From the yield condition (4.3) and the definitions (4.4) and (3.7), we can deduce the following criterion of loading

$$f = \varkappa \qquad \text{and} \qquad \mathrm{tr}(\partial_{\mathbf{T}_{\mathscr{R}}} f \dot{\mathbf{T}}_{\mathscr{R}}) + \partial_{\vartheta} f \dot{\vartheta} > 0. \tag{4.5}$$

Similarly, the criteria

$$f = \varkappa \qquad \text{and} \qquad \mathrm{tr}(\partial_{\mathbf{T}_{\mathscr{R}}} f \dot{\mathbf{T}}_{\mathscr{R}}) + \partial_{\vartheta} f \dot{\vartheta} \leqslant 0 \tag{4.6}$$

define the unloading and neutral state, respectively.

To satisfy the condition that the point representing the actual state of loading and temperature lies on the yield surface, it suffices to fulfil $\dot{f} = \dot{\varkappa}$, i.e.,

$$\mathrm{tr}(\partial_{\mathbf{T}_{\mathscr{R}}} f \dot{\mathbf{T}}_{\mathscr{R}}) + \partial_{\vartheta} f \dot{\vartheta} + \mathrm{tr}(\partial_{\mathbf{P}} f \dot{\mathbf{P}}) + \sum_{k=1}^{n-1} \mathrm{tr}(\partial_{\mathbf{\Gamma}^{(k)}} f \dot{\mathbf{\Gamma}}^{(k)}) = \mathrm{tr}(\mathbf{N}\dot{\mathbf{P}}). \tag{4.7}$$

Using (3.7) and (4.2), we have from (4.7)

$$\Lambda = \lambda[\mathrm{tr}(\partial_{\mathbf{T}_{\mathscr{R}}} f \dot{\mathbf{T}}_{\mathscr{R}}) + \partial_{\vartheta} f \dot{\vartheta}], \tag{4.8}$$

where

$$\lambda = \left\{\mathrm{tr}\left[\left(\mathbf{N} - \partial_{\mathbf{P}} f - \sum_{k=1}^{n-1} \partial_{\mathbf{\Gamma}^{(k)}} f \mathbf{L}^{(k)}\right) \mathbf{M}\right]\right\}^{-1}. \tag{4.9}$$

We shall assume the condition

$$\lambda > 0. \tag{4.10}$$

The relation for Λ (4.8), and the criteria of loading, unloading, and neutral state have shown that the differential equation determining the internal state tensor **P** for an elastic–plastic material (4.2) can be written as follows:

$$\dot{\mathbf{P}} = \lambda\langle \operatorname{tr}(\partial_{\mathbf{T}_{\mathscr{R}}} f \dot{\mathbf{T}}_{\mathscr{R}}) + \partial_\vartheta f \dot{\vartheta}\rangle\, \mathbf{M}(\mathbf{T}_{\mathscr{R}}, \vartheta, \mathbf{g}, \varkappa, \mathbf{P}, \mathbf{\Gamma}^{(k)}), \tag{4.11}$$

where the symbol $\langle \operatorname{tr}(\partial_{T_{\mathscr{R}}} f \dot{\mathbf{T}}_{\mathscr{R}}) + \partial_\vartheta f \dot{\vartheta}\rangle$ is defined by

$$\langle[\operatorname{tr}(\partial_{\mathbf{T}_{\mathscr{R}}} f \dot{\mathbf{T}}_{\mathscr{R}}) + \partial_\vartheta f \dot{\vartheta}]\rangle = \begin{cases} [\;] & \text{if} \quad f = \varkappa \quad \text{and} \quad [\;] > 0, \\ 0 & \text{if} \quad f = \varkappa \quad \text{and} \quad [\;] \leqslant 0, \\ & \text{or if} \quad f < \varkappa. \end{cases} \tag{4.12}$$

The full system of constitutive equations describing the behavior of an elastic–plastic material during the thermodynamic process at a material point X in $\mathscr{B}$ has the form*

$$\psi = \hat{\psi}(\mathbf{C}, \vartheta, \varkappa, \mathbf{P}, \mathbf{\Gamma}^{(k)}), \tag{4.13a}$$

$$\mathbf{T}_{\mathscr{R}} = 2\rho_{\mathscr{R}} \partial_{\mathbf{C}} \hat{\psi}, \tag{4.13b}$$

$$\eta = -\partial_\vartheta \hat{\psi}, \tag{4.13c}$$

$$\mathbf{q}_{\mathscr{R}} = \hat{\mathbf{q}}_{\mathscr{R}}(\mathbf{C}, \vartheta, \mathbf{g}, \varkappa, \mathbf{P}, \mathbf{\Gamma}^{(k)}), \tag{4.13d}$$

$$\dot{\mathbf{P}} = \mathbf{H}_1(\mathbf{C}, \vartheta, \mathbf{g}, \varkappa, \mathbf{P}, \mathbf{\Gamma}^{(k)})[\dot{\mathbf{C}}] + \mathbf{H}_2(\mathbf{C}, \vartheta, \mathbf{g}, \varkappa, \mathbf{P}, \mathbf{\Gamma}^{(k)})\dot{\vartheta}, \tag{4.13e}$$

$$\dot{\varkappa} = \operatorname{tr}\{\mathbf{N}_1(\mathbf{C}, \vartheta, \mathbf{g}, \varkappa, \mathbf{P}, \mathbf{\Gamma}^{(k)})\dot{\mathbf{C}}\} + \beta_1(\mathbf{C}, \vartheta, \mathbf{g}, \varkappa, \mathbf{P}, \mathbf{\Gamma}^{(k)})\dot{\vartheta}, \tag{4.13f}$$

$$\dot{\mathbf{\Gamma}}^{(k)} = \mathbf{L}_1^{(k)}(\mathbf{C}, \vartheta, \mathbf{g}, \varkappa, \mathbf{P}, \mathbf{\Gamma}^{(k)})[\dot{\mathbf{C}}] + \mathbf{L}_2^{(k)}(\mathbf{C}, \vartheta, \mathbf{g}, \varkappa, \mathbf{P}, \mathbf{\Gamma}^{(k)})\dot{\vartheta}, \tag{4.13g}$$

where the new tensor functions $\mathbf{H}_1$, $\mathbf{H}_2$, $\mathbf{N}_1$, β_1, $\mathbf{L}_1^{(k)}$ and $\mathbf{L}_2^{(k)}$ are determined by substitution of the relation (4.13b) into the equations (4.11), (4.4) and (3.7).

Equations (4.13e–g) are the definitions of the internal state parameters for an elastic–plastic material. It should be pointed out that all equations (4.13) are invariant under a change of time scale.

To satisfy the thermodynamic postulate of an elastic–plastic material, the constitutive equations (4.13) should fulfill the general dissipation inequality

$$\partial_\varkappa \hat{\psi}[\operatorname{tr}(\mathbf{N}_1\dot{\mathbf{C}}) + \beta_1\dot{\vartheta}] + \operatorname{tr}[\partial_{\mathbf{P}}\hat{\psi}(\mathbf{H}_1[\dot{\mathbf{C}}] + \mathbf{H}_2\dot{\vartheta})]$$
$$+ \sum_{k=1}^{n-1} \operatorname{tr}\{\partial_{\mathbf{\Gamma}^{(k)}}\hat{\psi}(\mathbf{L}_1^{(k)}[\dot{\mathbf{C}}] + \mathbf{L}_2^{(k)}\dot{\vartheta})\} + (1/\rho_{\mathscr{R}}\vartheta)\, \hat{\mathbf{q}}_{\mathscr{R}} \cdot \mathbf{g} \leqslant 0. \tag{4.14}$$

* This result may be compared with the constitutive equations presented by Green and Naghdi (1965a,b, 1968); Green *et al.* (1968).

The internal dissipation function $\hat{\sigma}$ for an elastic–plastic material is determined by the relation

$$\hat{\sigma} = -(1/\vartheta)\Big\{\partial_\varkappa\hat{\psi}[\mathrm{tr}(\mathbf{N}_1\dot{\mathbf{C}}) + \beta_1\dot{\vartheta}] + \mathrm{tr}[\partial_{\mathbf{P}}\hat{\psi}(\mathbf{H}_1[\dot{\mathbf{C}}] + \mathbf{H}_2\dot{\vartheta})]$$

$$+ \sum_{k=1}^{n-1} \mathrm{tr}[\partial_{\mathbf{\Gamma}^{(k)}}\hat{\psi}(\mathbf{L}_1^{(k)}[\dot{\mathbf{C}}] + \mathbf{L}_2^{(k)}\dot{\vartheta})]\Big\}.$$

V. Discussion of Particular Cases

Let us discuss the special case of the constitutive equations (3.7)–(3.13) for a rate-sensitive plastic material with the internal state variables as follows

$$\{\alpha^{(i)}, \mathbf{A}^{(j)}\} \equiv \{\varkappa, \mathbf{P}\}. \tag{5.1}$$

The full system of the constitutive equations for such a material has the following form (cf. Perzyna and Wojno, 1968)

$$\psi = \hat{\psi}(\mathbf{C}, \vartheta, \varkappa, \mathbf{P}), \tag{5.2}$$

$$\mathbf{T}_{\mathscr{R}} = 2\rho_{\mathscr{R}}\partial_{\mathbf{C}}\hat{\psi}, \tag{5.3}$$

$$\eta = -\partial_\vartheta\hat{\psi}, \tag{5.4}$$

$$\mathbf{q}_{\mathscr{R}} = \hat{\mathbf{q}}_{\mathscr{R}}(\mathbf{C}, \vartheta, \mathbf{g}, \varkappa, \mathbf{P}), \tag{5.5}$$

$$\dot{\mathbf{P}} = \gamma(\vartheta)\langle\Phi(\mathscr{F})\rangle\, \mathbf{M}(\mathbf{T}_{\mathscr{R}}, \vartheta, \mathbf{g}, \varkappa, \mathbf{P}), \tag{5.6}$$

$$\dot{\varkappa} = \mathrm{tr}\{\mathbf{N}(\mathbf{C}, \vartheta, \varkappa, \mathbf{P})\dot{\mathbf{P}}\}, \tag{5.7}$$

where $\mathscr{F}$ is now defined by the function

$$\mathscr{F} = (1/\varkappa)f(\mathbf{T}_{\mathscr{R}}, \vartheta, \mathbf{P}) - 1. \tag{5.8}$$

To ensure the fulfillment of the thermodynamic postulate (2.8), the constitutive equations (5.2)–(5.7) should satisfy the general dissipation inequality

$$\sigma - (1/\rho_{\mathscr{R}}\vartheta^2)\, \hat{\mathbf{q}}_{\mathscr{R}} \cdot \mathbf{g} \geqslant 0, \tag{5.9}$$

where the internal dissipation σ is now determined by the relation

$$\sigma = -(1/\vartheta)\{\partial_\varkappa\hat{\psi}\dot{\varkappa} + \mathrm{tr}[\partial_{\mathbf{P}}\hat{\psi}\dot{\mathbf{P}}]\}. \tag{5.10}$$

For a rate-sensitive plastic material we have, by (5.3),

$$\mathbf{T}_{\mathscr{R}} = \hat{\mathbf{T}}_{\mathscr{R}}(\mathbf{C}, \vartheta, \varkappa, \mathbf{P}). \tag{5.11}$$

After material differentiation of (5.11) we get

$$\dot{\mathbf{C}} = \mathbf{U}_1[\dot{\mathbf{T}}_{\mathscr{R}}] + \mathbf{U}_2\dot{\vartheta} + \gamma(\vartheta)\langle\Phi(\mathscr{F})\rangle\, \mathbf{M}^*, \tag{5.12}$$

where

$$\begin{aligned} \mathbf{U}_1 &= [\partial_{\mathbf{C}}\hat{\mathbf{T}}_{\mathscr{R}}]^{-1}, \\ \mathbf{U}_2 &= [\partial_{\mathbf{C}}\hat{\mathbf{T}}_{\mathscr{R}}]^{-1}[\partial_{\vartheta}\hat{\mathbf{T}}_{\mathscr{R}}], \\ \mathbf{M}^* &= [\partial_{\mathbf{C}}\hat{\mathbf{T}}_{\mathscr{R}}]^{-1}\{\partial_{\varkappa}\hat{\mathbf{T}}_{\mathscr{R}}\, \mathrm{tr}[\mathbf{NM}] + \partial_{\mathbf{P}}\hat{\mathbf{T}}_{\mathscr{R}}\mathbf{M}\}. \end{aligned} \tag{5.13}$$

In the case $\gamma(\vartheta) \to \infty$ the dynamical yield criterion yields

$$f(\mathbf{T}_{\mathscr{R}}, \vartheta, \mathbf{P}) = \varkappa \tag{5.14}$$

and the material again behaves as an elastic–plastic material.

Thus, the full system of constitutive equations for the inviscid theory of plasticity can now be written as follows [cf. (4.15)]:

$$\psi = \hat{\psi}(\mathbf{C}, \vartheta, \varkappa, \mathbf{P}), \tag{5.15a}$$

$$\mathbf{T}_{\mathscr{R}} = 2\rho_{\mathscr{R}}\partial_{\vartheta}\hat{\psi}, \tag{5.15b}$$

$$\eta = -\partial_{\vartheta}\hat{\psi}, \tag{5.15c}$$

$$\mathbf{q}_{\mathscr{R}} = \hat{\mathbf{q}}_{\mathscr{R}}(\mathbf{C}, \vartheta, \mathbf{g}, \varkappa, \mathbf{P}), \tag{5.15d}$$

$$\dot{\mathbf{P}} = \lambda\langle \mathrm{tr}(\partial_{\mathbf{T}_{\mathscr{R}}} f\dot{\mathbf{T}}_{\mathscr{R}}) + \partial_{\vartheta} f\dot{\vartheta}\rangle\, \mathbf{M}(\mathbf{T}_{\mathscr{R}}, \vartheta, \mathbf{g}, \varkappa, \mathbf{P}), \tag{5.15e}$$

$$\dot{\varkappa} = \mathrm{tr}[\mathbf{NP}]. \tag{5.15f}$$

After material differentiation of the equation (5.15b) we obtain the result

$$\dot{\mathbf{C}} = \mathbf{U}_1[\dot{\mathbf{T}}_{\mathscr{R}}] + \mathbf{U}_2\dot{\vartheta} + \lambda\langle \mathrm{tr}(\partial_{\mathbf{T}_{\mathscr{R}}} f\dot{\mathbf{T}}_{\mathscr{R}}) + \partial_{\vartheta} f\dot{\vartheta}\rangle\, \mathbf{M}^*. \tag{5.16}$$

Equations (5.12) and (5.16) show that the assumption $\dot{\mathbf{C}} = \dot{\mathbf{E}} + \dot{\mathbf{P}}$, where $\mathbf{E}$ denotes the elastic part of the deformation tensor, is not in general justified for finite strain (cf. the comments of Perzyna, 1968b,c). The functions $\mathbf{U}_1$ and $\mathbf{U}_2$ depend on the internal tensor $\mathbf{P}$.

VI. Equilibrium State and the Relaxation Process

The set of functions $(\mathbf{C}^*, \vartheta^*, \varkappa^*, \mathbf{P}^*)$ which satisfies the conditions

$$\begin{aligned} \mathbf{G}(\mathbf{C}^*, \vartheta^*, \mathbf{0}, \varkappa^*, \mathbf{P}^*) &= 0 \\ \beta(\mathbf{C}^*, \vartheta^*, \mathbf{0}, \varkappa^*, \mathbf{P}^*) &= 0 \end{aligned} \tag{6.1}$$

is called the equilibrium state of a material at X. Since for an elastic–viscoplastic material the functions $\mathbf{G}$ and β are defined by (5.6) and (5.7), the conditions (6.1) can be satisfied if

$$\Phi(\mathscr{F}) = 0, \qquad \text{i.e.,} \quad \mathscr{F} = 0. \tag{6.2}$$

The tensor function $\mathbf{M}$ and $\mathbf{N}$ in the entire domain of variability of $\mathbf{C}$, ϑ, $\mathbf{g}$, $\varkappa$, and $\mathbf{P}$ cannot be equal to zero. This is implied by the physical nature of an elastic–viscoplastic material. For a rate-sensitive and work-hardening plastic material $\dot{\mathbf{P}} = \mathbf{0}$ and $\dot{\varkappa} = 0$ if, and only if, $\mathscr{F} \leqslant 0$, and $\dot{\mathbf{P}} \neq \mathbf{0}$ and $\dot{\varkappa} \neq 0$ for $\mathscr{F} > 0$.

Following Coleman and Gurtin (1967), we introduce the definition of the domain of attraction of the equilibrium state $(\mathbf{C}^*, \vartheta^*, \varkappa^*, \mathbf{P}^*)$ at constant strain and temperature as a set $\mathscr{D}(\mathbf{C}^*, \vartheta^*, \varkappa^*, \mathbf{P}^*)$ of all initial values $\mathbf{P}^0$ and $\varkappa^0$ such that the solutions $\mathbf{P} = \mathbf{P}(t)$ and $\varkappa = \varkappa(t)$ of the initial-value problems

$$\begin{aligned} \dot{\mathbf{P}} &= \mathbf{G}(\mathbf{C}^*, \vartheta^*, \mathbf{0}, \varkappa, \mathbf{P}), \qquad & \mathbf{P}(0) &= \mathbf{P}^0, \\ \dot{\varkappa} &= \beta(\mathbf{C}^*, \vartheta^*, \mathbf{0}, \varkappa, \mathbf{P}), \qquad & \varkappa(0) &= \varkappa^0, \end{aligned} \tag{6.3}$$

exist for all $t \geqslant 0$ and tend to $\mathbf{P}^*$ and $\varkappa^*$, respectively, i.e.,

$$\mathbf{P}(t) \to \mathbf{P}^*, \quad \varkappa(t) \to \varkappa^* \qquad \text{if} \quad t \to \infty. \tag{6.4}$$

The equilibrium state $(\mathbf{C}^*, \vartheta^*, \varkappa^*, \mathbf{P}^*)$ is said to be asymptotically stable at constant strain and temperature if the domain of attaction of the equilibrium state at constant strain and temperature $\mathscr{D}(\mathbf{C}^*, \vartheta^*, \varkappa^*, \mathbf{P}^*)$ contains a neighbourhood of P^* and $\varkappa^*$, or, states differently, if there exist $\xi > 0$ and $z > 0$ such that every tensor of internal state $\mathbf{P}^0$ and every scalar parameter of internal state $\varkappa^0$, satisfying the conditions

$$\begin{aligned} |\, \mathbf{P}^0 - \mathbf{P}^* \,| &< \xi, \\ |\, \varkappa^0 - \varkappa^* \,| &< z, \end{aligned} \tag{6.5}$$

are in $\mathscr{D}(\mathbf{C}^*, \vartheta^*, \varkappa^*, \mathbf{P}^*)$.

Let us assume that there exists a set of quantities ($\mathbf{C}^*$, ϑ^*, $\varkappa^*$, $\mathbf{P}^*$) defining the equilibrium state, and let us investigate the solutions of the initial-value problems (4.3) with the conditions (6.5).

The thermodynamic process considered is characterized by constancy of the deformation tensor $\mathbf{C} = \mathbf{C}^*$ and temperature $\vartheta = \vartheta^*$ in the time interval $[t_0, \infty)$. The inelastic deformation tensor and the work-hardening parameter in such a thermodynamic process are described by Eqs. (6.3) and the stress tensor by

$$\mathbf{T}_{\mathscr{R}} = 2\rho_{\mathscr{R}}\partial_{\mathbf{C}}\hat{\psi}(\mathbf{C}^*, \vartheta^*, \varkappa, \mathbf{P}). \tag{6.6}$$

The process described by (6.3) and (6.6) is the isothermal relaxation process for an elastic–viscoplastic material.

For such a process

$$\dot{\mathbf{P}} \to \mathbf{0} \quad \text{and} \quad \dot{\varkappa} \to 0 \quad \text{if} \quad t \to \infty, \tag{6.7}$$

and

$$\mathbf{P} \to \mathbf{P}^* = \hat{\mathbf{P}}^*(\mathbf{C}^*, \vartheta^*), \tag{6.8}$$

where the function $\hat{\mathbf{P}}^*$ is determined by the condition

$$\mathscr{F} = 0. \tag{6.9}$$

Thus, we have proved the following proposition:

The asymptotically stable equilibrium state ($\mathbf{C}^*$, ϑ^*, $\varkappa^*$, $\mathbf{P}^*$) for an elastic–viscoplastic material is reached only by an isothermal relaxation process.

For the isothermal relaxation process and for grad $\vartheta = \mathbf{0}$, the thermodynamic postulate (2.8) yields

$$\dot{\psi} \leqslant 0. \tag{6.10}$$

Thus, we can write

$$\hat{\psi}(\mathbf{C}^*, \vartheta^*, \varkappa, \mathbf{P}) \geqslant \hat{\psi}(\mathbf{C}^*, \vartheta^*, \varkappa^*, \mathbf{P}^*) \tag{6.11}$$

for all internal state tensor parameters $\mathbf{P}$ and all scalar internal parameters $\varkappa$ in a neighborhood of $\mathbf{P}^*$ and $\varkappa^*$. Hence,

$$\partial_{\mathbf{P}}\hat{\psi}\Big|_{\substack{\mathbf{P}=\mathbf{P}^* \\ \varkappa=\varkappa^*}} = \mathbf{0} \quad \text{and} \quad \partial_{\varkappa}\hat{\psi}\Big|_{\substack{\mathbf{P}=\mathbf{P}^* \\ \varkappa=\varkappa^*}} = 0. \tag{6.12}$$

Equations (6.12) are called the equations of the internal equilibrium of an elastic–viscoplastic material. This implies the following statement:

In the isothermal relaxation process for an elastic–viscoplastic material, the free energy has minimum value in the asymptotically stable equilibrium state.

VII. Isotropic Material

The constitutive equations (5.2)–(5.7) describing the properties of an elastic–viscoplastic material are valid for arbitrary initial anisotropy. We shall now assume that the material considered is initially isotropic and the work-hardening parameter is not an internal state variable (cf. Perzyna and Wojno, 1968). In this case, polynomial representations can exist for the response tensor functions (cf. Truesdell and Noll, 1965).

The fundamental response function $\hat{\psi}(\mathbf{C}, \vartheta, \mathbf{P})$ may have a polynomial representation in the ten invariants

$$\begin{gathered}\operatorname{tr}\mathbf{C}, \quad \operatorname{tr}\mathbf{C}^2, \quad \operatorname{tr}\mathbf{C}^3, \quad \operatorname{tr}\mathbf{P}, \quad \operatorname{tr}\mathbf{P}^2, \quad \operatorname{tr}\mathbf{P}^3 \\ \operatorname{tr}\mathbf{CP}, \quad \operatorname{tr}\mathbf{C}^2\mathbf{P}, \quad \operatorname{tr}\mathbf{CP}^2, \quad \operatorname{tr}\mathbf{C}^2\mathbf{P}^2.\end{gathered} \tag{7.1}$$

Thus we can write

$$\begin{aligned}\hat{\psi}(\mathbf{C}, \vartheta, \mathbf{P}) = {} & \hat{\psi}_0(\vartheta) + \hat{\psi}_1(\vartheta)\operatorname{tr}\mathbf{C} + \hat{\psi}_2(\vartheta)\operatorname{tr}\mathbf{C}^2 + \hat{\psi}_3(\vartheta)\operatorname{tr}\mathbf{C}^3 \\ & + \hat{\psi}_4(\vartheta)\operatorname{tr}\mathbf{P} + \hat{\psi}_5(\vartheta)\operatorname{tr}\mathbf{P}^2 + \hat{\psi}_6(\vartheta)\operatorname{tr}\mathbf{P}^3 \\ & + \hat{\psi}_7(\vartheta)\operatorname{tr}\mathbf{CP} + \hat{\psi}_8(\vartheta)\operatorname{tr}\mathbf{CP}^2 + \hat{\psi}_9(\vartheta)\operatorname{tr}\mathbf{C}^2\mathbf{P} + \hat{\psi}_{10}(\vartheta)\operatorname{tr}\mathbf{C}^2\mathbf{P}^2.\end{aligned} \tag{7.2}$$

For the stress tensor and entropy we deduce the following relations from (5.3), (5.4), and (7.2):

$$\begin{aligned}\mathbf{T}_{\mathscr{R}} = {} & 2\rho_{\mathscr{R}}[\hat{\psi}_1(\vartheta)\mathbf{I} + 2\hat{\psi}_2(\vartheta)\mathbf{C} + 3\hat{\psi}_3(\vartheta)\mathbf{C}^2 \\ & + \hat{\psi}_7(\vartheta)\mathbf{P} + \hat{\psi}_8(\vartheta)\mathbf{P}^2 + 2\hat{\psi}_9(\vartheta)\mathbf{CP} + 2\hat{\psi}_{10}(\vartheta)\mathbf{CP}^2]\end{aligned} \tag{7.3}$$

$$\begin{aligned}\eta = {} & -[\hat{\psi}_0'(\vartheta) + \hat{\psi}_1'(\vartheta)\operatorname{tr}\mathbf{C} + \hat{\psi}_2'(\vartheta)\operatorname{tr}\mathbf{C}^2 + \hat{\psi}_3'(\vartheta)\operatorname{tr}\mathbf{C}^3 \\ & + \hat{\psi}_4'(\vartheta)\operatorname{tr}\mathbf{P} + \hat{\psi}_5'(\vartheta)\operatorname{tr}\mathbf{P}^2 + \hat{\psi}_6'(\vartheta)\operatorname{tr}\mathbf{P}^3 \\ & + \hat{\psi}_7'(\vartheta)\operatorname{tr}\mathbf{CP} + \hat{\psi}_8'(\vartheta)\operatorname{tr}\mathbf{CP}^2 + \hat{\psi}_9'(\vartheta)\operatorname{tr}\mathbf{C}^2\mathbf{P} + \hat{\psi}_{10}'(\vartheta)\operatorname{tr}\mathbf{C}^2\mathbf{P}^2]\end{aligned} \tag{7.4}$$

where $\hat{\psi}_0'(\vartheta),\ldots, \hat{\psi}_{10}'(\vartheta)$ are the derivatives with respect to temperature ϑ of the temperature-dependent coefficients $\hat{\psi}_0(\vartheta),\ldots, \hat{\psi}_{10}(\vartheta)$.

The differential equation determining the internal state tensor **P**, (5.6), may be written in the form

$$\dot{\mathbf{P}} = \gamma(\vartheta)\langle\Phi(\mathscr{F})\rangle[\varphi_0\mathbf{I} + \varphi_1\mathbf{T}_\mathscr{R} + \varphi_2\mathbf{P} + \varphi_3\mathbf{T}_\mathscr{R}{}^2 + \varphi_4\mathbf{P}^2 + \varphi_5(\mathbf{T}_\mathscr{R}\mathbf{P} + \mathbf{P}\mathbf{T}_\mathscr{R})$$
$$+ \varphi_6(\mathbf{T}_\mathscr{R}{}^2\mathbf{P} + \mathbf{P}\mathbf{T}_\mathscr{R}{}^2) + \varphi_7(\mathbf{T}_\mathscr{R}\mathbf{P}^2 + \mathbf{P}^2\mathbf{T}_\mathscr{R}) + \varphi_8(\mathbf{T}_\mathscr{R}{}^2\mathbf{P}^2 + \mathbf{P}^2\mathbf{T}_\mathscr{R}{}^2)], \tag{7.5}$$

where the function $\mathscr{F}$ and $\varphi_0, \ldots, \varphi_8$ depend on the temperature ϑ and the ten invariants

$$\begin{array}{cccccc} \operatorname{tr}\mathbf{T}_\mathscr{R}, & \operatorname{tr}\mathbf{T}_\mathscr{R}{}^2, & \operatorname{tr}\mathbf{T}_\mathscr{R}{}^3, & \operatorname{tr}\mathbf{P}, & \operatorname{tr}\mathbf{P}^2, & \operatorname{tr}\mathbf{P}^3, \\ \operatorname{tr}\mathbf{T}_\mathscr{R}\mathbf{P}, & \operatorname{tr}\mathbf{T}_\mathscr{R}{}^2\mathbf{P}, & \operatorname{tr}\mathbf{T}_\mathscr{R}\mathbf{P}^2, & \operatorname{tr}\mathbf{T}_\mathscr{R}{}^2\mathbf{P}^2. & & \end{array} \tag{7.6}$$

VIII. Physical Foundations of Viscoplasticity

A modern theory of plastic flow must be based on microscopic investigations because plastic deformation changes not only the external shape of a body but also its internal structure. Another essential feature of plasticity is that this theory must be intrinsically dynamical.

It is common experience that during a mechanical test on any plastic material we can control independently only two quantities: total deformation and temperature, or stress and temperature (cf. Kratochvil and Dillon, 1968, 1969). This is why we have introduced in the general theory of an elastic–viscoplastic material the total deformation tensor and the temperature as thermodynamic state variables or independent variables.

The basic result of the microscopic theory is that the elementary process of plastic deformation is the motion of a line-shaped crystal defect called dislocation, and that the arrangement of dislocations in the body is characteristic for the internal mechanical state of the solid which has undergone some plastic deformation. In the construction of the macroscopic theory from the microscopic knowledge, one therefore has to consider the possibility of describing the internal state of the solid by giving certain average quantities which characterize the dislocation arrangement in a macroscopic way (cf. Kröner, 1962).

In the thermodynamic theory of an elastic–viscoplastic material we have chosen as such quantities the work hardening parameter $\varkappa$, the inelastic deformation tensor **P**, and the dislocation arrangement tensors $\mathbf{\Gamma}^{(k)}$. These internal state variables are responsible for the structural changes of a material.

First we shall discuss the physical basis for the determination of the inelastic deformation tensor **P**. There is not yet available a satisfactory formal basis for the thermodynamics of crystal defects, either in equilibrium or in processes of activation.*

Since plastic flow occurs by the motion of dislocation lines, the rate at which it takes place depends on how fast the dislocations move, how many dislocations are moving in a given volume of material, and how much displacement is carried by each dislocation. The theory of crystal dislocations shows that for a one-dimensional state of stress the inelastic strain rate is

$$\dot{\epsilon} = \alpha N b V, \tag{8.1}$$

where α is an orientation factor, N the mean density of mobile dislocations, b the displacement per dislocation line (the Burgers vector), and V the mean dislocation velocity.

It is generally agreed that the finite stress needed to cause plastic flow is due to obstacles impeding the motion of dislocations through a crystal (cf. Conrad, 1964). Obstacles can be conveniently divided into two groups according to the distance over which they interact with the glide dislocations: (1) those that possess long-range stress fields (of the order of 10 atomic diameters or greater), and (2) those that possess short-range stress fields (of less than about 10 atomic diameters). The energy required to overcome the former type of obstacle may be so large that the thermal fluctuations cannot assist the applied stress in the temperature range under consideration. Thermal activation thus plays no role in overcoming these long-range obstacles; hence they are termed athermal obstacles. Thermal fluctuations can assist the applied stress in overcoming short-range or thermal obstacles. Thus the thermal obstacles are responsible for the dynamic aspects of plastic deformation.†

The mechanism of overcoming the dislocation forest has been developed theoretically by Seeger (1954a,b, 1955).‡ In Seeger's theory it was assumed that the forest is distributed uniformly, and that the activation energy decreases linearly with increasing applied stress over the whole process of intersection of dislocations. Here, we shall not

* This fact has been noted by Nabarro (1967). For a thermodynamic description of the activation process, see Gibbs (1964, 1967) and Schoeck (1965); cf. also comments on this subject presented by Sesták (1966).

† Common thermal obstacles or mechanisms in pure metals are the Peierls–Nabarro stress, forest dislocation, the motion of jogs in screw dislocations, cross-slip of screw dislocations, and climb of edge dislocations.

‡ This mechanism appears in cph, fcc, and bcc crystals of metals in the various temperature ranges (cf. Conrad, 1964).

assume *a priori* the form of the variation of activation energy with stress, but rather derive relations by which it can be obtained from experimentally observable quantities (cf. Basinski, 1959).

When the deformation is controlled by a single thermal activated process we have

$$V = V_0 \exp[-U/k\vartheta], \tag{8.2}$$

where U is the energy that must be supplied by a thermal fluctuation for each successful activation, k is the Boltzmann constant, and

$$V_0 = A\nu \tag{8.3}$$

if A is the area swept out per activation and ν is the Debye frequency.

Equations (8.1)–(8.3) give

$$\dot{\epsilon} = \alpha NAb\nu \exp[-U/k\vartheta]. \tag{8.4}$$

Let us assume that

$$U = \varphi[(\tau - \tau^*) Lb] \tag{8.5}$$

where τ^* is the back stress not surmountable by a thermal fluctuation and L is the mean cord distance between neighboring points at which the dislocation is arrested. Expansion of the function φ gives*

$$U = \varphi\,|_{\tau=\tau^*} + \varphi'\,|_{\tau=\tau^*} (\tau - \tau^*) Lb + \varphi''\,|_{\tau=\tau^*} \frac{(\tau - \tau^*)^2}{2!} L^2b^2 + \cdots. \tag{8.6}$$

Let us denote by

$$v^* = -\varphi'\,|_{\tau=\tau^*} Lb = vLb, \qquad U_0 = \varphi\,|_{\tau=\tau^*}, \tag{8.7}$$

the activation volume and the activation energy for intersection at zero stress, respectively. The linear approximation to Eq. (8.4) now gives Seeger's relation

$$\dot{\epsilon} = \alpha NAb\nu \exp\{-U_0/k\vartheta + [(\tau - \tau^*)\, v^*/k\vartheta]\} \tag{8.8}$$

or

$$\tau = (\tau^* + U_0/v^*) + k\vartheta/v^* \ln(\dot{\epsilon}/\alpha NAb\nu). \tag{8.9}$$

When the activation energy U is a nonlinear function of the stress the relation (8.4) yields†

$$\dot{\epsilon} = \alpha NAb\nu \exp\{-\varphi[(\tau - \tau^*) Lb]/k\vartheta\} \tag{8.10}$$

* Similar procedures have been applied by Lindholm (1965).

† The full discussion of thermal influences on dislocation motion has been presented by Nabarro (1967).

or

$$\tau = \tau^* + (1/Lb)\, \varphi^{-1}[k\vartheta \ln(\alpha NAb\nu/\dot{\epsilon})]. \tag{8.11}$$

In the linear theory [cf. (8.8) and (8.9)] we have four internal material parameters, namely, the frequency parameter $\nu^* = \alpha NAb\nu$, the activation volume v^*, the internal athermal stress component opposing dislocation motion τ^*, and the total activation energy U_0. In the most general case, each of these four parameters may be considered a function of the four independent variables ϵ, $\dot{\epsilon}$, τ, and ϑ.

In the nonlinear theory [cf. (8.10) and (8.11)] there are only two internal material parameters ν^* and τ^* and, in addition, one response function φ. We will restrict the frequency parameter ν^* to be a function of ϑ only and the athermal stress τ^* to be a function of the four independent variables.

Let us compare the theoretical dynamical yield criterion (3.6) with the physical predictions based on the thermally activated process (8.11). From this comparison it is seen that the phenomenological yield criterion (3.6) may be treated as a simple generalization to polycrystals in a general stress state, and finite deformations the physically justified relation (8.11). In this generalization it has been assumed that the influence of the strain rate and the temperature on the yield limit is described by the nonlinear function $\Phi(\mathscr{F})$. Thus, the differential equation for the inelastic deformation tensor (3.4) may be compared with (8.10).

The work-hardening parameter $\varkappa$ defined by (3.2) may be compared with the athermal stress τ^* in (8.10) and (8.11).

IX. Comparisons with Experimental Results

The effect of stress and temperature on dislocation velocity has been determined by the etch-pit technique employed by Johnston and Gilman (1959) on lithium fluoride and by Stein and Low (1960) on silicon–iron.* The results for silicon–iron are given in Fig. 1. It is noted that approximately straight lines result when the logarithm of the velocity V is plotted against the logarithm of the applied stress τ. This suggests a power relationship between dislocation velocity and stress. In general, the velocity of dislocation motion was found to be a very sensitive function of stress. Different relationships between velocity and stress have been proposed by Johnston and Gilman (1959), Gilman (1965, 1966, 1968a,b), Gillis and Gilman (1965), and Stein and Low (1960). However, since there does not exist a theoretical interpretation of these relation-

* A different technique was developed by Gorman, Wood, and Vreeland (1969) and applied for pure aluminum single crystals.

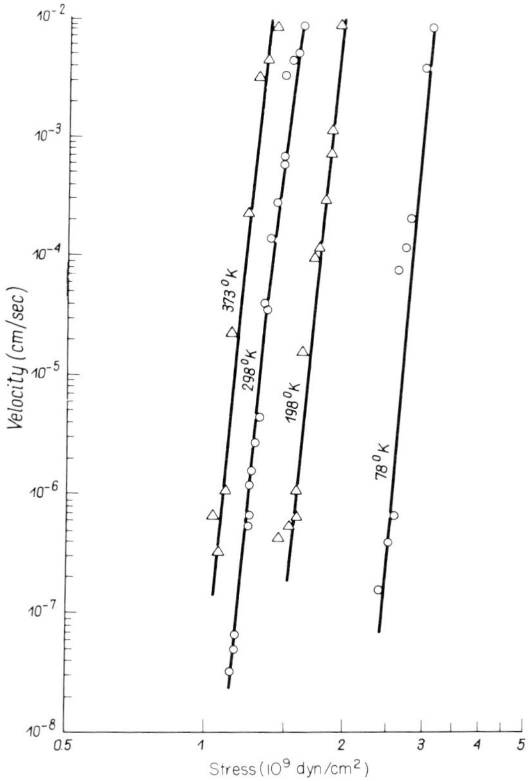

FIG. 1. Stress dependence of edge dislocation velocity at four different temperatures. After Stein and Low (1960).

ships, it seems more reasonable to start with (8.2) (cf. Conrad, 1964).

Lindholm and Yeakley (1965) investigated single crystal and polycrystalline specimens of high purity aluminum in compression at strain rates up to 500 $\sec^{-1}$ using the split Hopkinson pressure bar method. They obtained average stress–strain curves for the six orientations of a single crystal and similar curves for the polycrystalline material. Activation volume as a function of strain can be computed from the data obtained.* Results for the single and polycrystalline specimens of high

* From strain data at constant temperature the activation volume can be determined from the relation

$$v = k\vartheta \left.\frac{\partial \ln \dot{\epsilon}}{\partial \tau}\right|_{\vartheta=\text{const.}} \approx k\vartheta \left.\frac{\ln \dot{\epsilon}_2/\dot{\epsilon}_1}{\tau_2 - \tau_1}\right|_{\vartheta=278^\circ\text{K}}$$

purity aluminum are given in Fig. 2. The most interesting feature of these curves is that the activation volume for the polycrystalline material falls within the bounds and near the average of the single crystal data.

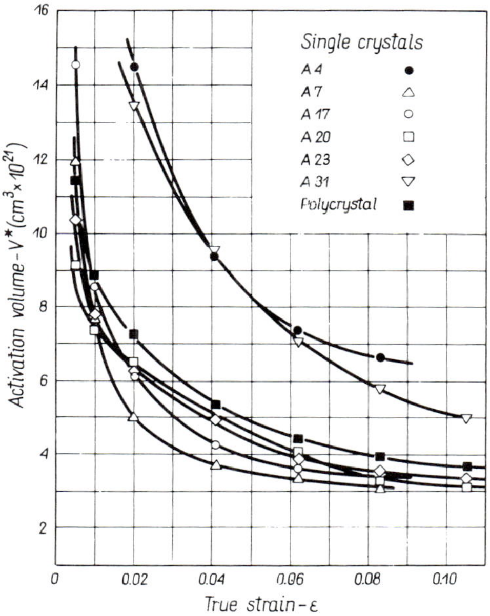

FIG. 2. Activation volumes for single and polycrystalline aluminum (99.995 %). After Lindholm and Yeakley (1965).

This implies that the same thermally activated mechanisms control the deformation in single and polycrystals and that the distribution of the activation barriers are essentially the same in both cases. This is in agreement with the previous results presented by Mitra and Dorn (1962)* for aluminum at low temperature and those of Conrad (1964) for iron and steel.

Ferguson, Kumar, and Dorn (1967) used impact shear tests of the Kolsky thin-wafer type to determine the effect of temperature and strain rate on the critical resolved shear stress for slip in aluminum single crystals at strain rates of 10^4 sec^{-1} and in the temperature range 20° to 500°K. They showed that if aluminum single crystals are deformed at a shear stress higher than the thermally activated stress range they behave

* Cf. also Mitra and Dorn (1963) for copper, and Klepaczko (1968) for polycrystalline aluminum.

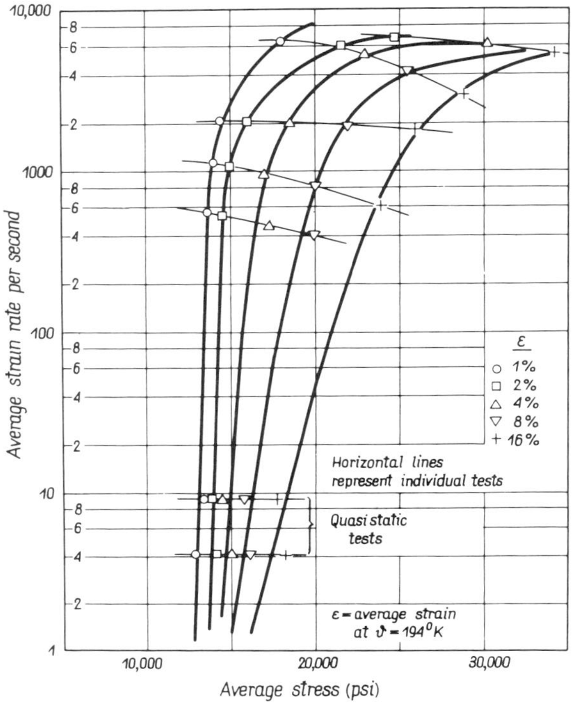

FIG. 3. Effect of stress and strain rate at constant strain and temperature $\vartheta = 194°K$. After Hauser *et al.* (1961).

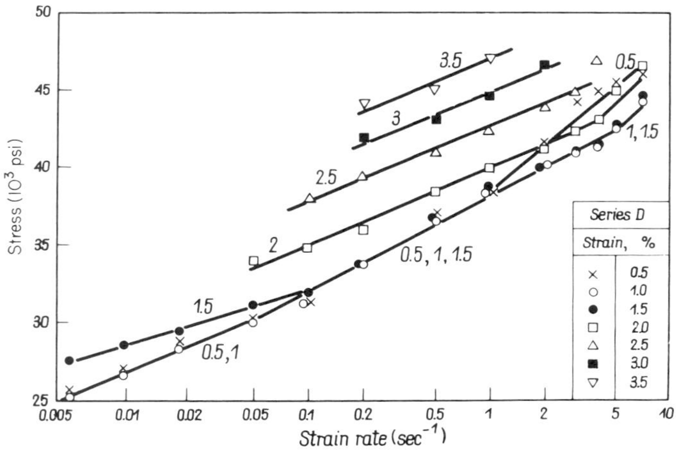

FIG. 4. Stress against strain rate (logarithmic scale) at constant strain (mean grain density, 2033 grains/mm²). After Marsh and Campbell (1963).

in a viscous manner, in that the stress is proportional to the shear–strain rate. At stresses higher than the thermally activated stress range, the stress–strain rate behavior is temperature-dependent for the temperature range investigated (cf. Larsen *et al.*, 1964; Rajnak *et al.*, 1962).

Hauser, Simmons, and Dorn (1961) presented the stress–strain versus strain rate properties of high purity aluminum for 295, 194, and 77.4°K over strain rates from 2 to 12×10^3 sec^{-1}. An example of these experimental results for 194°K is shown in Fig. 3.

Marsh and Campbell (1963) obtained results of constant stress tests on mild steel specimens of different mean ferrite grain sizes (cf. Fig. 4).

Chiddister and Malvern (1963) investigated aluminum specimens at strain rates from 300 to 200 sec^{-1} at six temperatures from 30° to 550°C. The stress–strain rate curves at each temperature were plotted on both semilogarithmic and log–log coordinate systems in order to evaluate the validity of the logarithmic and the power flow laws. The semi-logarithmic curves are shown in Fig. 5.

Lindholm (1964) investigated three annealed face centered cubic metals—lead, aluminum, and copper—both dynamically and at lower strain rates to determine their rate sensitivity. His results are plotted in Figs. 6–8.

To assess the thermal dependence of the constitutive relations Lindholm (1968) performed elevated temperature tests for aluminum in tension and compression. These results are presented in plots of stress versus strain rate in Fig. 9 and stress versus temperature in Fig. 10.

Campbell and Cooper (1967) described experimental results obtained in tensile tests on annealed low-carbon steel at mean plastic strain rates in the range 10^{-3} to 10^2 sec^{-1} (cf. Fig. 11).

Lindholm (1968) presented experimental results on the plastic flow of aluminum suject to a wide spectrum of loading conditions. This includes stress states of pure compression, tension, and torsion as well as combined stress states. The rate of loading is varied to produce strain rates within the range of 10^{-3} sec^{-1} to 10^3 sec^{-1}. Elevated temperature data from approximately 300°K to 700°K were obtained for compression and tension.

In Fig. 12, the data of a large number of tests are plotted to show the relation between the stress and strain rate invariants at constant strain amplitude and temperature. According to the linearized theory (Seeger, 1955), this should be a linear relationship on the semilogarithmic plot. The straight line through the experimental points at each strain amplitude is a best least-squares fit. The standard estimate of error of the data about the mean is 5.4 % for the lowest strain and less than 3 % for the two higher strain amplitudes.

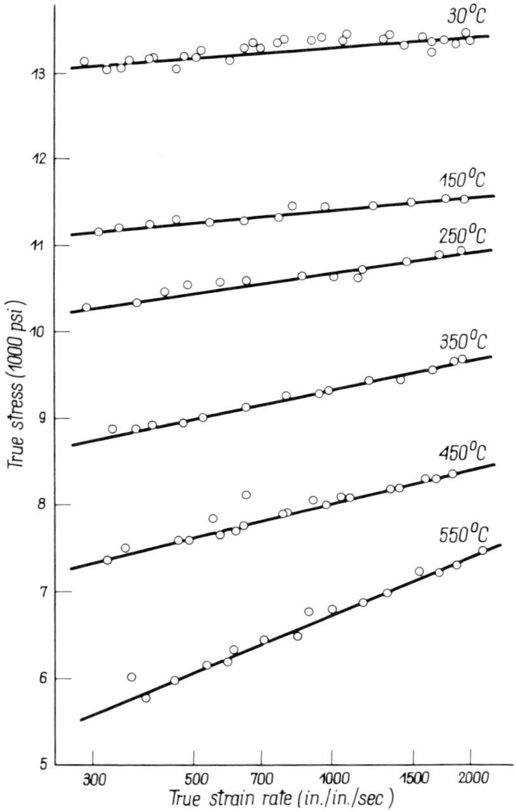

FIG. 5. Semilogarithmic plot of stress versus true strain rate. After Chiddister and Malvern (1963).

All the preceding experimental data of Lindholm were obtained with the stress ratio, shear stress–tension stress, and the rate of plastic deformation nearly constant. Several tests of an exploratory nature were performed by Lindhom (1968) to determine the material response under sudden changes in the direction or the rate of loading. These tests indicate that within the normal scatter of the data, there does not appear to be any marked evidence of the effect of previous loading history, or rate or direction of loading, on the subsequent material response.

The tensile stress–strain behavior of high purity iron over a range of temperatures from 0° to 900°C and strain rates from 0.04 min^{-1} to 40 min^{-1} have been examined by R. W. Evans and Simpson (1969).

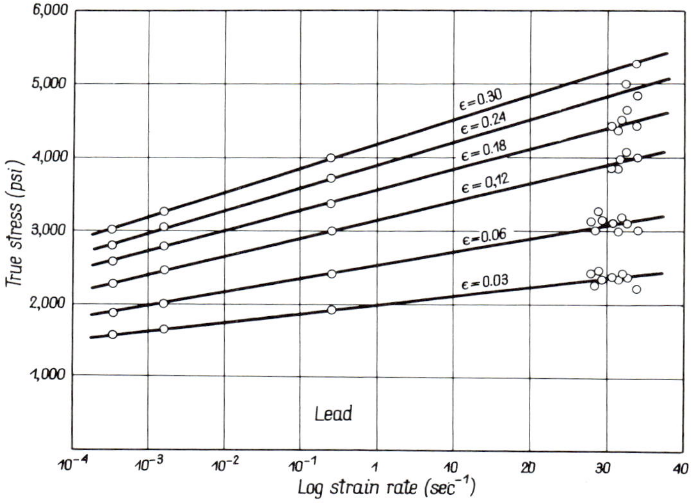

FIG. 6. Flow stress as a function of strain rate for lead. After Lindholm (1964).

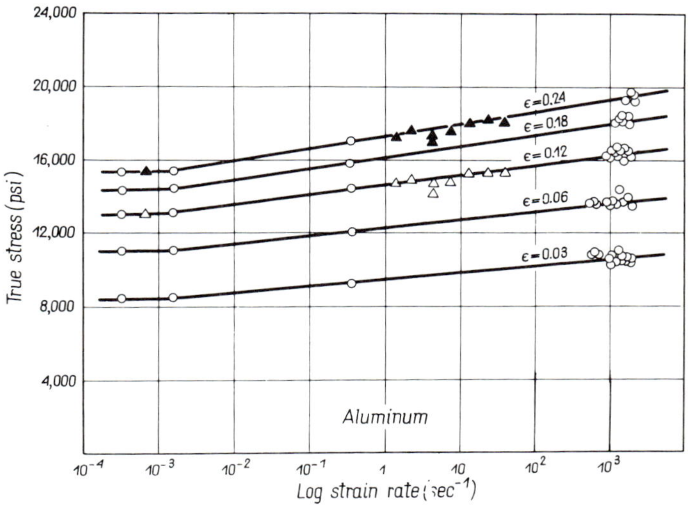

FIG. 7. Flow stress as a function of strain rate for aluminum. After Lindholm (1964). $\epsilon = 0.10$, △; $\epsilon = 0.20$, ▲. After Alder and Phillips (1954-1955).

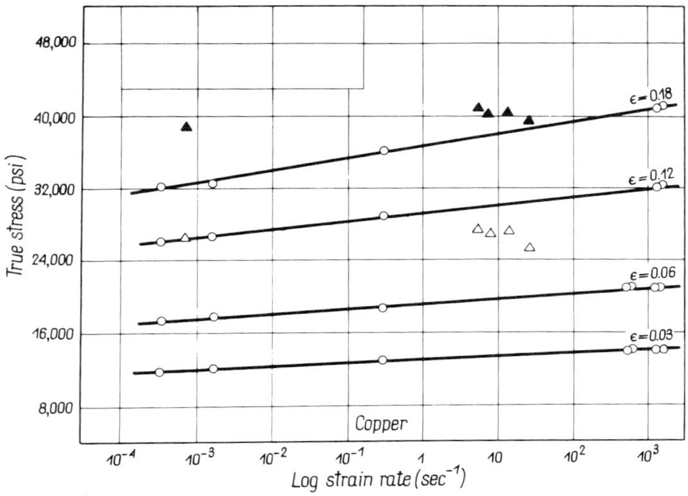

FIG. 8. Flow stress as a function of strain rate for copper. After Lindholm (1964). $\epsilon = 0.10$, △; $\epsilon = 0.20$, ▲. After Alder and Phillips (1954-1955).

These results have shown that the strain rate sensitivity of the flow stress of α-iron decreases with increasing temperature and decreasing strain rate [similar results have been obtained for carbon steel by Maiden and Campbell (1958); cf. also K. R. Evans and Flanagan (1968) for copper crystals]. The activation volume is not a unique function of the effective stress but depends on the strain rate and temperature at which the test is conducted. The results can best be explained by the fact that the thermally activated component of the flow stress is determined by the nonconservative motion of jogs in screw dislocations.

The basic assumptions of the theory of viscoplasticity for complex states of stress have also been discussed by Klepaczko (1969) in the light of an experimental study of the strain rate behavior of technically pure iron in pure shear. These results have shown that the iron is more strain rate sensitive in pure shear than in tension or compression. This implies that the assumption about isotropic strain rate sensitivity used in the theory of viscoplasticity does not seem to be exactly satisfied.

The comparison of theoretical predictions with experimental results shows that the hypothesis that the strain rate and temperature dependence in metals is due primarily to a thermally activated process described by the linearized relation (8.9) is valid for certain metals and for certain ranges of strain rate only (cf. Alder and Phillips, 1954–1955; Chiddister and Malvern, 1963; Krafft *et al.*, 1954; Lindholm, 1964, 1968;

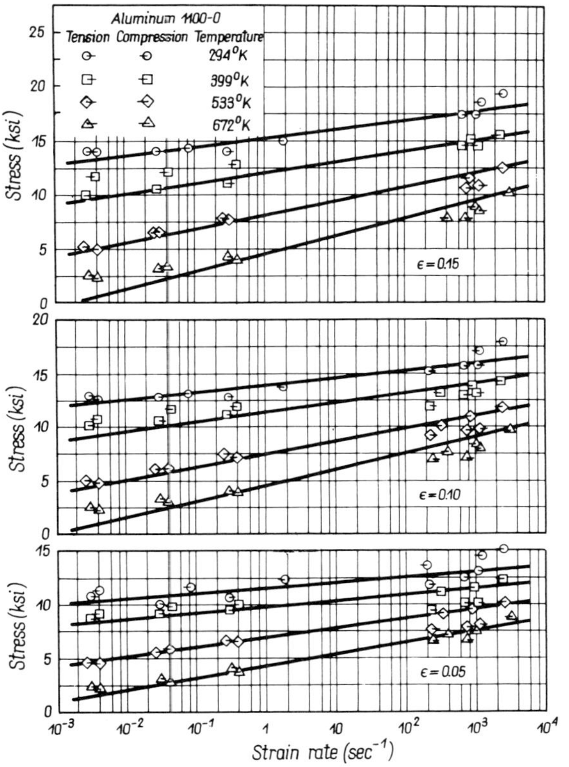

FIG. 9. Semilogarithmic plot of stress versus true strain rate for aluminum. After Lindholm (1968).

Trozera *et al.*, 1957). On the other hand the nonlinear relation (8.11) may describe the strain rate and temperature phenomenon throughout the entire range of the strain rate (cf. Campbell and Cooper, 1967; Hauser *et al.*, 1961; Marsh and Campbell, 1963).

Thus, this analysis indicates that the introduction of the nonlinear function $\Phi(\mathscr{F})$ into the constitutive equations (3.8)–(3.14) and its choice on the basis of experimental results may be considered a very well-founded hypothesis.

The constitutive equations (4.13) for an elastic–plastic material were obtained as a limiting case of the strain rate sensitive constitutive equations (3.8)–(3.14), and therefore they describe the quasistatic behavior of a material.

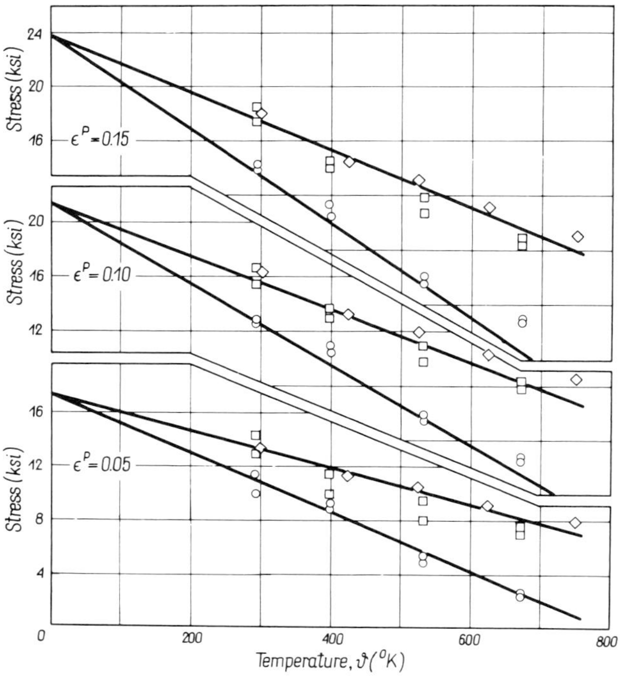

FIG. 10. Plot of stress versus temperature for aluminum. After Lindholm (1968). $\dot{\epsilon}^p = 10^{-2}$ sec^{-1}, ○; $\dot{\epsilon}^p = 10^3$ sec^{-1}, □ Lindholm (1968), ◇ Chiddister and Malvern (1963).

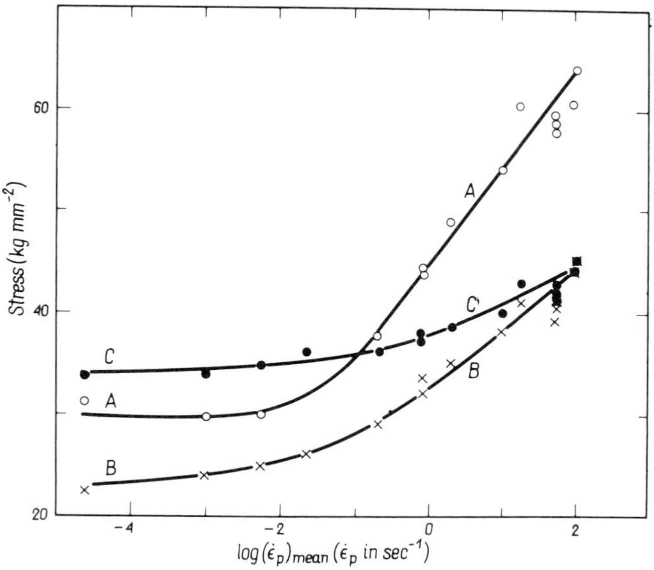

FIG. 11. Variation of strength with strain rate. A—Upper yield stress; B—lower yield stress; C—ultimate tensile stress. After Campbell and Cooper (1967).

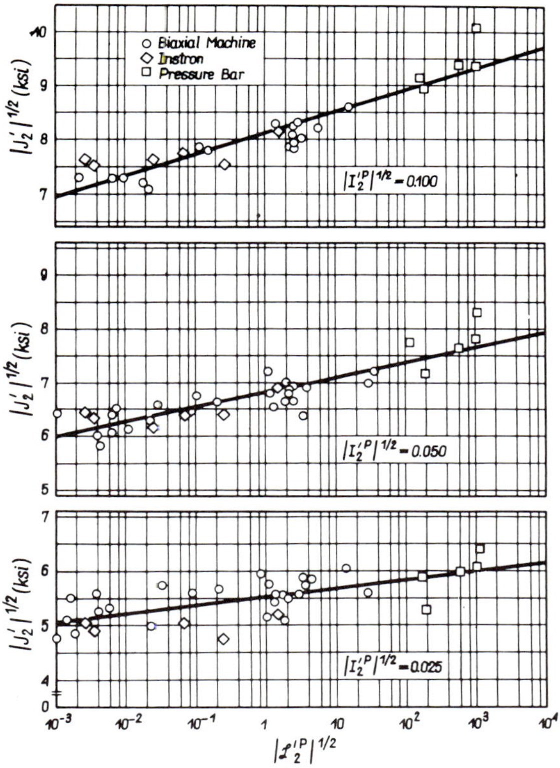

FIG. 12. Semilogarithmic stress–strain plots for aluminum in terms of the invariants at constant temperature $\vartheta = 294°K$ and strain rate. After Lindholm (1968). J_2' = second invariant of stress deviation; $I_2^{P'}$ = second invariant of inelastic strain rate deviation; $\mathscr{L}_2^{P'}$ = second invariant of inelastic strain deviation.

REFERENCES

ALDER, J. F., and PHILLIPS, V. A. (1954–1955). The effect of strain rate and temperature of aluminium, copper and steel to compression. *J. Inst. Metals* **83**, 80–86.

BASINSKI, Z. S. (1959). Thermally activated glide in face-centred cubic metals and its application to the theory of strain hardening. *Phil. Mag.* [8] **4**, No. 37, 393–432.

BECKMAN, M. E. (1964). Form for the relation between stress and finite elastic and plastic strains under impulsive loading. *J. Appl. Phys.* **35**, 2524–2533.

BESSELING, J. F. (1968). A thermodynamic approach to rheology. *IUTAM Symp. Irreversible Aspects Continuum Mech.*, 1966 pp. 16–53.

BIOT, M. A. (1954). Theory of stress-strain relations in anisotropic viscoelasticity and relaxation phenomena. *J. Appl. Phys.* **25**, 1385–1391.

BIOT, M. A. (1956). Thermoelasticity and irreversible thermodynamics. *J. Appl. Phys.* **27**, 240–253.

BIOT, M. A. (1958). Linear thermodynamics and the mechanics of solids. *Proc. U.S. Nat. Congr. Appl. Mech., 3rd*, 1958 pp. 1–17.

BODNER, S. R. (1968). Constitutive equations for dynamic material behavior. *Symp. Mech. Behav. Mater. Dyn. Loads*, 1967 pp. 176–190.

BRIDGMAN, P. W. (1950). The thermodynamics of plastic deformation and generalized entropy. *Rev. Mod. Phys.* **22**, 56–63.

CAMPBELL, J. D., and COOPER, R. H. (1967). Yield and flow of low-carbon steel at medium strain rates. *Proc. Conf. Phys. Basis Yield Fracture*, 1967, pp. 77–87.

CASIMIR, H. B. G. (1945). On Onsager's principle of microscopic reversibility. *Rev. Mod. Phys.* **17**, 343–350.

CHIDDISTER, J. L., and MALVERN, L. E. (1963). Compression-impact testing of aluminium at elevated temperature. *Exp. Mech.* **3**, 81–90.

CHRISTENSEN, R. M., and NAGHDI, P. M. (1967). Linear non-isothermal viscoelastic solids. *Acta Mech.* **3**, 1–12.

COLEMAN, B. D. (1964a). Thermodynamics of materials with memory. *Arch. Ration. Mech. Anal.* **17**, 1–46.

COLEMAN, B. D. (1964b). On thermodynamics, strain impulses, and viscoelasticity. *Arch. Ration. Mech. Anal.* **17**, 230–254.

COLEMAN, B. D., and GURTIN, M. E. (1967). Thermodynamics with internal state variables. *J. Chem. Phys.* **47**, 597–613.

COLEMAN, B. D., and MIZEL, V. J. (1964). Existence of caloric equations of state in thermodynamics. *J. Chem. Phys.* **40**, 1116–1125.

COLEMAN, B. D., and NOLL, W. (1963). The thermodynamics of elastic materials with heat conduction and viscosity. *Arch. Ration. Mech. Anal.* **13**, 167–178.

CONRAD, H. (1964). Thermally activated deformation of metals. *J. Metals* **16**, 582–588.

DE GROOT, S. R. (1966). "Thermodynamics of Irreversible Processes." North-Holland Publ., Amsterdam.

DE GROOT, S. R., and MAZUR, P. (1962). "Non-Equilibrium Thermodynamics." North-Holland Publ., Amsterdam.

DILLON, O. W. (1963). Coupled thermoplasticity. *J. Mech. Phys. Solids* **11**, 21–33.

DILLON, O. W. (1967). A thermodynamic basis of plasticity. *Acta Mech.* **3**, 182–195.

DRUCKER, D. C. (1964). Stress-strain-time relations and irreversible thermodynamics. *Proc. Int. Symp. Second Order Effects Elasticity, Plasticity Fluid Dyn.*, 1962 pp. 331–351.

ECKART, C. (1940). The thermodynamics of irreversible processes. I. The simple fluid. II. Fluid mixtures. III. Relativistic theory of simple fluid. *Phys. Rev.* **58**, 267–269, 269–275, and 919–924.

ECKART, C. (1948). IV. The theory of elasticity and anelasticity. *Phys. Rev.* **73**, 373–382.

ERINGEN, A. C. (1960), Irreversible thermodynamics and continuum mechanics. *Phys. Rev.* **117**, 1174–1183.

EVANS, K. R., and FLANAGAN, W. F. (1968). The nature of obstacles to dislocation motion in Cu and Cu–Si solid solutions. *Phil. Mag.* [8] **17**, 535–551.

EVANS, R. W., and SIMPSON, L. A. (1969). The strain-rate sensitivity of the plastic properties of α-iron at high temperatures. *Phil. Mag.* [8] **18**, 809–819.

FERGUSON, W. G., KUMAR, A., and DORN, J. E. (1967). Dislocation damping in aluminium at high strain rate. *J. Appl. Phys.* **38**, 1863–1869.

GIBBS, G. B. (1964). The thermodynamics of creep deformation. *Phys. Status Solidi* **5**, 693–696.

GIBBS, G. B. (1967). The activation parameters for dislocation glide. *Phil. Mag.* [8] **16**, 97–102.

GILES, R. (1964). "Mathematical Foundations of Thermodynamics." Pergamon Press, Oxford.

GILLIS, P. P., and GILMAN, J. J. (1965). Dynamical dislocation theory of crystal plasticity. I. The yield stress. II. Easy glide and strain hardening. *J. Appl. Phys.* **36**, 3370–3380 and 3380–3386.

GILMAN, J. J. (1965). Microdynamics of plastic flow at constant stress. *J. Appl. Phys.* **36**, 2772–2777.

GILMAN, J. J. (1966). Progress in the microdynamical theory of plasticity. *Proc. U.S. Nat. Congr. Appl. Mech., 5th*, 1966 pp. 385–403.

GILMAN, J. J. (1968a). Dynamic behavior of dislocations. *Symp. Mech. Behav. Mater. Dyn. Loads*, 1967 pp. 152–175.

GILMAN, J. J. (1968b). Dislocation dynamics and the response of materials to impact. *Appl. Mech. Rev.* **21**, 767–783.

GORMAN, J. A., WOOD, D. S., and VREELAND, T., JR. (1969). Mobility of dislocation in aluminium. *J. Appl. Phys.* **40**, 833–841.

GREEN, A. E. (1956a). Hypo-elasticity and plasticity. *J. Ration. Mech. Anal.* **5**, 725–734.

GREEN, A. E. (1956b). *Proc. Roy. Soc., Ser. A* **234**, 46–59.

GREEN, A. E., and ADKINS, J. E. (1960). "Large Elastic Deformations and Non-Linear Continuum Mechanics." Oxford Univ. Press, London and New York.

GREEN, A. E., and NAGHDI, P. M. (1965a). A general theory of an elastic-plastic continuum. *Arch. Ration. Mech. Anal.* **18**, 251–281.

GREEN, A. E., and NAGHDI, P. M. (1965b). Plasticity theory and multipolar continuum mechanics. *Mathematica* **12**, 21–26.

GREEN, A. E., and NAGHDI, P. M. (1967). A class of viscoelastic-plastic media. *Acta Mech.* **4**, 288–295.

GREEN, A. E., and NAGHDI, P. M. (1968). A thermodynamic development of elastic-plastic continua. *Proc. IUTAM Symp. Irreversible Aspects Continuum Mech.*, 1966 pp. 117–131.

GREEN, A. E., and ZERNA, W. (1968). "Theoretical Elasticity," 2nd ed. Oxford Univ. Press, London and New York.

GREEN, A. E., MCINNIS, B. C., and NAGHDI, P. M. (1968). Elastic-plastic continua with simple force dipole. *Int. J. Eng. Sci.* **6**, 373–394.

GURTIN, M. E. (1965). Thermodynamics and the possibility of spatial interaction in elastic materials. *Arch. Ration. Mech. Anal.* **19**, 339–352.

GURTIN, M. E., and PIPKIN, A. C. (1968). A general theory of heat conduction with finite wave speeds. *Arch. Ration. Mech. Anal.* **31**, 113–126.

GURTIN, M. E., and WILLIAMS, W. O. (1966). On the Clausius-Duhem inequality. *Z. Angew. Math. Phys.* **17**, 626–633.

GURTIN, M. E., and WILLIAMS, W. O. (1967a). An axiomatic foundation for continuum thermodynamics. Report. Carnegie Institute of Technology.

GURTIN, M. E., and WILLIAMS, W. O. (1967b). *Arch. Ration. Mech. Anal.* **26**, 83–117.

HAUSER, F. E., SIMMONS, J. A., and DORN, J. E. (1961). Strain rate effects in plastic wave propagation. *In* "Response of Metals to High Velocity Deformation," pp. 93–114. Wiley (Interscience), New York.

HOHENEMSER, K., and PRAGER, W. (1932). Über die Ansätze der Mechanik isotroper Kontinua. *Z. Angew. Math. Mech.* **12**, 216–226.

ILYUSHIN, A. A. (1954). On the law between stresses and small deformations in the mechanics of continua. *Appl. Math. Mech. (USSR)* **18**, 641–666 (in Russian).

ILYUSHIN, A. A. (1961). On the postulate of plasticity. *Appl. Math. Mech. (USSR)* **25**, 507 (in Russian).

JOHNSTON, W. G., and GILMAN, J. J. (1959). Dislocation velocities, dislocation densities, and plastic flow in lithinum fluoride crystals. *J. Appl. Phys.* **30**, 129–144.

KESTIN, J. (1966). On the application of the principles of thermodynamics to strained solid materials. Report. Brown University.

KESTIN, J. (1968). *IUTAM Symp. Irreversible Aspects Continuum Mech.*, 1966 pp. 177–212.

KLEPACZKO, J. (1968). Strain rate history effects for polycrystalline aluminium and theory of intersections. *J. Mech. Phys. Solids* **16**, 255–266.

KLEPACZKO, J. (1969). The strain rate behavior of iron in pure shear. *Int. J. Solids Struct.* **5**, 533–548.

KLUITENBERG, G. A. (1962a). Thermodynamical theory of elasticity and plasticity. *Physica* **28**, 217–232.

KLUITENBERG, G. A. (1962b). A note on the thermodynamics of Maxwell bodies, Kelvin bodies (Voigt bodies), and fluids. *Physica* **28**, 561–568.

KLUITENBERG, G. A. (1962c). On rheology and thermodynamics of irreversible processes. *Physica* **28**, 1173–1183.

KLUITENBERG, G. A. (1963). On the thermodynamics of viscosity and plasticity. *Physica* **29**, 633–652.

KLUITENBERG, G. A. (1964). A unified thermodynamic theory for large deformations in elastic media and in Kelvin (Voigt) media, and for viscous fluid flow. *Physica* **30**, 1945–1972.

KLUITENBERG, G. A. (1966). Application of the thermodynamics of irreversible processes to continuum mechanics. *Proc. Symp. Non-Equilibrium Thermodyn., Variation. Tech. Stability*, 1965 pp. 91–99.

KLUITENBERG, G. A. (1971). On heat dissipation due to irreversible mechanical phenomena in continuous media. *Physica* (to be published).

KOH, S. L., and ERINGEN, A. C. (1963). On the foundations of non-linear thermo-viscoelasticity. *Int. J. Eng. Sci.* **1**, 199–229.

KRAFFT, J. M., SULLIVAN, A. M., and TIPPER, C. F. (1954). The effect of static and dynamic loading and temperature on the yield stress of iron and mild steel in compression. *Proc. Roy. Soc., Ser. A* **221**, 114–127.

KRATOCHVIL, J., and DILLON, O. W. (1968). Thermodynamics of elastic-plastic materials as a theory with internal state variables. Report. University of Kentucky.

KRATOCHVIL, J., and DILLON, O. W. (1969). *J. Appl. Phys.* **40**, 3207–3218.

KRÖNER, E. (1960). Allgemeine Kontinuumstheorie der Versetzungen und Eigenspannungen. *Arch. Ration. Mech. Anal.* **4**, 273–334.

KRÖNER, E. (1962). Dislocation: A new concept in the continuum theory of plasticity. *J. Math. Phys.* **42**, 27–37.

KRÖNER, E. (1966). Dislocation field theory. *In* "Theory of Crystal Defects" (B. Gruber, ed.), pp. 231–256. Academic Press, New York.

LANGDON, T. G., and DORN, J. E. (1968). Low temperature dislocation mechanisms in ordered and disordered Cu_3Au. *Phil. Mag.* [8] **17**, 999–1015.

LARSEN, T. L., RAJNAK, S. L., HAUSER, F. E., and DORN, J. E. (1964). Plastic stress/strain-rate/temperature relations in H.C.P. Ag-Al under impact loading. *J. Mech. Phys. Solids* **12**, 361–376.

LEE, E. H., and LIU, D. T. (1967). Finite-strain elastic-plastic theory with application to plane-wave analysis. *J. Appl. Phys.* **38**, 19–27.

LINDHOLM, U. S. (1964). Some experiments with the split Hopkinson pressure bar. *J. Mech. Phys. Solids* **12**, 317–335.

LINDHOLM, U. S. (1965). Dynamic deformation of metals. *In* "Behavior of Materials under Dynamic Loading," pp. 42–61. Am. Soc. Mech. Eng., New York.

LINDHOLM, U. S. (1968). Some experiments in dynamic plasticity under combined stress. *Symp. Mech. Behav. Mater. Dyn. Loads*, 1967 pp. 77–95.

LINDHOLM, U. S., and YEAKLEY, L. M. (1965). Dynamic deformation of single and polycrystalline aluminium. *J. Mech. Phys. Solids* **13**, 41–53.

MAIDEN, C. J., and CAMPBELL, J. D. (1958). The static and dynamic strength of a carbon steel at low temperature. *Phil. Mag.* [8] **3**, 872–885.

MARSH, K. J., and CAMPBELL, J. D. (1963). The effect of strain rate on the post-yield flow of mild steel. *J. Mech. Phys. Solids* **11**, 49–63.

MEIXNER, J. (1941a). *Ann. Phys. (Leipzig)* [5] **39**, 333.

MEIXNER, J. (1941b). *Ann. Phys. (Leipzig)* [5] **40**, 165.

MEIXNER, J. (1942). *Ann. Phys. (Leipzig)* [5] **41**, 409.

MEIXNER, J. (1943). *Ann. Phys. (Leipzig)* [5] **43**, 244 and 470.

MEIXNER, J. (1964). On the theory of linear passive systems. *Arch. Ration. Mech. Anal.* **17**, 278–296.

MEIXNER, J. (1965). On the theory of linear viscoelastic behavior. *Rheol. Acta* **4**, 77–85.

MEIXNER, J. (1966). Consequences of an inequality in non-equilibrium thermodynamics. *J. Appl. Mech.* **33**, 481–488.

MEIXNER, J. (1968a). TIP has many faces. *IUTAM Symp. Irreversible Aspects Continuum Mech.*, 1966 pp. 237–249.

MEIXNER, J. (1968). Entropie im Nichtgleichgewicht. *Rheol. Acta* **7**, 8–13.

MEIXNER, J. (1968c). Beziehungen zwischen Netzwerktheorie und Thermodynamik. *Sitzungsber. Arbeitsgemeinschaft Forsch. Landes Nordrhein-Westfalen*, 1968 pp. 7–35.

MEIXNER, J. (1969a). Thermodynamik der Vergänge in einfachen fluiden Medien und die Charakterisierung der Thermodynamik irreversibler Prozesse. *Z. Phys.* **219**, 79–104.

MEIXNER, J. (1969b). On the foundation of thermodynamics of processes. *Int. Symp. Critical Rev. Found. Relativistic Classical Thermodyn.*, 1969, pp. 37–47.

MEIXNER, J. (1969c). Processes in simple thermodynamic materials. *Arch. Ration. Mech. Anal.* **33**, 33–53.

MEIXNER, J., and REIK, H. G. (1959). Thermodynamik der irreversiblen Prozesse. *In* "Handbuch der Physik" (S. Flügge, ed.), Vol. 3, Part II, pp. 413–523. Springer, Berlin.

MITRA, S. K., and DORN, J. E. (1962). On the nature of strain hardening in face-centred cubic metals. *Trans. AIME* **224**, 1062–1071.

MITRA, S. K., and DORN, J. E. (1963). On the nature of strain hardening in polycrystalline aluminium and aluminium-magnesium alloys. *Trans. AIME* **227**, 1015–1024.

NABARRO, F. R. N. (1967). "Theory of Crystal Dislocations." Oxford Univ. Press, London and New York.

NAGHDI, P. M., and MURCH, S. A. (1963). On the mechanical behavior of viscoelastic/plastic solids. *J. Appl. Mech.* **30**, 321–328.

NOLL, W. (1958). A mathematical theory of the mechanical behavior of continuous media. *Arch. Ration. Mech. Anal.* **2**, 117–226.

NOLL, W. (1959). The foundations of classical mechanics in the light of recent advances in continuum mechanics. *Proc. Int. Symp. Axiomatic Method Spec. Ref. Geom. Phys.*, 1957–1958, pp. 226–281.

OLSZAK, W., and PERZYNA, P. (1966). The constitutive equations of the flow theory for a non-stationary yield condition. *Appl. Mech., Proc. Int. Congr., 11th*, 1964, pp. 545–553.

ONAT, E. T. (1968). The notion of state and its implications in thermodynamics of inelastic solids. *IUTAM Symp. Irreversible Aspects Continuum Mech.*, 1966, pp. 292–314.

ONSAGER, L. (1931a). Reciprocal relations in irreversible processes. I. *Phys. Rev.* **37**, 405–426.

ONSAGER, L. (1931b). II. *Phys. Rev.* **38**, 2265–2279.

OWEN, D. R. (1968). Thermodynamics of materials with elastic range. *Arch. Ration. Mech. Anal.* **31**, 91–112.

OWEN, D. R., and WILLIAMS, O. W. (1968). On the concept of rate-independence, *Quart. Appl. Math.* **26**, 321–329.

PERZYNA, P. (1963a). The constitutive equations for rate sensitive plastic materials. *Quart. Appl. Math.* **20**, 321–332.

PERZYNA, P. (1963b). The constitutive equations for work-hardening and rate sensitive plastic materials. *Proc. Vibr. Probl.* **4**, 281–290.

PERZYNA, P. (1964). *Bull. Acad. Pol. Sci., Ser. Sci. Tech.* **12**, 199–206.

PERZYNA, P. (1966a). Fundamental problems in viscoplasticity. *Advan. Appl. Mech.* **9**, 243–377.

PERZYNA, P. (1966b). On thermodynamics of the rate type material. *Bull. Acad. Pol. Sci., Ser. Sci. Tech.* **14**, 397–407.

PERZYNA, P. (1966c). On thermodynamics of elastic-viscoplastic material. *Bull. Acad. Pol. Sci., Ser. Sci. Tech.* **14**, 409–416.

PERZYNA, P. (1968a). On thermodynamic foundations of viscoplasticity. *Symp. Mech. Behav. Mater. Dyn. Loads*, 1967 pp. 61–76.

PERZYNA, P. (1968b). On the physical foundations of viscoplasticity. Report. Institute of Fundamental Technical Research.

PERZYNA, P. (1968c). *Int. Congr. Appl. Theor. Mech., 12th*, 1968.

PERZYNA, P., and WIERZBICKI, T. (1964). On temperature dependent and strain rate sensitive plastic materials. *Bull. Acad. Pol. Sci., Ser. Sci. Tech.* **12**, 225–232.

PERZYNA, P., and WOJNO, W. (1966). On the constitutive equations of elastic/viscoplastic materials at finite strain. *Arch. Mech. Stosowanej* **18**, 85–100.

PERZYNA, P., and WOJNO, W. (1968). Thermodynamics of a rate sensitive plastic material. *Arch. Mech. Stosowanej* **20**, 499–511.

PIPKIN, A. C., and RIVLIN, R. S. (1965). Mechanics of rate-independent materials. *Z. Angew. Math. Phys.* **16**, 313–327.

RAJNAK, S. L., LARSEN, T., HAUSER, F. E., and DORN, J. E. (1962). The effect of strain rate and temperature on the yield strength in Ag-Al single crystals. Tech. Rep. MRL, University of California, Berkeley, California.

SCHAPERY, R. A. (1964). Application of thermodynamics to thermodynamics to thermomechanical fracture and birefringent phenomenon in viscoelastic media. *J. Appl. Phys.* **35**, 1451–1465.

SCHOECK, G. (1965). The activation energy of dislocation movement. *Phys. Status Solidi* **8**, 499–507.

SEEGER, A. (1954a). Theorie der Kristallplastizität. I. Grundzüge der Theorie. II. Die Grundstruktur der dichtest gepackten Metalle und ihr Einfluss auf die plastische Verformung. III. Die Temperatur- und Geschwindigkeitsabhängigkeit der Kristallplastizität. *Z. Naturforsch. A* **9**, 758–775, 856–869, and 870–881.

SEEGER, A. (1954b). The temperature dependence of the critical shear stress and of work-hardening of metal crystals. *Phil. Mag.* [7] **45**, 771–773.

SEEGER, A. (1955). The generation of lattice defects by moving dislocations and its application to the temperature dependence of the flow-stress of f.c.c. crystals. *Phil. Mag.* [7] **46**, 1194–1217.

SESTÁK, B. (1966). Glide and work-hardening in b.b.c. metals. *In* "Theory of Crystal Defects" (B. Gruber, ed.), pp. 357–402. Academic Press, New York.

STEIN, D. F., and LOW, J. R., JR. (1960). Mobility of edge dislocations in silicon-iron crystals. *J. Appl. Phys.* **31**, 362–369.

TIETZ, T. E., and DORN, J. E. (1948). The effect of strain histories on the work hardening of metals. *In* "Cold Working of Metals," pp. 163–179. Am. Soc. Metals, Cleveland, Ohio.

TROZERA, T. A., SHERBY, O. D., and DORN, J. E. (1957). Effect of strain rate and temperature on the plastic deformation of high purity aluminium. *Trans. Amer. So Metals* **49**, 173–188.

TRUESDELL, C. (1966a). "Six Lectures on Modern Natural Philosophy." Springer, Berlin.

TRUESDELL, C. (1966b). Letter to the participants. *IUTAM Symp. Irreversible Aspects Continuum Mech.*, 1966, pp. 1–3.

TRUESDELL, C. (1966c). Comments on Professor Meixner's Lecture. 1966.

TRUESDELL, C., and NOLL, W. (1965). The non-linear field theories of mechanics. *In* "Handbuch der Physik" (S. Flügge, ed.), Vol. 3, Part III, pp. 1–602. Springer, Berlin.

TRUESDELL, C., and TOUPIN, R. (1960). The classical field theories. *In* "Handbuch der Physik" (S. Flügge, ed.), Vol. 3, Part I, pp. 226–793. Springer, Berlin.

VAKULENKO, A. A. (1958). On stress-strain relations for inelastic bodies. *Dokl. Akad. Nauk SSSR* **118**, 665–668 (in Russian).

VAKULENKO, A. A. (1959). Thermodynamic investigation of stress-strain relations in isotropic elasto-plastic bodies. *Dokl. Akad. Nauk SSSR* **126**, 736–739 (in Russian).

VALANIS, K. C. (1966). Thermodynamics of large viscoelastic deformations. *J. Math. Phys.* **45**, 197–212.

VALANIS, K. C. (1967). Entropy, fading memory and Onsager's relations. *J. Math. Phys.* **46**, 164–174.

VALANIS, K. C. (1968). Unified theory of thermomechanical behavior of viscoelastic materials. *Symp. Mech. Behav. Mater. Dyn. Loads*, 1967 pp. 343–364.

WANG, C.-C., and BOWEN, R. M. (1966). On the thermodynamics of non-linear materials. *Arch. Ration. Mech. Anal.* **22**, 79–99.

WEHRLI, C., and ZIEGLER, H. (1962). Einige mit dem Prinzip von der grössten Dissipationsleistung verträgliche Stoffgleichungen. *Z. Angew. Math. Phys.* **13**, 372–393.

WOJNO, W. (1967). On thermodynamics of elastic/viscoplastic materials. Thesis, Inst. Basic Tech. Res., Polish Academy of Sciences.

ZIEGLER, H. (1957). Thermodynamik und rheologische Probleme. *Ing.-Arch.* **25**, 58–70.

ZIEGLER, H. (1958). An attempt to generalize Onsager's principle, and its significance for rheological problems. *Z. Angew. Math. Phys.* **9b**, 748–763.

ZIEGLER, H. (1961). Zwei Extremalprinzipien der irreversiblen Thermodynamik. *Ing.-Arch.* **30**, 410–416.

ZIEGLER, H. (1962a). Über ein Prinzip der grössten specifischen Entropieproduktion und seine Bedeutung für die Rheologie. *Rheol. Acta* **2**, 230–235.

ZIEGLER, H. (1962b). Die Statistischen Grundlagen der irreversiblen Thermodynamik. *Ing.-Arch.* **31**, 317–342.

ZIEGLER, H. (1963a). Some extremum principles in irreversible thermodynamics, with application to continuum mechanics. *Progr. Solid Mech.* **4**, 91 and 193.

Ziegler, H. (1963b). Thermodynamics aspects of continuum mechanics. *CIME Colloq.*, p. 1.

Ziegler, H. (1964). Thermodynamic considerations in continuum mechanics. Minta Martin Lecture. Dept. Aeronaut. Astronaut., Massachusetts Institute of Technology, Cambridge, Massachusetts.

Ziegler, H. (1966). Thermodynamik der Deformationen. *Appl. Mech. 11th, Proc. Int. Congr.* 1964, pp. 99–108.

Author Index

Numbers in italics refer to the pages on which the complete references are listed.

I

J

K

N

O

P

T

U

V

W

Subject Index

D

E

F

O

P

R

S

T

U

V

W

Y